A Practical Guide to Optimization in Engineering and Data Science

Wellington Rodrigo Monteiro ·
Gilberto Reynoso Meza

A Practical Guide to Optimization in Engineering and Data Science

PUCPR
GRUPO MARISTA
PUCPRESS
Springer

Wellington Rodrigo Monteiro
Pontifícia Universidade Católica do Paraná
Curitiba, Paraná, Brazil

Gilberto Reynoso Meza
Pontifícia Universidade Católica do Paraná
Curitiba, Paraná, Brazil

ISBN 978-3-032-04632-1 ISBN 978-3-032-04633-8 (eBook)
https://doi.org/10.1007/978-3-032-04633-8

Jointly published with Associação Paranaense de Cultura

This Springer imprint is published by the registered company Springer Nature Switzerland AG
The registered company address is: Gewerbestrasse 11, 6330 Cham, Switzerland

Acknowledgements The authors wish to express their sincere gratitude to their families for their unwavering support and understanding throughout the process of bringing this work to completion. They are also deeply indebted to the Pontifical Catholic University of Paraná (PUCPR) for the invaluable assistance and resources that were instrumental in bringing this book to fruition.

Furthermore, the authors wish to disclose that Generative AI technologies were utilized in a limited capacity during the preparation of this manuscript. Specifically, Large Language Models (LLMs) were employed to assist with refining English grammar and vocabulary, as English is not the authors' native language. This use was aimed at enhancing the clarity and precision of the text. The authors retained full editorial control, and all outputs from these LLMs were subject to their careful review, refinement, and final approval.

Competing Interests The authors have no competing interests to declare that are relevant to the content of this manuscript.

Contents

Chapter 1
Grokking Optimization

Abstract This chapter introduces the core concept of optimization by helping to build an instinctive understanding (or "grokking") of its principles. Here, we explore optimization as a natural human and environmental process, using examples from daily life and nature. The goal of this chapter is to define and demonstrate key optimization terms, such as objectives, decision variables, and constraints. This chapter also highlights the real-world importance and wide use of optimization for both engineers and data scientists through various optimization problem types, such as single objective versus multi-objective, linear versus non-linear, and deterministic versus stochastic. The aim of this chapter is to be an interesting and easy read for beginners, preparing them for more technical discussions and definitions in later chapters through real-world optimization examples.

Optimization is an art. Nature breathes and inspires optimization, and you've been seeking to optimize things for as long as you can remember. We could very well start explaining optimization with mathematical equations or by using some technical terms. However, what purpose would that serve? You could understand all these formal definitions without major problems, but what would be the value if you didn't understand where to apply them, or how to apply them? Therefore, we believe it's important to first think about examples and analogies.

We don't know if this is your case, but the life of many university students is demanding, to say the least. You have to absorb the maximum amount of content as quickly as possible. You need to submit assignments, take exams, and manage a social life. You have to juggle deadlines, grades, relationships, and your own health. It's no accident that, in many countries, students suffer from anxiety and stress.

But, let's suppose this situation *isn't* your case. Even so, you have your own personal life concerns to take care of. You have your priorities, dreams, and desires. And, still, there are constraints. You wouldn't be able to achieve all your plans and dreams within the 24 hours of a day. Other people might have financial limitations. Or, perhaps, they have space restrictions—after all, you can't be in two places at the same time. Still, many strive to make the most of their available resources.

These examples, in practice, mean that you and others are trying to *optimize* your lives with the limitations you have. No one has an infinite brain, or infinite time, or infinite space. Despite this, they try to achieve the best possible outcome given the

W. R. Monteiro and G. R. Meza, *A Practical Guide to Optimization in Engineering and Data Science*, https://doi.org/10.1007/978-3-032-04633-8_1

Fig. 1.1 An example of a desire path (on the right). *Source* The author, 2025

circumstances. And look: this way of thinking doesn't only apply to grand plans, like your academic education or your career. This even applies to things that might seem trivial, like the *desire path*.

A "desire path" is a path preferred by humans that becomes visible through erosion or soil use. Observe Fig. 1.1, for example. Even though there's a paved path that someone designed at some point, people preferred to cut across the lawn. And this shortcut is used so frequently that the grass ceased to exist on that path. In other words, people preferred to *optimize* their time and the distance traveled by creating that extra path. Perhaps it wasn't a conscious decision by a crowd, but rather a natural one.

Similarly, nature also performs its optimizations. A small plant that sprouted in the shade grows toward the sun. A flock of birds collectively decides where to fly to feed, reproduce, and protect themselves. A swarm of bees jointly chooses where the new hive will be for similar reasons. Speaking of which, have you ever seen a honeycomb in a beehive? It's composed of several hexagons built in a way that approaches symmetrical perfection. This shape is not the result of an engineering directive to the bees—in fact, this shape is *optimized* to reduce the effort and materials needed to build the comb (Fig. 1.2).

Fig. 1.2 A honeycomb. *Source* Wikimedia Commons

Tip
Many innovative techniques and processes in industry were inspired by nature. So, how about taking a walk and getting some fresh air when you're facing a problem at work? Several renowned professionals and academics had ideas that were born outside the classroom but were refined within it.

Did you notice that terms like *best*, *largest*, and *shortest* are embedded in these examples? We constantly seek to make decisions that are the best for us considering our objectives and available resources. This decision-making process is called **optimization**, and it's the focus of this book.

Important
Optimization is the art of doing the best you can with what you have available at the moment.

It's common for engineering students to learn about optimization during their studies. After all, there are several processes in the industry that require you to *do more with less*. That's optimizing, isn't it? Whether it's maximizing the quantity of products transported with fewer vehicles, minimizing electrical energy consumption in an electrical circuit, or designing engines with the highest energy efficiency.

Data scientists in industry also use optimization, but its role is less explicit. Meaning: although these professionals seek to produce Artificial Intelligence (AI) algorithms with the best possible effectiveness given the data available at a certain time, much of their effort is dedicated to producing a quality dataset and, also, to preparing (or, as we call it in the field, *training*) algorithms that work well. The optimization step itself transforms into an additional parameter or configuration in an algorithm. There's rarely a dedication to understanding and exploring optimization as a field of study—it's relegated from a method to a configuration line. On the other hand, this is understandable: as we mentioned earlier, it's common for engineers to learn about optimization, but not necessarily for those in other fields. And, in practice, data scientists can come from various courses besides engineering.

Thus, you might be an engineer who will work with AI in some way in the future. Or, perhaps, a data scientist who has never set foot in an engineering classroom. Regardless of the case, it's important that you know what optimization is so you can apply it correctly in your career. After all, memorizing something is different from knowing it. And we wouldn't want you to just skim through this book to memorize something for a test, but not know how or where to use it in your career. Instead, we want you to master optimization for your life. A good performance on exams and assignments would merely be a consequence.

1.1 Concepts

That said, how do we build an optimization algorithm? Notice that we've given some simple examples, but we haven't examined any specific case so far. So, how about we think about donuts?

> **Attention**
> Just as different mathematical problems have different strategies for their resolution, different people think (and learn) in different ways. Therefore, in this first part, we will use analogies and other non-mathematical resources to explain the concepts of optimization. From there, we will deep dive into the mathematical concepts in Chap. 2.

Let's suppose you own "Optimal Donut."[1] Now, any self-respecting donut shop offers several flavors, and each flavor has a price (Fig. 1.3). Customers might be willing to pay a higher price for a chocolate-filled donut than a plain cinnamon donut due to the ingredients, for example. We could also offer even more expensive

[1] The name of this place is a pun on the word *optimization*, and is purely fictional. Any similarity to a real place is purely coincidental.

Fig. 1.3 Our donut shop could sell products like these. *Source* Wikimedia Commons

products due to their specificities—a well-filled coffee roll or cinnamon roll could have a special price, indeed.[2]

Well, given this context, let's understand how we could apply optimization here. First, let's start with the **objective**.

1.1.1 Objective

Every optimization problem begins with defining its objective, and this greatly depends on the context. Companies usually pay us salaries to **solve problems**. When we solve a problem, we first invest time to understand, define, and delimit it. In optimization, this is no different. So, the first point is to define *what the objective is*. And, if you can't explain this objective using the words *minimize* or *maximize*, we are sorry, but it's probably not an optimization problem. Let's look at some examples.

- ***Maximize*** *factory profit*: This phrase makes sense—after all, many industries wouldn't want to produce at a loss, right?
- ***Minimize*** *production cost*: This also makes sense. If the sales department is interested in profit, the production department would like to reduce its own costs.
- ***Minimize*** *waste*: This also makes sense. Whether due to sustainability concerns or because raw material is expensive, some companies may focus on reducing waste of time, equipment, or raw material.

[2] Would a cinnamon roll or a coffee roll be special types of donut?

- ***Maximize*** *the accuracy of an AI algorithm*: We also think this makes sense. After all, nobody wants an AI algorithm that doesn't work, right? Naturally, there are different mathematical definitions of what "accuracy" would be, but the fact is that companies would like to have the algorithm that gets more things right (or fewer things wrong).

Now, in what cases would it *not* be an optimization problem? Let's see some examples:

- ***Forecast*** *factory profit for the coming months*: This is an interesting problem. In fact, many data scientists work on problems like this. However, this is a *time series* problem. We use optimization here indirectly: we optimize a time series forecasting model, which is responsible for providing this type of prediction. Notice that we don't use optimization to find the *highest* (maximization) or *lowest* (minimization) possible profit.
- ***Classify*** *products into groups that generate more and less profit*: This is a job for clustering algorithms, not optimization. Note that we are only dividing into groups. In fact, we wouldn't be maximizing or minimizing anything in this challenge.
- ***Analyze*** *the main contributors to waste*: Reread this sentence. Notice that we seem to be talking about something that has already occurred, right? Meaning: Our challenge is to understand what is contributing most to waste, and we don't want to find out what we can do to reduce this in the future. This is an interesting job for a data analyst, and *Business Intelligence* (BI) tools are often used for this. However, it wouldn't be an optimization case.
- ***Improve*** *the quality of the database for training an AI algorithm*: To have a good AI algorithm, it's important to have a good database. It's the concept of *garbage in, garbage out*: If we have bad data, an AI's predictions will be bad. In this case, a data engineer or scientist cleans and removes erroneous information from the database, handles null values, and deals with outliers. Perhaps after this, we could also use some optimization strategy to *minimize* data multicollinearity, but this would only be one (and not the only) step in this process. Thus, optimization becomes a subproblem here instead of the sole problem.

Important
So, notice that the objective is a **goal you need to achieve**. This goal can be defined either by something you'd like to maximize or as something you'd like to minimize.

With that, let's return to the donut shop: a valid objective we could have might be **maximizing profits**—after all, you wouldn't want a donut shop that operates at a loss. Now, if you're the owner, it would be very good for you if all your customers always bought the most expensive product. But not everyone wants that donut. Or, as our friends from Business School define it more elegantly, the *law of supply and demand*. Therefore, even if you don't profit as much from the cheaper donuts, you'll need to produce them anyway. We also have a second problem: even if you have the only donut shop on the planet, you can't absorb an infinite demand for donuts. In a single day, you'll be able to produce a limited quantity of donuts due to your infrastructure, the number of employees, and the amount of available ingredients.

So, how would we define this on paper? What do you think of the definition below?

Defining the donut shop problem
We want to **maximize** the **profit** from donut sales. The profit is **subject to** the availability and cost of ingredients, donut demand, and the time available to make the donuts.

Notice that we've clearly established what we want to do: we don't want to "*improve*" profit (after all, this is subjective), nor "*increase*" profit (how much of an increase would be good? 5% more? 10%?). We want the **maximum** possible. Of course, it would be great if this maximum were infinite, but we are subject to **constraints** such as, for example, ingredient availability and customer demand. And, to achieve this maximum profit, we would have an **ideal combination** of donuts to be produced.

At first, it seems simple, doesn't it? How about we better understand the definitions that every optimization problem can have?

1.1.2 Objective Function, Variables, and Constraints

Now, how do we clearly define this type of objective? That's where the objective function comes in. If the objective is the goal we'd like to achieve with optimization, then the objective function is the *formalization of what* we are trying to maximize or minimize.

Important
If the objective is a goal you need to achieve, the objective function is the quantitative (mathematical) representation of this goal. That is, if the objective is *what* you need to achieve, the objective function represents *how* you should achieve it.

Let's go back to the donut case: if our objective was to **maximize** donut profit, now would be the time to formalize how we could arrive at this value. To simplify things, let's assume our shop offers five types of donuts: plain, coconut, cinnamon, chocolate, and hazelnut cream. Each donut would have a price, something like what's shown in Table 1.1.

Table 1.1 Prices of donuts at the Optimal Donut shop

Donut type	Selling price
Plain	$1.00
Coconut	$1.25
Cinnamon	$1.25
Chocolate	$1.50
Hazelnut cream	$2.00

Think with us for a moment: if we want to maximize profit and the price of each donut has already been defined, we practically want to find out *what quantity of each type of donut we need to sell*, right?

The objective function, in this case, would be something like:

Defining the donut shop problem

Maximize the profit of the Optimal Donut shop, where the profit will be
1.00 * quantity of plain donuts sold +
1.25 * quantity of coconut donuts sold +
1.25 * quantity of cinnamon donuts sold +
1.50 * quantity of chocolate donuts sold +
2.00 * quantity of hazelnut cream donuts sold.

It got a bit wordy, didn't it? In this case, let's shorten the names a bit. For example, instead of saying *quantity of cinnamon donuts sold*, we'll just call it `cinnamon`. So, we'll have

(Re)defining the donut shop problem

Maximize profit, defined by
1.00 * plain + 1.25 * coconut + 1.25 * cinnamon + 1.50 * chocolate + 2.00 * hazelnut

Simple, right? If we sell one of each, our profit would be $7.00. If we only sell hazelnut cream donuts, but sell five of them, our profit would be even higher: $10.00. As you can imagine, the secret here would be to sell as many of the most expensive donuts as possible. If we sold infinite donuts, our profit would be infinite! However, life isn't an endless sea of donuts.

So, in real problems, we also have constraints. Let's start with the easiest: common sense. We can't sell negative donuts, right? Thus, the minimum quantity of donuts we can sell is zero. We might also have some raw material constraints: let's suppose we have enough dough to make only 20 donuts in total. So, we can already set some rules—one per line:

Defining sales constraints for the donut shop, part I

plain ≥ 0
coconut ≥ 0
cinnamon ≥ 0
chocolate ≥ 0
hazelnut ≥ 0
plain + coconut + cinnamon + chocolate + hazelnut ≤ 20

Now things are starting to get more interesting. We can't sell infinite donuts, not even 50 donuts if we wanted to. So, think with us: there must be a combination that respects these rules in order to achieve the maximum profit we were looking for, right? And, as you may have inferred from these examples, the answer would be 20 donuts of the most expensive type (hazelnut cream). Thus, if we produce 20 hazelnut cream donuts, we would have the maximum profit of $100 while respecting these rules.

But real life still isn't that simple. Let's suppose there's a *demand* from your customers for the simpler donuts. If you don't sell these simpler donuts, you'll have fewer customers and, over time, your business will fail. Therefore, imagine the new rules are

Defining sales constraints for the donut shop, part II

plain ≥ 3
cinnamon ≥ 5
chocolate ≥ 1

Notice that some rules have changed: the minimum limit for these three types of donuts is no longer zero. Therefore, we will need to produce at least this minimum quantity of these donuts. Furthermore, let's make it a bit more difficult, and suppose we only have enough raw material to produce four hazelnut cream donuts. So, we also need to insert the following rule:

Defining sales constraints for the donut shop, part III

hazelnut ≤ 4

How about we put all the rules together? It will look like this:

Putting all the donut shop rules together

Maximize profit, defined by
1.00 * plain + 1.25 * coconut + 1.25 * cinnamon + 1.50 * chocolate + 2.00 * hazelnut

Respecting the following constraints:
plain ≥ 0
plain ≥ 3

coconut ≥ 0

cinnamon ≥ 0
cinnamon ≥ 5

chocolate ≥ 0
chocolate ≥ 1

hazelnut ≥ 0
hazelnut ≤ 4

plain + coconut + cinnamon + chocolate + hazelnut ≤ 20

You might notice that some rules overlap. So, if we summarized all these rules, we would have

Defining the donut shop problem, final part

Maximize profit, defined by
1.00 * plain + 1.25 * coconut + 1.25 * cinnamon + 1.50 * chocolate + 2.00 * hazelnut

Respecting the following constraints:
plain ≥ 3
coconut ≥ 0
cinnamon ≥ 5
chocolate ≥ 1
0 ≤ hazelnut ≤ 4
plain + coconut + cinnamon + chocolate + hazelnut ≤ 20

And now, we have the definition of our objective function ready. Here we can extract some important components for defining any objective function:

- **Decision Variables**: These are the variables that influence the outcome of the objective function. Let's consider the problem above: what would influence the result (profits) would be the quantity of donuts sold, right? In this case, each type of donut would be a decision variable. We'll explore this a bit more later, but it's possible to have different types of variables such as integers, real numbers, and binary variables.
- **Constraints**: These are the limitations that might affect (or delimit) the decision variables. In our case, we mentioned that we can't sell a negative quantity of donuts because it wouldn't make much sense. We also said that some types of donuts have a maximum quantity that could be produced. These would be the constraints.

Not every objective function will have constraints, but every objective function will have decision variables. That said, let's identify these different components in the definition above:

- **Objective**: Maximize the profit defined by the objective function.
- **Objective Function**: 1.00 * plain + 1.25 * coconut + 1.25 * cinnamon + 1.50 * chocolate + 2.00 * hazelnut.
- **Variables**: plain; coconut; cinnamon; chocolate; hazelnut.
- **Constraints**:
  plain $\geq$ 3;
  coconut $\geq$ 0;
  cinnamon $\geq$ 5;
  chocolate $\geq$ 1;
  0 $\leq$ hazelnut $\leq$ 4;
  plain + coconut + cinnamon + chocolate + hazelnut $\leq$ 20.

With the objective function is ready, it's time to find the best possible combination. We could even take a spreadsheet and write down all possible combinations to achieve a result by brute force, without using mathematical techniques. However, this can take a very long time or, depending on the problem, be impossible to calculate given its complexity. After all, in real problems, we'll have much more than just five variables and six constraints.[3]

That said, since our problem is simple, we decided to create some solutions for you to visualize in Table 1.2. Notice that the first solution is invalid, as it doesn't respect all constraints: remember we had a minimum quantity for some donuts. The second solution respects all constraints, but we could still have an even better combination. Thus, you can see that from the third solution onward, we tested some other perfectly valid combinations until we reached the best possible combination—that is, with the *highest possible profit*, which was our initial objective.

---

[3] We will explore the main optimization techniques in the upcoming chapters. In fact, there will be different techniques for different types of optimization problems starting from Chap. 2. Just as someone working in cuisine or construction chooses the best equipment for each case, we will also choose the best optimization technique to be applied to each type of problem.

Table 1.2 Some combinations for the donut shop optimization problem. The best combination is in bold

Plain	Coconut	Cinnamon	Chocolate	Hazelnut	Constraints respected?	Profit
1	1	1	1	1	No	$7.00
3	0	5	1	0	Yes	$10.75
3	0	5	1	4	Yes	$18.75
3	0	5	5	4	Yes	$24.75
3	0	8	5	4	Yes	$28.50
3	**0**	**5**	**8**	**4**	**Yes**	**$29.25**

In the industry

To begin with, we don't have a magical ready-made solution, nor do we create problems in a vacuum. That is, developing an optimization problem involves investigating data and talking to people. Usually, we start with a **business problem** proposed by an area. For example, the Logistics department might have some issues with high transportation costs. Or, perhaps, the Production department in an industry might have problems with manufacturing costs.

After understanding the business problem, we start translating it into a **mathematical problem**. Or, perhaps, we could call this **mathematical modeling**. It's in this step that we define, first and foremost, *what* we want to optimize. The Logistics business problem could become "minimize transportation costs," and the Production department's problem would transform into "minimize manufacturing costs." This is where we start fitting in the words "minimize" and "maximize," and understanding what the objectives are. There are cases where there might be more than one objective we want to optimize simultaneously.

Still in this step, we also define *which* variables matter for that problem. Perhaps for Logistics, transportation routes, vehicles, time, and fuel are important. For production, perhaps production lines, raw material, and the products that would come out of there. Notice that the variables need to be linked to the problem—perhaps the question to ask those in these departments who are interested in optimizing would be: "*how would we calculate the* `$PROBLEM`*?*". That is: how would we calculate transportation costs, or how would we calculate manufacturing costs?

It's also in this mathematical modeling step that we define the *constraints*, such as the minimum and maximum capacity of the variables. In our examples, we would have a minimum and maximum quantity of vehicles, or raw material, or time. Notice that each element—be it a variable or a constraint—has a story and a meaning for solving this problem.

After modeling, it's natural to validate this problem with other people. Preferably, the people who work in that area so they can evaluate if the mathematical modeling makes sense. After all, if your results don't make sense to them, it's natural that they won't trust your model's results and, consequently, won't use it. In other words, it would be a waste of time for you, and we wouldn't want that to happen. Also, it's worth remembering that mathematical modeling is a *abstraction* of real life; therefore, it will hardly be able to reproduce all the nuances and uncertainties of real life. And, in most cases, this wouldn't be necessary. Thus, it might be interesting for you to focus on generalizing the real-world problem using mathematical variables: something that represents it with a certain (but not total) fidelity. In the case of Logistics, it wouldn't make much sense to think about all possible combinations of scenarios that could occur, such as a traffic accident, a flat tire, or a detour due to road work. Or, in the case of production, perhaps you wouldn't need to consider in your modeling what would happen if a factory employee was three minutes late for the start of their shift, or if the raw material yielded 0.02% more than expected.

That said, after validating the problem, we implement and test it using software. That is, we convert the problem into computer code. It's common to use programming languages like Python; code libraries like PuLP, Pymoo, and OR-Tools; or platforms like CPLEX, Gurobi, and MATLAB for this. It's also common for it not to work on the first try: we test different models, run simulations, and adjust parameters until we consider it ready to run periodically.

In fact, this whole process can take from a few days to a few weeks, depending on the complexity of the problem and the free time to implement and test these algorithms. Still, we might face some challenges such as, for example, incomplete data, uncertainties in the parameters, or even high computational complexity (that is, the problem is so complex that we might need a better computer to run it).

1.2 Practical Importance: Is It Still Relevant?

At this point, we assume you already understand the fundamentals of optimization—that is, you know what an objective function, variables, and constraints are. But, if you still have doubts, don't worry: we'll explore more examples throughout the next sections.

It's equally important to understand if learning these concepts is *worth anything*. Depending on your exposure to technology news, you probably haven't heard much about optimization. In contrast, topics like generative artificial intelligence (Gen AI), blockchains, cryptocurrencies, quantum computing, and even electric or autonomous cars frequently dominate the discussion. I'll bet a donut that "Optimization" doesn't appear among the most talked-about technical skills.

Therefore, it's quite fair to question whether it's worth dedicating your precious time to study this. And this is an excellent question, as optimization is a peculiar field. To illustrate this, let's look at another example: computers. How many job openings related to your course explicitly require you to know how to use a computer? Probably very few, even though this information is implicit in most cases.

The Millennial generation grew up in a kind of technological "Goldilocks zone"[4]—a period when personal computers and the Internet were becoming accessible. Many middle-class families managed to buy their first computer, and this generation was exposed to a variety of technologies from an early age. In some countries, it was common for children to have access to computer labs, where they learned to perform tasks like creating folders, compressing files, and typing documents. In college, many of these students who enrolled in IT courses already mastered these skills, and it was a premise that they already knew them.

However, younger generations haven't necessarily had the same experience. Not everyone grew up using desktops or laptops; the cellphone became the main communication tool. Ironically, some current Information Technology students may not be familiar with basic computer tasks, like creating folders or compressing files.

And why is this relevant in this book? You'll rarely see job openings requiring "computer skills." However, you're expected to know how to use computers, including creating spreadsheets and presentations. Similarly, you might use optimization in your work, even if it's not explicitly listed in a job's requirements. Therefore, yes, optimization is relevant.

The daily life of good data scientists and engineers is built on optimization. In engineering, it's impossible to design electrical circuits, production lines, sustainable constructions, or high-performance engines without optimization. Nor can you create an efficient data architecture without optimizing it. And the same goes for AI: a good machine learning (ML) model fundamentally depends on optimization strategies to recognize patterns and "learn" from data. A good data scientist selects algorithms; an excellent data scientist masters optimization strategies to maximize results.

Now, you might be wondering: "what if I don't end up working with this?" Perhaps your immediate focus might involve social activities, personal well-being, or academic success. But if you observe the job market, the requirements for positions in different regions, or even the paths taken by those who graduated before you, you'll notice two things. First: successful careers are born from interdisciplinarity. Thus, knowing a little about different things will make a difference in enabling you to converse with people of various profiles and solve problems. Second, not everyone is interested in the same subjects or has the same aptitude. One of the authors of this book, for example, graduated in Computer Engineering, and his colleagues followed

[4] The Goldilocks zone is that ideal region around a star for life as we know it to exist. Not too cold, and not too hot.

diverse paths—some went into hardware and embedded systems, others into academic research, others into ML or data engineering. And there are colleagues who don't even use the title "engineer" but have built great careers.

The key lies in the competencies you acquire. Even if you can't visualize it now, optimization concepts can be useful in your endeavor. Therefore, *optimization matters*.

1.3 Different Types of Optimization

Alright, if optimization matters and we use it in various fields, what are its main types? To be honest, depending on the researcher or professor you ask, it's extremely likely you'll receive different answers. Therefore, our idea here is to present the main types to you.

Just as we could divide the different types of sandwiches in the world by the type of bread, the region where they were created, the main ingredient, or whether they are sweet or savory, something similar happens in optimization. Thus, we can divide optimization by how we work with variables, constraints, objectives, or even depending on the types of data we are working with. Each of these professionals could categorize optimization depending on the characteristic that is most important to them.

1.3.1 Number of Objectives

Let's start with the **objectives**. Remember that donut shop example we gave earlier? Well, that was an optimization problem where we were trying to improve only *one* thing: profit.

In fact, it's common to work with problems similar to the donut shop one when we're learning about optimization. After all, the challenge seems relatively simple: find the largest (or smallest) possible value considering the available variables and, also, respecting the constraints.

Problems like that—that is, when you need to maximize or minimize only one value—are called **Single-Objective Optimization** (SOO) problems. Data scientists work with such problems almost exclusively. When training a neural network or some other ML model, we want to reduce the model's error so that its predictions are more reliable. For this, we use error metrics like Root Mean Squared Error (RMSE) or F1-score.

Although this explanation might seem to make sense, not everything in life has only one objective to achieve, or one dimension to analyze. Let's go back to the donut shop case: the combination that would yield the highest possible profit isn't necessarily the most suitable for a specific situation. Sometimes, the solution with the highest profit also generates the most waste, or has the highest risk, or is the solution whose raw material might expire the quickest. Let's think of another example—an investment manager at a large bank could use these optimization concepts to find the best asset allocation in an investment portfolio that yields the highest returns. However, the result might be so aggressive that the risk becomes impractical. Therefore, **there are situations where we need to consider more than one objective simultaneously**.

For these cases, there's **Multi-Objective Optimization** (MOO).[5] In this case, we work with two or more conflicting objectives (for example, every time profit increases (good), waste also increases (bad). Or, every time financial return increases (good), risk also increases (bad)). A classic example in Civil Engineering is structural design. Suppose we are designing a bridge and want to minimize both construction cost and the total weight of the structure. Now, reducing cost might imply using cheaper materials, but this can lead to an increase in the bridge's total weight, which, in turn, requires a more robust foundation and can generate additional problems and costs. On the other hand, to reduce weight, we can use more advanced materials, which impacts the cost.

Thus, there's a trade-off between these two objectives, and the "best" solution (or optimal solution) isn't a single point, but a set of solutions known as the **Pareto front**. In the problem we just mentioned, each solution would be a different bridge design. In the case of our donut shop, each solution would be a combination of products with distinct profits and waste levels. We'll see this in more detail in Chap. 5.

The cellphone example As we've discussed before, we believe a good introduction to MOO would be an example closer to your own life. After all, not everyone has worked on a factory floor, right? So, imagine your cellphone is quite old, and you'd like to buy a new one (Fig. 1.4). On the other hand, you also have some bills and responsibilities, and your budget is tight. Thus, notice that your challenge is to buy the *best cellphone your money can buy*. We don't know if you noticed, but we have two conflicting objectives here.

Let's see:

- **Objective 1**: Minimize price. You have a certain budget you wouldn't want to exceed, but you also wouldn't want to use it to the limit. After all, if you found two devices with the same performance but different prices, you'd likely opt for the cheaper one, right?
- **Objective 2**: Maximize performance. Some people want a good processor, or a good camera, a good battery, or perhaps a good screen resolution. Or, maybe, a weighted average of all these characteristics. In this case, imagine we are considering all these factors in a single value called "performance,"

5 Within MOO, we also have *Many-Objective* (MaOO) problems. MOO problems usually involve two to four objectives, while many-objective problems deal with five or more objectives simultaneously. The more objectives we have, the greater the computational complexity to generate and test different solutions. Therefore, some optimization algorithms are specifically built for *many-objective* problems.

Fig. 1.4 You (probably), choosing a cellphone. *Source* Unsplash

represented by a score from 0 to 100, where a higher score indicates better performance. In this case, you'd want the cellphone with the highest possible performance.

Since you're an intelligent person, you know that the cheapest cellphone certainly won't be the best on the market. On the other hand, the cellphone with the best performance will have a hefty price tag. So, notice that you intuitively already know there's a trade-off between these two conflicting objectives.

Now, let's suppose you're researching on an e-commerce site that offers several cellphones. Of these, ten models caught your attention—each with a different price and performance. And, when comparing each model, you realize some are much better than others. For example, you noticed that Cellphone X is much faster than Cellphone Y and, yet, is cheaper. So, it wouldn't even make sense for you to consider buying Cellphone Y, right? Since Cellphone X is better than Cellphone Y in those two objectives, we discard Cellphone Y from our comparison. Or, technically speaking, Cellphone Y is a "**dominated**" solution.

Let's say that, of those ten models that caught your attention, six are dominated solutions. This means there's at least one other cellphone that is better in both price and performance. Therefore, we can remove these six devices from our decision-making process. Thus, we end up with four devices—or, technically speaking, "**non-dominated**" solutions. These are cellphones that are not strictly better than the others on this list. For example:

- Cellphone A: Costs $166, and has a performance score of 22/100.
- Cellphone B: Costs $281, and has a performance score of 40/100.
- Cellphone C: Costs $410, and has a performance score of 59/100.
- Cellphone D: Costs $618, and has a performance score of 88/100.

Notice that none of these options can be removed because they are better in at least one objective without compromising the other. Let's look at cellphone A, for example, it has the best price of all. Since there's no other cellphone with a better price, it remains on the list. On the other hand, Cellphone D has the best performance among all other options. Therefore, it remains on the list because there's no other option with better performance.

Now, let's consider Cellphone B: Cellphone A is cheaper but has worse performance compared to it. Similarly, Cellphones C and D have better performance but a worse price. The same logic applies to Cellphone C, where Cellphones A and B are cheaper than it but have worse performance. Likewise, Cellphone D is faster but more expensive.

That said, which of these four cellphones would we choose? The answer is one that every quantitative scientist or engineer hates to hear: *it's subjective*. The final choice depends on each person's preferences and needs. If you have a lot of money at the moment and really care about having a fast cellphone in hand, you'll likely choose Cellphone D. Now, if you were making this decision a few months ago when you had many bills to pay, you'd probably opt for Cellphone B or C. Or, if you really had a lot of bills, you'd probably opt for Cellphone A.

Thus, notice that the decision-making is based on the context and specific needs you might have. Similarly, a company's board could opt for production plan A, B, C, or D when analyzing conflicting objectives for that type of problem. Or, perhaps, a data scientist could choose between ML model A, B, C, or D when analyzing the model's accuracy metrics and its response time. Sometimes, it's worth opting for a model that has a marginally higher error in exchange for an exceptionally better response speed.

1.3.2 Variables

Another way to categorize optimization problems is by the type of variables we have. If you're reading this book, it's very likely you've already taken some basic Mathematics and Statistics courses, and you know what a continuous or discrete variable is.[6]

Thus, it's possible to have optimization problems of a few different types considering the nature of the variables. The first are problems that use **discrete variables**, where all variables are discrete. Our donut shop optimization problem is an example of this—after all, we can't produce 2.3491 donuts, but only a whole number of donuts of each type. Here, it's worth remembering that a discrete variable can be an interval of integers, a set of categories, or even binary variables.

The second type deals with **continuous variables**, where all variables are, as the name suggests, continuous. Here, we'd like you to think of examples like the amount of fuel a car consumes per kilometer, the height of a building, or the time it would take you to finish an exam. All these examples can take values at an infinite number of points within an interval. For example, a building's height can be 40, 40.1, 40.01 m, and so on, without limits. Or, if we're considering the duration of a task, it could be 3.7, 3.77, 3.777 h, and so forth.

The third type considers **mixed variables**, where we can mix discrete and continuous variables in a single optimization problem. Think, for example, of a transportation problem, where the number of vehicles to be used would be a discrete variable (since we can only have a whole number of vehicles), while travel time or distance covered could be treated as a continuous variable. Or, consider a factory that needs to decide how many units of each product type to produce (discrete variable) and, at the same time, determine the amount of raw material needed, which could be a continuous variable.

1.3.3 Scale

One common complaint within some Universities is the gap between Industry and Academia. This occurs more frequently in Information Technology courses, and a bit less in others. In such cases, it's common to hear phrases like "*you don't even need a degree to work in the field*," or "*I never used anything I learned in college at work*," or "*everything I learned was only because of my work experience, not college*." We're not here to discuss the merit of these statements, but to think a bit more about them.

[6] So, in short: continuous variables are those that have infinite values within a numerical interval. For example, in an interval from 25 °C to 28 °C, we can have infinite values, such as 26.5 °C, 27.4591 °C, or 27.99995 °C. A discrete variable, however, is one whose possible values are finite or a countably infinite set. An example of this would be if we counted temperature only in whole numbers or in 0.5 °C intervals. In the case of whole numbers, the *only* possible values in the closed interval between 25 °C and 28 °C would be 25 °C, 26 °C, 27 °C, and 28 °C.

Now, if these complaints exist, we understand they likely contain some element of truth.

Perhaps one of the many reasons behind phrases like these is the distance between the problems presented in the classroom and real-life problems. You see: in the classroom, it's common to study overly simple problems, even just to better understand the concept. In Engineering or Statistics and Data Science courses, this is practically a rule: we see simple, already studied and treated cases. Problems that can be easily applied in an exam, without the noise and uncertainties of real life. You receive an exercise where the professor already knows the answer, not a problem that another area of the company sent (and that they probably put together incorrectly, or with missing information).

With this, we want to get to **scale**. We can have **small-scale problems**, like the donut shop one. As you noticed, we only have a few variables and constraints, easily printed in a book or on a few lines of paper. For these cases, the search scope is smaller: there are fewer combinations and comparisons that a computer (or, in rare cases, a person) needs to make to find the best solution.

A classic example of small-scale problems that we and other professors often use in the classroom is the beam sizing problem. In this problem, the objective is to determine the ideal dimensions of a beam (i.e., width, height, and length) to support a specific load. We have strength and deformation constraints, and the objective is cost minimization. We like to work with problems like this for two reasons: first, it's easy to describe on a sheet of paper, in a book, or on a classroom board. Second, because it uses a real-life example, and it's easier to think about the applicability of optimization problems with such examples. In fact, small-scale problems also exist in real life, especially in contexts where the scope is limited and decisions are localized. So, it's not as if we're wasting time by studying this type of problem in the classroom.

On the other hand, when there are many variables and constraints at play, such that a computer could run for extended periods (hours or even days) to find a solution, we would have a **large-scale problem**. In these cases, it might be worthwhile to use some methods where we lose a bit of accuracy and precision to gain execution time to produce some valid result.

You know that beam sizing problem we mentioned? Well, what if it were the design of something more complete, like a building or a bridge? In that case, we would have multiple beams, columns, and connections, and each would have its own requirements and constraints, right? Here, you certainly wouldn't have the time, patience, or enough paper to do it manually. There are so many possible combinations that it might even be unfeasible for your computer to find the best possible solution. Therefore, in these large-scale problems, it's common to use **approximate methods**, such as genetic algorithms and other metaheuristics. These techniques allow finding solutions that are good enough, even if they aren't necessarily the best possible. Practically speaking, sometimes an approximate method can find a very good solution in 30 minutes. On the other hand, an exact method would find the best possible solution in a few *days*, even if this best possible solution is only marginally better than the other. Can you see the cost-benefit in these approximate methods?

1.3.4 Uncertainty

Another way to categorize problems is by the uncertainty involved in them. One of the most interesting (or most frustrating, depending on your point of view) things in real life is precisely the presence of uncertainty, in its different levels and degrees. That is, you can be *almost* certain that tomorrow will be colder than today, based on weather forecasts and climate patterns already calculated by your phone's weather app. However, this degree of uncertainty is significantly lower than, for example, trying to predict the stock prices of major tech companies next month. While weather follows more predictable trends, the financial market is influenced by a multitude of unpredictable factors, such as political decisions, technological changes, or even unexpected global events.

In an industrial environment, like a factory, we also see this difference in the levels of (un)certainty. The factory might have reasonably precise control over its own production capacity, knowing how many units of a product it can manufacture in a given period. On the other hand, it faces much more uncertainty when it comes to market demand. And, in turn, the factory might know more about market demands than about its competitors' actions in the coming months, or the problems its suppliers are facing. These layers of uncertainty are an intrinsic part of decision-making in real environments, and that's precisely what makes the problems so interesting for us.

Thus, there are problems where we know in advance the behavior of all variables involved, and all elements are known and controlled. In these cases, there's greater predictability, and we can model the problem with more precision. Classic linear programming problems are like this. If all variables (such as raw material costs, machine capacity, and demand) are known and fixed, we are dealing with a **deterministic problem**. And, as you can imagine, these problems are frequently addressed in the classroom because they allow for clear analysis, without the additional complexity introduced by uncertainty.

On the other hand, there are cases where uncertainties or variations are present, and the idea isn't necessarily to hit the exact value, but rather to have a general notion of the values produced.[7] Let's go back to the donut production example: variations in oven temperature, environmental humidity, or even the preparation method can lead to a produced quantity that is different from what was originally expected. Raw material might end up yielding more or fewer of our donuts, depending on factors that aren't entirely under control. Or, in the case of a factory, we might have a certain degree of tolerance in the products it produces, such as small variations in the size or weight of an item. These cases are typical of a **stochastic problem**, where uncertainty is a fundamental part of the problem.

In stochastic problems, mathematical modeling needs to incorporate this uncertainty, often using probability and statistics concepts. Let's consider an example: instead of assuming that the demand for a product will be exactly 100 units per day,

[7] Agroindustry is a traditional case of this. After all, not all plants and animals have the same weight and size, and not all batches have the same yield.

we could model it as a random variable with a probability distribution, such as a normal distribution with a mean of 100 and a standard deviation of 10. This allows the problem's solution to take variability into account and provide results that are robust even in the face of uncertainties.

We have different tools and methods for solving deterministic and stochastic problems—so, knowing how to differentiate whether a problem has uncertainties or not by analyzing its context is very important. While deterministic problems allow for exact solutions, stochastic problems require more sophisticated approaches, such as simulations, Monte Carlo methods, or robust optimization. These techniques help deal with the unpredictability of the real world, providing solutions that, although not perfect, are good enough to support decision-making in complex and dynamic environments.

The pandemic example

The clearest example of uncertainty was the COVID-19 pandemic. During the pandemic, one of us worked in the agroindustry. For administrative teams, managing the pandemic was relatively easy considering the circumstances—after all, it was just a matter of sending employees to operate remotely with their computers. However, the same cannot be said for factories that couldn't operate remotely. After all, workers could contract the virus and be absent from work for a few weeks due to quarantine. Depending on the time and waves of contagion, entire areas could be impacted by this. And, as you can imagine, this affects the availability of labor within factories and outside as well: after all, people also operate vehicles transporting raw materials, people operate distribution centers for industrialized products, and people buy food in supermarkets and restaurants, which also had their operations limited. Therefore, it was impossible to predict what might happen in the coming days or weeks.

This chain of uncertainties generated complex optimization problems for all companies in the sector. Planning production in scenarios where the availability of inputs, workers, and transportation varied constantly required adapting traditional optimization models. Therefore, it was necessary to adapt optimization problems and adopt techniques that incorporated uncertainties and risks, allowing for more robust decisions in this scenario of constant unpredictability.

1.3.4.1 Robustness

We think a special subsection to talk about robustness is worthwhile. We just told you that we could divide optimization problems regarding the uncertainty of conditions in two ways: deterministic and stochastic problems. In reality, there's a third type: **robust optimization** problems.

Both robust optimization and stochastic optimization deal with uncertainties, but in different ways. Perhaps the keyword to determine one type or the other is **probability**: if you take probability into account and consider the *average* of possible scenarios, then you likely have a stochastic problem on your hands. Otherwise, if you're working with the *worst* possible scenario, you might have a robust problem.

Let's think of some examples to clarify these differences. First suppose you're working for a water supply company for a large city. Specifically, you'd be working on an optimization problem to minimize the risk of water shortage. For this problem, we'd have two alternatives:

- Thinking realistically, you might conclude that there have only been two periods of extreme drought in the last century. So, you design your supply system considering the *average* water demand and storage from recent years. Your system will work well (cool), save the company a good amount of money (very cool), and maybe you'll even get a promotion (awesome). On the other hand, if there's a drought in the coming years, the city will run out of water (not cool at all).
  In this scenario, we'd have a **stochastic optimization** problem. We thought about the problem according to the city's recent behavior, and not the worst-case scenario.
- Thinking pessimistically, you know that climate change will mean more frequent and longer droughts. Therefore, you plan your supply system for a drought period so long it has never happened in the city's history, but it's not impossible. Your supervisor might question the high costs of your system (not cool), and there might be more bureaucracy to approve your project (absolutely not nice), but your idea could save many people in your city in cases of public calamity (very cool).
  In this case, we'd have a **robust optimization** problem. We prepared for the worst-case scenario.

Let's consider a second example. Data scientists frequently work on creating predictive models using ML. So, let's imagine a scenario where a data scientist is tasked with training an ML model using a neural network architecture to predict fraud on an e-commerce website. As you know from common sense, fraud doesn't happen often, but it causes losses. Meaning, even if 0.5% of all transactions are fraudulent (and not a number like 50.0%), this would still represent a good amount of money lost by the company. Therefore, it's in the company's interest to detect this in advance.

Now, training an ML algorithm involves an optimization problem where we try to minimize a cost function. There are several different error metrics, and it's usually the data scientist who defines this. Well then:

- We could consider that the data scientist is training the model to have good performance, overall. That is: to work very well and in a balanced way for both fraudulent and legitimate transactions. Since it would be a balanced model, the algorithm might make some errors here and there, but still, *overall*, it would have acceptable performance.
  This would be a case of **stochastic optimization**. We try to combine metrics and techniques (like oversampling and undersampling, or combining different ML models) to arrive at a good result. Notice that the phrasing is very much directed toward having something that works well, *in general*.

- On the other hand, we could consider a much greater focus on fraudulent transactions. We could think of techniques to penalize errors and that work well even in cases of uncertainty.
  This would be a case of **robust optimization**. We could use techniques like adversarial training and regularization. With this, very similar transactions will have a higher chance of having the same prediction. Notice that the focus of these phrases is almost always on *working robustly even in cases of uncertainty*.

1.3.5 Linearity (or Lack Thereof)

Another way to think about the optimization problem is through linearity. However, it might seem overly simplistic to just say that problems can be divided into **linear optimization** and **non-linear optimization**, as this distinction is rather apparent.

Linear optimization problems (also known as **linear programming**, or LP) are widely studied and have an enormous amount of available resources. To give you an idea, there are dozens of excellent books dedicated exclusively to linear programming, in addition to courses, specialized software, and a vast range of practical applications. The popularity of LP is no accident: it's one of the most versatile tools in optimization, capable of solving problems in areas like logistics, production, economics, and even healthcare.

For example, our donut shop problem can be modeled as an LP problem. Let's suppose you need to decide how many donuts of each type to produce to maximize profit, considering constraints like raw material availability, production capacity, and market demand. This could be modeled as a classic LP problem, where the objective function (maximize profit) and constraints (resource limits) are expressed by linear equations. Another example would be a logistics distribution problem, where the objective is to minimize transportation costs while meeting the demand of several customers. Or, even, a problem of allocating people and equipment in a factory, where you need to efficiently distribute limited resources. Even problems in Economics, such as maximizing a company's profit or allocating investments, can be adapted as linear programming problems.

Furthermore, LP is an excellent starting point for anyone beginning to learn optimization, as linear relationships between variables are easier to interpret and visualize. For example, in an LP problem, the objective function is usually a simple sum of variables, weighted by coefficients representing costs, profits, or other factors. Constraints are also expressed clearly and directly, as linear inequalities that limit the value of variables. This simplicity allows even someone without much familiarity with Algebra to understand and formulate LP problems.

On the other hand, we would have a non-linear optimization problem when the relationships between variables are not directly proportional. Let's return to the donut case: it would be an LP problem if the objective function was simply the sum of the different types of donuts to be produced, and if the constraints depended on linear factors, such as raw material capacity or market demand as a static value. However, what if the problem wasn't that simple? What if the relationships between variables were more complex and couldn't be expressed by linear equations?

It's in cases like these that we work with non-linear optimization. Let's look, for example, at a problem involving economies of scale, a common concept in business and production. Economies of scale occur when costs decrease as the scale of production increases. For example, imagine that to make a single chocolate donut, you spend $0.40. However, if you decide to produce 100 chocolate donuts, you can negotiate discounts with your suppliers and reduce the unit cost to $0.28. In this case, the cost is no longer a linear function of the quantity produced: it varies non-linearly, depending on the production volume. This transforms the problem into a non-linear optimization case, where the objective function and/or constraints involve quadratic, exponential, or other more complex mathematical forms.

Another example of a non-linear problem would be optimizing a production function that depends on multiple interrelated factors, such as machine efficiency, raw material quality, and worker experience. These factors can interact in complex ways, creating relationships that cannot be expressed by linear equations. To solve these problems, non-linear methods, such as Newton's method, genetic algorithms, or gradient descent techniques, are useful. These methods allow dealing with the complexity and nuances that arise when variables don't behave proportionally or predictably.

1.3.6 Combining Everything

You may have seen that we've given a few different ways to think about an optimization problem. That said, we can combine them for each problem. For example, we could have problems that are

- Single objective, with continuous variables, small-scale, deterministic, and linear.
- Single objective, with mixed variables, small-scale, deterministic, and non-linear.
- Multi-objective, with continuous variables, large-scale, deterministic, and linear.
- Multi-objective, with discrete variables, small-scale, stochastic, and non-linear.
- Multi-objective, with mixed variables, large-scale, stochastic, and non-linear.

Therefore, consider the table below. In it, we've placed all the most common combinations[8] considering the division we made above (i.e., number of objectives, variables, scale, uncertainty, and linearity) (Table 1.3). Notice that some combinations are more rare or specialized, such as:

- Stochastic linear problems. Stochastic Linear Programming (SLP) is a well-established field within optimization, and used in areas like finance, logistics, and

[8] Notice that this is somewhat subjective—you might have worked only with some combination that is *not* in the table. And that's okay: after all, the table aims to show you only the most common situations. Don't interpret it as dogma, but rather as a sample of combinations you might encounter, some rarer than others.

Table 1.3 Some common combinations of characteristics of optimization problems

Objective	Variables	Scale	Nature	Linearity
Single objective	Continuous	Small	Deterministic	Linear, Non-linear
Single objective	Continuous	Small	Stochastic	Non-linear
Single objective	Continuous	Large	Deterministic	Linear, Non-linear
Single objective	Continuous	Large	Stochastic	Non-linear
Single objective	Mixed	Small	Deterministic	Linear, Non-linear
Single objective	Mixed	Small	Stochastic	Non-linear
Single objective	Mixed	Large	Deterministic	Linear, Non-linear
Single objective	Mixed	Large	Stochastic	Non-linear
Single objective	Discrete	Small	Deterministic	Linear, Non-linear
Single objective	Discrete	Small	Stochastic	Non-linear
Single objective	Discrete	Large	Deterministic	Linear, Non-linear
Single objective	Discrete	Large	Stochastic	Non-linear
Multi-objective	Continuous	Large	Deterministic	Non-linear
Multi-objective	Continuous	Large	Stochastic	Non-linear
Multi-objective	Mixed	Large	Deterministic	Non-linear
Multi-objective	Mixed	Large	Stochastic	Non-linear
Multi-objective	Discrete	Large	Deterministic	Non-linear
Multi-objective	Discrete	Large	Stochastic	Non-linear

energy systems. Many real-world systems involving uncertainty can be effectively modeled with linear objectives and constraints, where only certain parameters are stochastic.

- Small-scale multi-objective problems. Outside the classroom setting, it's normal to have situations where multi-objective problems are complex enough to require many variables and constraints. However, we can still use them in component-level design, preliminary studies, and specialized research areas to understand fundamental trade-offs with fewer variables.
- Large-scale linear multi-objective problems. The greater the complexity, the higher the chances of having non-linearity. And, in this case, we have two factors contributing to this complexity: the large scale and the fact that we have a MOO problem. However, we can still find them in specialized areas such as public policy modeling and network planning.

1.4 Engineering Applications of Optimization

Of course, the example with a donut shop helps to understand the basic concepts of optimization, but it might still be somewhat distant from reality. After all, we understand you're not reading this book thinking only about working with donuts, right? If your interest is Engineering, we'll provide some examples of optimization use in the three major areas: Electrical, Mechanical, and Civil. Of course, within these areas, we also have other Engineering fields such as Electronics Engineering, Aerospace Engineering, Computer Engineering, or Production Engineering.

1.4.1 Mechanical Engineering

Mechanical Engineering is a field full of problems that can be addressed with optimization techniques. A classic example is the design of heat exchangers, an equipment that allows thermal energy transfer between fluids without them coming into direct contact. Refrigeration systems, for example, depend on these devices. Although it's tempting to choose the largest and most expensive heat exchanger available to avoid cooling headaches, this decision might not be the most efficient or economical. Therefore, optimization algorithms are used to design cooling systems that maximize heat transfer while minimizing energy consumption and operational costs.

Furthermore, aerodynamic optimization is important when designing vehicles. After all, aerodynamics influences factors like fuel consumption and vehicle stability. And, although the shape of a teardrop is the most aerodynamic possible, other factors also need to be considered during design, such as safety, engine cooling, and weight distribution. Optimization algorithms help find the ideal balance between these requirements.

1.4.2 Electrical, Electronics, and Computer Engineering

Electrical Engineering also benefits enormously from optimization techniques. A relevant example is economic dispatch, which aims to minimize operating cost by distributing the energy load among different power plants, respecting generation and demand constraints. This problem is fundamental to ensure the electrical grid operates without service interruptions or overloads at any point, especially in systems with multiple energy sources, such as thermal and hydroelectric plants.

Another classic problem is optimal power flow, which seeks to optimize energy flow in electrical grids. The objective can be to minimize costs, energy losses, or maximize grid stability. This type of optimization is essential to ensure that the electrical infrastructure supports growing demands and operates reliably.

Another case is antenna optimization, where the objective is to design antennas with the best possible performance in terms of gain, directionality, and bandwidth. Here, we optimize parameters like the antenna's shape and size, ensuring it meets technical specifications. We also use optimization for problems in embedded systems, such as those found in smartphones, autonomous cars, and IoT devices. In this case, the challenge is to maximize system performance while minimizing energy consumption and cost. Optimization techniques are used to adjust hardware and software parameters, like clock frequency, operating voltage, and memory allocation.

Furthermore, on a smaller scale, integrated circuit optimization is also an important application. In this case, the objective is to minimize energy consumption, the size of the integrated circuit, processing time, and its errors, while maximizing its efficiency. It's also common to use Optimization to position components on a chip to reduce connection lengths, improve performance, and minimize overheating points.

1.4.3 Civil Engineering

In Civil Engineering, the use of optimization is no different. For example, when designing a bridge or a building, the objective is to minimize the weight or cost of materials, while respecting technical standards and safety constraints. And, by using optimization algorithms, we can find the ideal balance between these factors, resulting in structures that are both safe and economically viable.

It's also common to use optimization algorithms in water resource management projects, such as water distribution in urban networks or the operation of dams and reservoirs. In these cases, the objective might be to maximize system efficiency, minimize losses, or ensure continuous supply, even in situations of scarcity. We also apply optimization in drainage and stormwater management systems in urban projects. In cities, soil sealing and high construction density can lead to problems like floods and waterlogging. Therefore, optimization algorithms are used to design drainage networks that minimize these risks, considering factors like flow capacity, installation cost, and environmental impact. Thus, these algorithms help determine the ideal size and location of manholes, galleries, and retention reservoirs, for example.

Similarly, optimization is widely applied in the design of sustainable structures. With growing concern about the environmental impact of civil construction, engineers seek solutions that minimize the use of natural resources and reduce carbon emissions. For example, optimization algorithms can be used to determine the ideal combination of materials, such as concrete, steel, and wood, that meets strength and durability requirements while minimizing environmental cost. In practice, this involves selecting recycled or low-impact materials, like concrete with fly ash or steel produced with renewable energy.

1.4.4 Industrial and Production Engineering

Production Engineering is one of the areas that benefits most from optimization, especially in production planning and inventory management problems. Consider this: one of the classic problems in the field is job shop scheduling. In a factory, various machines and production lines need to be organized to maximize efficiency. The objective is to find the best possible combination of machines and tasks, considering constraints like production time, resource availability, and order priorities. This type of problem is common in automotive, aeronautical, and general manufacturing industries.

Another important application is vehicle routing problems (VRPs). In logistics teams, it's common to face the challenge of distributing a fleet of vehicles among various destinations, ensuring all deliveries are made within a predetermined period. This type of problem can also be applied to public transportation planning or traffic engineering, where the objective is to maximize vehicle flow and minimize congestion in large cities.

We also use optimization in other problems of designing and operating industrial processes, such as oil refineries, chemical plants, and water treatment systems. A classic example is its application in separation processes, like distillation and extraction, where we need to maximize product purity while minimizing energy consumption and operational costs. Or, in the optimization of chemical reactors, where we need to maximize the conversion of reactants into products, considering constraints like temperature, pressure, and reaction time. Here, we optimize parameters such as the feed rate of reactants and temperature control.

1.5 Data Science Applications of Optimization

If it was easy to think of optimization applications for Engineering, could we perform the same exercise for the Data Science field? The answer is yes, although Computer Science courses, technical books, or study programs involving AI rarely dedicate significant time to discussing optimization in depth. It's common to find terms like "optimize," "maximize," or "minimize" in materials about ML, but there's hardly the same level of detail seen in Operations Research, Optimization, or Control Systems disciplines. In practice, optimization is often treated as an automated step, delegated to some algorithm, without much explanation of its fundamentals or methods. It ends up being seen more as a parameter or configuration than as a field of study in itself.

Thus, it's much more common to see data scientists testing ML techniques and libraries, like random forest, LightGBM, XGBoost, and neural networks, without explicitly discussing the role of optimization in these models. It's normal for the conversation to revolve around feature selection to improve model performance or the analysis of metrics to evaluate its effectiveness. However, there's rarely talk about how optimization is intrinsically linked to the functioning of these algorithms. Thus, we run the risk of just accepting that the model "learns" on its own, using some "magic" embedded in the libraries, and the discussion ends there. In Data Science, we find professionals with a deep understanding of the mathematical foundations of models as well as those who limit themselves to applying them. Regardless of the case, Optimization often remains in the background, even though it's essential for the success of these techniques.

Therefore, it might be useful to highlight some basic cases where optimization plays a central role in Data Science. The first example is regularization, a technique that allows controlling model complexity and reducing the risk of overfitting. Algorithms like neural networks, random forest, support vector machines (SVM), and linear and logistic regression methods use regularization. Remember the objective function we discussed earlier? Well, these models have an objective function called a "loss function," which measures how well the model is performing by comparing its predictions with the actual values. Regularization adds a term to this function, penalizing overly complex models and encouraging simpler, more generalizable solutions. This process is, in essence, an optimization problem, where the objective is to minimize the loss function while controlling model complexity.

Another popular application of optimization concepts is **hyperparameter optimization**. Each ML technique has several hyperparameters that need to be adjusted to obtain the best performance. For example, a simple neural network has hyperparameters like the learning rate, the number of layers, the number of neurons per layer, the type of activation function, and the batch size of data used in each training iteration. Although there are default values for these hyperparameters, exploring other combinations can lead to significantly better results. However, since there are many hyperparameters and each can take a wide range of values, the number of possible combinations is enormous. This is clearly an optimization problem: we want to find the best combination of hyperparameters (variables) considering the constraints (possible values for each variable) to minimize the ML model's error.

And, to solve this type of problem, we naturally use optimization techniques. Tools like Optuna, Hyperopt, Keras Tuner, and Ray Tune allow exploring the hyperparameter space using methods like random search, grid search, or more sophisticated algorithms, such as Bayesian optimization. In some cases, the process of testing and evaluating different combinations is performed implicitly,[9] and data scientists don't necessarily know or have control over the optimization methods in these cases.

Furthermore, optimization is also present in techniques like principal component analysis (PCA), where the objective is to reduce data dimensionality while preserving as much information as possible. PCA solves an optimization problem to find the directions (components) that maximize data variance. Similarly, in clustering algorithms, like K-means, optimization is used to minimize the distance between data points and their respective cluster centroids.

[9] This is especially common in automated ML (AutoML) market tools, where data scientists provide the dataset, the problem to be solved, and some additional configurations like the model evaluation metric.

1.6 Conclusion

We believe that learning is different from getting a high grade on an exam or in a company performance review. That is, someone who memorizes contexts and gets top marks in everything at a University isn't necessarily fully equipped to work in a certain field. That's why we believe it's important for you to know the "heart of the problem": that is, to know the general idea behind it, so we can advance a bit further in explanations and concepts.

That's why in this chapter, we preferred to steer clear of mathematical definitions for a bit because it's preferable to assume that not everyone has the same theoretical background (or the same mood) to talk about it at this initial moment. Sometimes, you're a person who learns by application, or you're simply too tired to talk about more complex equations right now. Therefore, for us, it's critical that you first understand the utility of optimization and why it's relevant so that, in the upcoming chapters, we can deep dive into mathematical formalizations.

As a suggestion, we'd like you to be able to *design a system* at this point: just as we gave an example of the donut shop, why don't you try taking a piece of paper and thinking about another optimization problem you're familiar with? If you can think in terms of objective function, variables, and constraints based on what we've discussed in the last few pages, we'll be very happy to take another step forward in the next chapter.

Chapter 2
Essential Mathematics for Optimization

Abstract This chapter moves from the intuitive ideas of Optimization presented in the first chapter to the formal mathematical language required for its detailed study and application. Therefore, this chapter provides the fundamentals required to work with Optimization problems in academic and professional environments. Here, we systematically define the main building blocks of optimization problems, including variables (scalar and vector, covering continuous, discrete, and mixed types), equality and inequality constraints, and functions, detailing their properties such as linearity, non-linearity, and convexity. We also cover the difference between Single-Objective and Multi-Objective Optimization and discuss the formal definition of solutions, including concepts like the feasible region, feasible and infeasible points, local and global optima, Pareto dominance, and the Pareto front.

If you're reading this chapter, we assume you already have a basic understanding of what Optimization is, what it's for, and how it generally works. That is, we expect you to at least know the difference between an objective function, a variable, or a constraint, as we explained in Chap. 1. If you skipped directly here and these terms seem strange to you, we suggest you go back to that chapter.

Indeed, up to this point, we've been somewhat concerned with avoiding mathematical concepts to explain a bit about the relevance and applicability of Optimization. In Chap. 1, we started by identifying a real-life problem to be solved that needs to be minimized or maximized. From there, we mapped the variables and constraints to have a mathematical formalization of a real-life problem. It's upon this formalization that we execute Optimization algorithms to arrive at the best result.

We can only go so far with analogies. Therefore, in this chapter, we will try to present and explain the mathematical definitions of the concepts. It will be from these definitions that we can understand how Optimization algorithms work in practice and, also, in which cases one algorithm might work better than another. Shall we begin?

W. R. Monteiro and G. R. Meza, *A Practical Guide to Optimization in Engineering and Data Science*, https://doi.org/10.1007/978-3-032-04633-8_2

2.1 Variables

In Sect. 1.3.2, we mentioned that we have some types of variables, such as scalar, discrete, and mixed. Let's understand this a bit better.

Let's start with a "single" (scalar) variable. It refers to an individual numerical value denoted simply by x. See the definition in Eq. 2.1.

$$\begin{aligned} x &\in \mathbb{R}, \quad \text{for real numbers as a single continuous variable,} \\ x &\in \mathbb{Z}, \quad \text{for integers as a single discrete variable.} \end{aligned} \tag{2.1}$$

Notice that the value of x can be a continuous value in the domain of $\mathbb{R}$. Most introductory problems in universities are continuous: after all, we have more Optimization algorithms available for these problems, and it's simpler to implement these variables in most software. Here, we could include variables like mass, height, temperature, speed, or volume.

On the other hand, we can also have discrete variables, which are in the domain of $\mathbb{Z}$. Here, we consider variables whose representation needs to be in the domain of $\mathbb{Z}$ for practical reasons,[1] binary variables (i.e., where $x \in \{0, 1\}$), or even categorical variables (i.e., where $x \in \{0, 1, 2, \ldots, n\}$).

Important
In Data Science, it's common to work with *category encoding*: that is, converting textual values into a numerical value. In this case, *no/yes* or *failed/approved* become the values 0 and 1. Or, perhaps, the categories "low," "medium," and "high" become the codes 0, 1, and 2. The same logic applies here: these cases become variables $x \in \mathbb{Z}$.

Similarly, a group of scalar variables x are treated in an n-dimensional vector called $\mathbf{x}$, as seen in Eq. 2.2.

$$\mathbf{x} = [x_1, x_2, \ldots, x_n], \quad \text{where} \quad x_i \in \mathbb{R} \tag{2.2}$$

In reality, this is the simplest definition of a vector variable, where all variables are continuous. Similarly, Eq. 2.3 shows a vector variable where all its variables are discrete.

$$\mathbf{x} = [x_1, x_2, \ldots, x_n], \quad \text{where} \quad x_i \in \mathbb{Z} \tag{2.3}$$

And, we also have Eq. 2.4. This equation has *mixed* variables: that is—some are discrete, and others are continuous.

[1] For example: we cannot produce 3.14159265 chocolate donuts—we either produce 3, or we produce 4. Or, perhaps, we can allocate five trucks to a factory, but it wouldn't make sense to allocate 5.21 trucks. After all, we either send five or six trucks.

$$\mathbf{x} = [x_1, x_2, \ldots, x_n], \quad \text{where} \quad x_i \in \begin{cases} \mathbb{R}, & \text{if } x_i \text{ is continuous,} \\ \mathbb{Z}, & \text{if } x_i \text{ is discrete.} \end{cases} \tag{2.4}$$

Regardless of the individual characteristics of each variable x_i in a vector, they must always be ordered within our problem. That is: the meaning of x_1 doesn't change within the same problem, and the same applies to the other variables $x_2, \ldots, x_n$. That said, the person developing the problem is usually responsible for defining what each variable means.

Attention
In Sect. 1.1.2 we defined the variables representing the donut flavors as:

$$[plain, coconut, cinnamon, chocolate, hazelnut] \tag{2.5}$$

However, it's much more usual to standardize the variables this way in our problems:

$$[x_1, x_2, x_3, x_4, x_5] \tag{2.6}$$

Of course, we can later convert these names back to something humanly intelligible when presenting to someone who isn't from the field. So, instead of telling the manager "*We should produce 6* x_1 and 3 x_4." We would say "*We should produce 6 plain donuts and 3 chocolate donuts*."

2.2 Functions

Now that we've talked about variables, it's time to talk about functions. Functions are the heart of our Optimization problems. They represent the mathematical relationship between the variables and the objective we want to achieve. That said, the term can cause a bit of confusion depending on your level of familiarity with the topic of your field of study. After all, the term *function* is used in various fields and disciplines.

Therefore, let's take it one step at a time: a mathematical function, in its most basic form, is a relation that associates each element of an input set to exactly one element of an output set. In Optimization, we are typically interested in functions that map a vector of variables **x** to a real value. And this real value can represent a real-life concept like cost, profit, distance, or any other metric we wish to minimize or maximize.

As an example, let's consider the function shown in Eq. 2.7 below.

$$f(\mathbf{x}) = x_1^2 + x_2^2 \tag{2.7}$$

This function takes a vector $\mathbf{x} = [x_1, x_2]$ and returns the sum of the squares of its components (i.e., the two variables x_1 and x_2). If we are trying to minimize this function, we are essentially trying to find the values of x_1 and x_2 that result in the smallest possible value of $f(\mathbf{x})$. Functions like this are common in Optimization problems with continuous variables, where the variables can take any value within a real interval. However, in problems with discrete variables, functions can take different forms.

Let's look, for example, at a resource allocation problem in Eq. 2.8. In this case, we might have a function that counts the number of allocated resources, which is a discrete function.

$$f(\mathbf{x}) = \sum_{i=1}^{n} x_i, \quad \text{where} \quad x_i \in \{0, 1\} \tag{2.8}$$

Notice that, in this case, the function $f(\mathbf{x})$ sums the values of x_i, which are binary.[2] In fact, it's common to have this type of function with binary variables in combinatorial Optimization problems.

2.2.1 Linear Functions

Additionally, functions can be linear or non-linear. We talked a bit about linearity in Sect. 1.3.5, although at that time we hadn't explored the mathematical part. Linear functions are those that can be expressed as a linear combination of the variables, as you can see in Eq. 2.9 below.

$$f(\mathbf{x}) = a_1x_1 + a_2x_2 + \cdots + a_nx_n \tag{2.9}$$

Notice that it's a linear combination of the variables $x_1, x_2, \ldots, x_n$, where each variable x_i is multiplied by a numerical coefficient a_i. Thus, the result will be a value that depends on the values of the variables and the coefficients.

Some instant noodles to help you

As you may have noticed in previous sections, we like analogies to explain some concepts to you. So, to explain linearity, we'd like to bring another real-life example. For instance, let's say you're short on cash and need to make some instant noodles for dinner, but you'd like to supplement it with some other ingredients. And, in this case, let's assume you like eggs, carrots, shredded chicken, and cheese. The ideal would be to mix these ingredients so it doesn't

[2] That is, mathematically represented by 0 and 1. As we commented in this same chapter, these variables represent binary decisions, like no/yes or off/on.

become too bland or nutritionally poor. In this scenario, imagine each variable x_i in Eq. 2.9 represents an ingredient. Something like:

- x_1: number of eggs
- x_2: a 100g package of shredded carrots
- x_3: a 100g package of shredded chicken
- x_4: a 100g package of cheese

In this case, each coefficient a_i would be the price of each product. Following the example:

- a_1: \$0.20/egg
- a_2: \$0.30/100g of shredded carrots
- a_3: \$0.90/100g of shredded chicken
- a_4: \$1.00/100g of cheese

So, in practice, the function would be represented by $f(\mathbf{x}) = 0.2x_1 + 0.3x_2 + 0.9x_3 + 1.0x_4$.

2.2.2 *Non-linear Functions*

If linear functions are relatively simple and involve only linear combinations of variables (i.e., sums and multiplications by constants), non-linear functions are more complex and can include other mathematical operations, such as products between variables, exponentials, logarithms, trigonometric functions, or any other relationship that isn't a linear combination. Seems simple, doesn't it?

For example, let's consider Eq. 2.10. Here, we have some types of non-linearity with variables having powers, square roots, rational functions, exponentials, logarithms, trigonometric functions, products between variables, and fractions with polynomials.

$$f(\mathbf{x}) = a_1x_1^2 + x_2^3 \qquad f(\mathbf{x}) = \sqrt{x_1} + x_2 \qquad f(\mathbf{x}) = \frac{1}{x_1 + a_2x_2} \tag{2.10a}$$

$$f(\mathbf{x}) = e^{x_1} + x_2 \qquad f(\mathbf{x}) = \log(x_1) - x_2^4 \qquad f(\mathbf{x}) = \sin(x_1 + x_2) \tag{2.10b}$$

$$f(\mathbf{x}) = a_1x_1 \cdot a_2x_2 \qquad f(\mathbf{x}) = \frac{x_1}{1 + x_2^2} \qquad f(\mathbf{x}) = \cos(x_1) \cdot e^{x_2} \tag{2.10c}$$

Tips

If the function has any non-linear operation, it is non-linear. That is: if there are products between variables or if any of the variables has an exponent other than 1, or if it has logarithms, trigonometric functions, or any non-linear mathematical functions, the entire function will be non-linear.

And, of course: the idea isn't for you to "force" your function to be either linear or non-linear, but rather for you to create a mathematical function that well represents the problem you're trying to depict. For example, we use linear functions in problems where the relationship between variables and the objective is proportional and direct. For instance, in resource allocation or production planning problems, the total cost is a weighted sum of individual costs. On the other hand, we use non-linear functions where the relationship between variables and the objective is more complex, such as in engineering problems where one seeks to minimize potential energy or maximize system efficiency.

Many real-life problems have non-linearities because we need to capture more complex relationships between variables. Deformations of structures, turbulence in fluids, and electrical circuits have non-linearities in their characteristics, for example. Or, when we're training machine learning (ML) models like neural networks, it's common to use non-linear activation functions (such as the sigmoid or ReLU function) to capture complex patterns in the data.

2.2.3 Convexity

Functions can also be divided into convex and non-convex, as this directly influences the complexity and how easy (or not) it would be to find good solutions. A function or set is considered **convex** if it satisfies specific mathematical properties that guarantee the absence of local minima distinct from the global minimum, making optimization problems more tractable.[3]

Thus, a function $f(\mathbf{x})$ is convex if, for any two points x_1 and x_2 and $\lambda \in [0, 1]$, the following inequality condition is satisfied:

$$f(\lambda x_1 + (1 - \lambda)x_2) \leq \lambda f(x_1) + (1 - \lambda) f(x_2) \tag{2.11}$$

---

[3] We'll explain in more detail what a "local minimum" versus a "global minimum" is in Sect. 2.3.2. Until then, allow us to use another analogy. Let's say you've always liked Margherita pizza because it was the best they made at the pizzeria near your house, and it was the best pizza *you knew*. However, you've never tried a mushroom pizza because they didn't offer that flavor near your house. Even though the Margherita is really good and the best they make near your house, it wouldn't be the best in the world—the mushroom one would be in this hypothetical example. In this case, the "local minimum" would be the Margherita: good, but not *the best*. The "global minimum" would be the mushroom pizza.

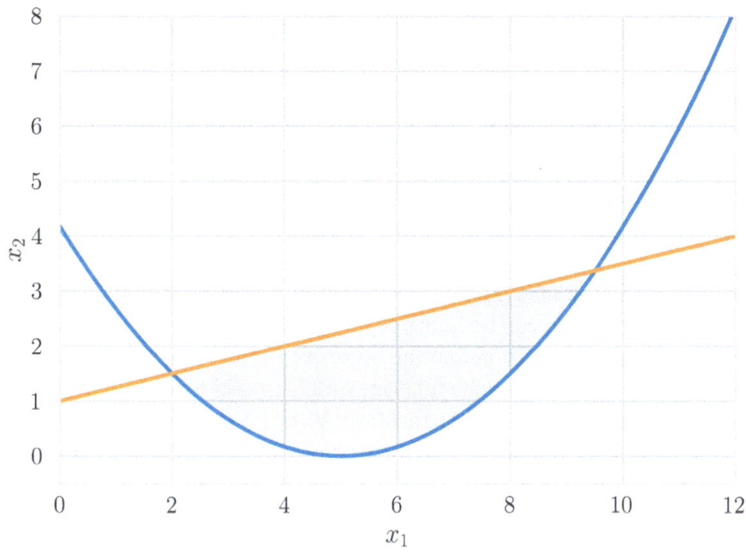

Fig. 2.1 Example of convexity of two variables. The feasible region is shaded. *Source* The author, 2025

This property implies that the line segment connecting any two points on the graph of $f(\mathbf{x})$ lies above or on the graph itself.

Now, we could make a small change to this equation. We would have a *strictly* convex function if we have the following strict inequality for $x_1 \neq x_2$ and $\lambda \in (0, 1)$:

$$f(\lambda x_1 + (1-\lambda)x_2) < \lambda f(x_1) + (1-\lambda) f(x_2) \tag{2.12}$$

Now, we don't expect everyone to read the equations above and react with something like "*Oh, it's obvious that's it! I got it right away!*" because different people understand the same content in different ways. Therefore, allow us to provide an additional explanation.

First, let's look at Fig. 2.1. It illustrates convexity in an optimization problem with two variables (x_1 and x_2). This problem also has two constraints. The parabola represents one of the constraints, while the straight line represents the other.

The shaded region between these curves defines the feasible set of solutions.[4] Since this region satisfies the definition of a convex set (any line segment between two points within it remains entirely contained in the set), it shows how convexity ensures well-behaved problems.

[4] We'll see more about this in Sect. 2.3.1. Until then, imagine that we could only have acceptable solutions for an optimization problem if the values of x_1 and x_2 are in this area.

2.2.4 Objective

Now that we have our variables (and we know what type of variables we can have) and have organized them within a function (and we know what a function is in our context), we also need to explicitly state what the objective is.

The objective function is a specific type of function we use in Optimization problems. It represents the goal we are trying to achieve—be it minimization or maximization.

The objective is the core of any Optimization problem. It's represented by the objective function, which can be linear or non-linear, and defines what we're trying to maximize or minimize. To define the objective correctly, it's essential to understand the real problem, identify the relevant variables, and formulate a mathematical function that reflects what we want to achieve. With a well-defined objective, we can apply Optimization techniques to find the best possible solution. To make it clearer, see the equations below, where:

- Equation 2.13a is a function that can describe any mathematical relationship, and doesn't necessarily describe an Optimization problem.
- Equations 2.13b and 2.13c are objective functions. In them, we are explicitly indicating that we want to **minimize**. For this, we use the notation min (Eq. 2.13b) or $minimize$ (Eq. 2.13c).
- Similarly, Eqs. 2.13d and 2.13e are maximization objective functions. Notice that we explicitly use the notation max (Eq. 2.13d) or $maximize$ (Eq. 2.13e).

$$f(\mathbf{x}) = a_1x_1 + a_2x_2 + \cdots + a_nx_n \quad (2.13a)$$

$$\min f(\mathbf{x}) = a_1x_1 + a_2x_2 + \cdots + a_nx_n \quad (2.13b)$$

$$\text{minimize } f(\mathbf{x}) = a_1x_1 + a_2x_2 + \cdots + a_nx_n \quad (2.13c)$$

$$\max f(\mathbf{x}) = a_1x_1 + a_2x_2 + \cdots + a_nx_n \quad (2.13d)$$

$$\text{maximize } f(\mathbf{x}) = a_1x_1 + a_2x_2 + \cdots + a_nx_n \quad (2.13e)$$

Tips

Depending on the professor or context, you won't necessarily see the notations $min, minimize, max$, or $maximize$ in the equation. However, this information might be described in the text, and it's good practice to make it explicit to avoid misunderstandings.

2.2.5 Multi-objective Functions

In Sect. 1.3.1, we discussed the existence of problems with multiple objectives. Indeed, some real-world situations have multiple objectives that can conflict with each other, and we'll need to find a balance between them. In Sect. 1.3.1, we gave the example of buying a cellphone, but we could also think of a product design problem, where we need to maximize product durability while simultaneously minimizing production cost. If you've ever bought very cheap socks or t-shirts, you might have noticed that cost and durability are conflicting.

Or, perhaps, an ML algorithm that we need to have the lowest possible error but, at the same time, be as fast as possible. Now, ML algorithms that are fast usually have a simpler architecture which, in turn, translates into a higher error. Therefore, these two objectives (error and speed) can also be conflicting.

Thus, in cases like this, we use multi-objective functions. Look at Eq. 2.14. In it, we represent a vector of functions $\mathbf{f}(\mathbf{x})$. Notice that it's composed of several functions—in this case, $f_1(\mathbf{x}), f_2(\mathbf{x}), \ldots, f_k(\mathbf{x})$ represent k objectives. Also, observe that all these objectives share the same vector of decision variables $\mathbf{x}$.

$$\mathbf{f}(\mathbf{x}) = \begin{bmatrix} f_1(\mathbf{x}) \\ f_2(\mathbf{x}) \\ \vdots \\ f_k(\mathbf{x}) \end{bmatrix} \tag{2.14}$$

However, let's make an adjustment—after all, we just mentioned in Sect. 2.2.4 that there's a difference between a function and an objective function, right? So, a multi-objective minimization function could be described as illustrated by Eq. 2.15.

$$\min \mathbf{f}(\mathbf{x}) = \begin{bmatrix} f_1(\mathbf{x}) \\ f_2(\mathbf{x}) \\ \vdots \\ f_k(\mathbf{x}) \end{bmatrix} \tag{2.15}$$

Or, if it were a multi-objective maximization function, we could describe it as illustrated by Eq. 2.16.

$$\max \mathbf{f}(\mathbf{x}) = \begin{bmatrix} f_1(\mathbf{x}) \\ f_2(\mathbf{x}) \\ \vdots \\ f_k(\mathbf{x}) \end{bmatrix} \tag{2.16}$$

Fig. 2.2 Let's return to the idea of comparing cellphones, but now as if we were responsible within a company for producing new models. *Source*: Unsplash

2.2.5.1 Returning to the Cellphone Example

Do you remember we exemplified what a Multi-objective Optimization (MOO) problem would look like with a hypothetical scenario where you were trying to buy a cellphone, right?

And, if you remember correctly, we mentioned in Sect. 1.3.1 that in this case, we would have two main objectives:

- Minimize the cellphone's price (i.e., spend as little as possible).
- Maximize the cellphone's performance (i.e., have the best possible performance).

These two objectives are conflicting, as cellphones with better performance tend to be more expensive. Now, we'd like you to take a moment to think with us about how we could model this problem (Fig. 2.2). First, let's think about the variables. Any industry responsible for manufacturing cellphones and its commercial departments don't arrive at a selling price randomly—there are several factors (or, following the Optimization nomenclature, *variables*) that are taken into consideration. So, let's assume that only the variables shown in Eq. 2.17 influence those two objectives. Of course, in real life, there are many more variables, but our intention is to simplify the example.

$$\mathbf{x} = \begin{bmatrix} x_1 = \text{Battery life, in hours} \\ x_2 = \text{Processor benchmark score (0-100)} \\ x_3 = \text{RAM, in GB} \\ x_4 = \text{Storage capacity, in GB} \\ x_5 = \text{Battery capacity, in mAh} \\ x_6 = \text{Camera quality (0-10)} \end{bmatrix} \tag{2.17}$$

With the variables in hand, we next need to model the functions that represent the interaction of these variables. As a simplified example, we could say that the two objectives could be described as a linear equation. Thus, we would have Eq. 2.18, where $f_1(\mathbf{x})$ represents the price and $f_2(\mathbf{x})$ represents performance.

$$\mathbf{f}(\mathbf{x}) = \begin{bmatrix} f_1(\mathbf{x}) = a_1x_1 + a_2x_2 + \cdots + a_6x_6 \\ f_2(\mathbf{x}) = b_1x_1 + b_2x_2 + \cdots + b_6x_6 \end{bmatrix} \tag{2.18}$$

Each variable could be multiplied by a coefficient (weight) a_i for $f_1(\mathbf{x})$ and a coefficient b_i for $f_2(\mathbf{x})$. We don't assign a random value to each coefficient. In reality, an expert in the field[5] develops, based on their knowledge, what the weight should be for each of these variables and for each of the objectives. Let's assume that marketing, production, and product design experts concluded that the weights follow what is shown in Eq. 2.19.

$$\mathbf{f}(\mathbf{x}) = \begin{bmatrix} f_1(\mathbf{x}) = 5x_1 + 2x_2 + 10x_3 + 0.500x_4 + 0.010x_5 + 2.000x_6 \\ f_2(\mathbf{x}) = 0.100x_1 + 0.500x_2 + 2x_3 + 0.050x_4 + 0.001x_5 + 0.500x_6 \end{bmatrix} \tag{2.19}$$

Tips

Now, remember that we also need to specify if we're trying to minimize or maximize these objectives, right? We could even do something like min $f_1(\mathbf{x})$ and max $f_2(\mathbf{x})$, but it's common to standardize all objectives to be either all minimization or all maximization. Since Optimization algorithms typically minimize all objectives by default, we could just invert $f_2(\mathbf{x})$ to transform it into a minimization objective. Thus, we would have the notation in Eq. 2.20 as a result.

$$\min \mathbf{f}(\mathbf{x}) = \begin{bmatrix} f_1(\mathbf{x}) \\ -f_2(\mathbf{x}) \end{bmatrix} \tag{2.20}$$

[5] Also called a decision-maker.

2.2.6 A General Notation for Constraints

In Optimization problems, we usually express constraints as a set of inequalities and equalities involving the decision variables.

So, let's start with Eq. 2.21. Here, we have a function called $g_i(\mathbf{x})$ that defines an equality constraint—that is: the function must evaluate **exactly** to the value c_i to be valid. An Optimization problem can have as many equality constraints as necessary. In Eq. 2.21, we would have m equality constraints. Constraints of this form ensure we are strictly following norms and laws, like those in Engineering and Physics.

$$g_i(\mathbf{x}) = c_i \text{ for } i = 1, \dots, m \tag{2.21}$$

Let's use an example of a product distribution problem in a supply chain. In a distribution center, the total quantity of products sent to stores must be exactly equal to the quantity available in the central stock. Here, the equality constraint of Eq. 2.21 ensures there are no inconsistencies in inventory control—neither shortages nor surpluses of goods. Let's see: if the distribution center has a quantity c_i units of a product (for example, 1000 units), the sum of units sent to all stores must be exactly 1000. It wouldn't make sense for stores to receive more than what's available at the distribution center, right?

Similarly, we also have inequality constraints. They are represented by inequalities and are responsible for defining a limit for the variables of the objective function. Consider Eq. 2.22, for example. Here, we have a function called $h_i(\mathbf{x})$ whose result must be less than or equal to the coefficient d_i for the solution to be valid. Here, we would have p inequality constraints.

$$h_i(\mathbf{x}) \leq d_i \text{ for } i = 1, \dots, p \tag{2.22}$$

To better understand the application of inequalities, we'd like to invoke another example from logistics: as you may know, trucks that transport cargo have a maximum load capacity—a maximum weight d_i. In this case, $h_i(\mathbf{x})$ would represent the total weight of the loaded goods, which must not exceed this maximum weight.

We could also modify inequality constraints so that the variable values are obligatorily within an interval. For example, in Eq. 2.23, we establish that the values of variables x must necessarily be between the minimum value l_i and the maximum value u_i.[6]

$$l_i \leq x_i \leq u_i, \forall i \in \{1, 2, \dots, n\} \tag{2.23}$$

Let's go back to the example of the maximum cargo capacity of trucks: if we think carefully, there isn't just a maximum limit, but also a minimum limit, right? Or, to be fairer, there are two minimum limits: the limit of Physics (where the minimum limit is zero, since we can't transport cargo with negative weight), and the limit imposed

[6] The letters l and u stand for *lower bound* and *upper bound*, respectively.

by the company (where the minimum limit is a capacity established by management to avoid losses. After all, no manager would like to see a huge truck crossing the country idly and uneconomically carrying only a 5kg box in its cargo compartment). Here, we take all these factors into account to define a minimum and maximum limit for each variable. Similarly, for discrete variables, we can define that they can only have values defined within a predetermined set of values **S**, as illustrated by Eq. 2.24. As an example, in binary problems, we would have $\mathbf{S} = \{0, 1\}$ since only two values would be allowed in these cases.

$$x_i \in \mathbf{S}, \forall i \in \{1, 2, \ldots, n\} \tag{2.24}$$

It's very common to see this type of condition established by Eq. 2.24 in resource allocation problems. In fact, this is quite common for those of us who work in universities or schools. In a problem of assigning classes to teachers in a school, each teacher can be assigned or not to a particular subject, represented by 1 (assigned) or 0 (not assigned). The constraint $x_i \in \{0, 1\}$ would help ensure that no invalid solution is considered in the planning.

Tips
We can have Optimization problems without any constraints, as well as problems containing any combination of constraints. For example, we can have only inequality constraints, only equality constraints, or a combination of both. Additionally, it's common for the problem to be harder to converge (i.e., find valid solutions and be more complex to calculate) the more constraints we have.

2.2.7 Constraints in the Cellphone Example

That said, how would we adapt the cellphone purchase problem with the inclusion of constraints? Recalling our example, we have six variables representing cellphone characteristics, as shown in Eq. 2.17. Let's now apply the different types of constraints discussed earlier.

First, we can establish inequality constraints according to Eq. 2.22. For example, the total price of the cellphone should not exceed a maximum value that the consumer is willing to pay. If this maximum value were \$1000, we could establish a constraint $h_1(\mathbf{x})$ as:

$$h_1(\mathbf{x}) = 5x_1 + 2x_2 + 10x_3 + 0.500x_4 + 0.010x_5 + 2.000x_6 \leq 1000 \tag{2.25}$$

Table 2.1 Some fictitious cellphone models

Cellphone	x_1	x_2	x_3	x_4	x_5	x_6	f_1	f_2
A	10	20	2	32	3000	5	166	22.1
B	15	40	4	64	4000	7	281	40.2
C	20	60	6	128	5000	8	410	59.4
D	30	90	8	256	6000	10	618	87.8

We can also define limits for each variable, according to Eq. 2.23. As we saw in the examples of cellphones X and Y in Table 2.1, certain characteristics must be within realistic intervals:

$$\begin{aligned}
20 \leq x_1 \leq 72 \quad & \text{(Battery life between 20 and 72 h)} \\
0 \leq x_2 \leq 100 \quad & \text{(Benchmark score between 0 and 100)} \\
2 \leq x_3 \leq 32 \quad & \text{(RAM between 2 and 32 GB)} \\
32 \leq x_4 \leq 2048 \quad & \text{(Storage between 32 and 2048 GB)} \\
3000 \leq x_5 \leq 8000 \quad & \text{(Battery capacity between 3000 and 8000 mAh)} \\
0 \leq x_6 \leq 10 \quad & \text{(Camera quality score between 0 and 10)}
\end{aligned}$$

We can also include equality constraints according to Eq. 2.21. For example, if we're developing a cellphone for a specific line that requires exactly the best possible camera (score 10), we would have:

$$g_1(\mathbf{x}) = x_6 = 10 \tag{2.26}$$

Another equality constraint could be related to the balance between performance and battery consumption. The faster the cellphone, the more battery it consumes. Therefore, let's assume there's a technical relationship that needs to be maintained and is established by the following equation:

$$g_2(\mathbf{x}) = 0.05x_2 + 0.5x_3 - 0.001x_5 = 10 \tag{2.27}$$

Now, let's return to the variable intervals we just established. We usually find specific values for storage capacity and cellphone memory—meaning, it's common to find cellphones with values such as 2, 3, 4, 6, 8, 12, 16, 24, or 32 GB of RAM, but not continuous values like 7.3 GB, 3.9 GB, or 17.4 GB. Therefore, we can define that some discrete variables must respect specific values, as seen in Eq. 2.24. Thus, we would have:

$$x_3 \in \{2, 3, 4, 6, 8, 12, 16, 24, 32, \ldots\} \text{ GB}$$

$$x_4 \in \{32, 64, 128, 256, 512, 1024, 2048\} \text{ GB}$$

Notice that all these constraints define, *collectively*, the space of valid solutions for the optimization problem. Thus, the idea is to have solutions that are optimal (or close to optimal), but also realistic. So, we now have a complete formulation of the optimization problem for the cellphone design in a single structured equation in Eq. 2.28.

$$\min \mathbf{f}(\mathbf{x}) = \begin{bmatrix} f_1(\mathbf{x}) = 5x_1 + 2x_2 + 10x_3 + 0.500x_4 + 0.010x_5 + 2.000x_6 \\ -f_2(\mathbf{x}) = -(0.100x_1 + 0.500x_2 + 2x_3 + 0.050x_4 + 0.001x_5 + 0.500x_6) \end{bmatrix}$$

subject to

$$\begin{aligned}
&5x_1 + 2x_2 + 10x_3 + 0.500x_4 + 0.010x_5 + 2.000x_6 \leq 1000 \\
&0.05x_2 + 0.5x_3 - 0.001x_5 = 10 \\
&20 \leq x_1 \leq 72 \\
&0 \leq x_2 \leq 100 \\
&x_3 \in \{2, 3, 4, 6, 8, 12, 16, 24, 32\} \\
&x_4 \in \{32, 64, 128, 256, 512, 1024, 2048\} \\
&3000 \leq x_5 \leq 8000 \\
&x_6 = 10
\end{aligned} \tag{2.28}$$

Important
Notice the term *subject to* in Eq. 2.28, also sometimes referred to as *s.t.*. This means, in practice, that the solutions found according to our objective function are *subject to* all the constraints established below. That's why it's important to invest some time to review and refine these constraints—after all, they are the difference between generating solutions that are useful and representative of our real problem, or not.

2.2.8 *Thinking About Engines*

Sometimes, just one example still leaves us confused. So, how about we think about another Optimization problem? In this case, consider a problem of manufacturing electric motors, like the motor in Fig. 2.3, with the inclusion of constraints.

For brevity, let's also consider six variables. These variables represent different characteristics of the motor, namely:

Fig. 2.3 An illustrative example of an electric motor. *Source* Unsplash

$$\mathbf{x} = \begin{bmatrix} x_1 = \text{Nominal power, in kW} \\ x_2 = \text{Energy efficiency } (\%) \\ x_3 = \text{Maximum operating temperature, in°C} \\ x_4 = \text{Estimated lifespan, in hours} \\ x_5 = \text{Total weight, in kg} \\ x_6 = \text{Noise level, in dB} \end{bmatrix} \tag{2.29}$$

That said, we have two conflicting objectives: we need to minimize production cost (f_1) and maximize motor performance (f_2). As you may know, the cheapest motor isn't the one with the best performance, and vice versa. The chief engineer suggested that the equations defining these two objectives would be:

$$\mathbf{f}(\mathbf{x}) = \begin{bmatrix} f_1(\mathbf{x}) = 150x_1 + 20x_2 + 5x_3 + 0.01x_4 + 25x_5 - 10x_6 \\ f_2(\mathbf{x}) = 0.8x_1 + 1.5x_2 - 0.2x_3 + 0.005x_4 - 0.1x_5 - 0.3x_6 \end{bmatrix} \tag{2.30}$$

Now that we have the function defined, let's apply the different types of constraints we discussed earlier. As an example, consider that the total production cost should not exceed a maximum value due to market limitations. Thus, we use an inequality constraint according to Eq. 2.22. So, we have:

$$h_1(\mathbf{x}) = 150x_1 + 20x_2 + 5x_3 + 0.01x_4 + 25x_5 - 10x_6 \leq 15000 \tag{2.31}$$

Notice that we don't yet know how the factors will combine. We only have a definition of the maximum cost (\$15,000) and how much each variable contributes to this value. Here, we could conclude that for each kW of power (x_1), we have an increase of \$150 in production cost. Similarly, for each % increase in energy efficiency, we have an increase of \$20 in production cost, and for each °C of maximum operating temperature, we increase production cost by \$5,[7] and so on.

That said, we could add additional engineering constraints. For example, we could add constraints associated with the motor's thermal dissipation, which limits the relationship between power and temperature:

$$h_2(\mathbf{x}) = x_1 - 0.05x_3 \leq 50 \tag{2.32}$$

Also, notice that in Eq. 2.31, we didn't bother to establish a minimum production limit, like \$0. Most of the time, by defining the limits of each variable, we are already intrinsically defining this minimum limit. In fact, now we need to define limits for each variable, according to Eq. 2.23:

$$5 \leq x_1 \leq 250 \quad \text{(Power between 5 and 250 kW)} \tag{2.33}$$

$$80 \leq x_2 \leq 98 \quad \text{(Efficiency between 80\% and 98\%)} \tag{2.34}$$

$$60 \leq x_3 \leq 180 \quad \text{(Maximum temperature between 60°C and 180°C)} \tag{2.35}$$

$$10000 \leq x_4 \leq 100000 \quad \text{(Lifespan between 10,000 and 100,000 h)} \tag{2.36}$$

$$10 \leq x_5 \leq 500 \quad \text{(Weight between 10 and 500 kg)} \tag{2.37}$$

$$40 \leq x_6 \leq 90 \quad \text{(Noise level between 40 and 90 dB)} \tag{2.38}$$

We can also include equality constraints according to Eq. 2.21.[8] For example, to meet a specific industrial standard, energy efficiency must be precisely 92% for a certain line of motors:

$$g_1(\mathbf{x}) = x_2 = 92 \tag{2.39}$$

Or, perhaps, we could add a constraint to model power losses due to limited efficiency and the equipment's thermal dissipation. So, let's assume we have a simplified equation that establishes the relationship between power, efficiency, temperature, and weight within this Optimization problem:

$$g_2(\mathbf{x}) = x_1(1 - x_2/100) - 0.05x_3x_5 = 0 \tag{2.40}$$

[7] Of course, these values are purely illustrative.

[8] However, note that this is *not* mandatory: there are Optimization problems without constraints, or only with inequality constraints, or only with equality constraints.

All these constraints together define the space of valid solutions for our Optimization problem. Thus, we could describe the problem as shown in Eq. 2.41. When reading this equation, pay special attention to $f_2(\mathbf{x})$. Contrary to what we described in Eq. 2.30, here we established that this objective has a negative sign. After all, it was a maximization objective that is now described as minimization—hence the inversion.

$$\min \mathbf{f}(\mathbf{x}) = \begin{bmatrix} f_1(\mathbf{x}) = 150x_1 + 20x_2 + 5x_3 + 0.01x_4 + 25x_5 - 10x_6 \\ -f_2(\mathbf{x}) = -(0.8x_1 + 1.5x_2 - 0.2x_3 + 0.005x_4 - 0.1x_5 - 0.3x_6) \end{bmatrix}$$

subject to

$$\begin{aligned}
&150x_1 + 20x_2 + 5x_3 + 0.01x_4 + 25x_5 - 10x_6 \leq 15000 \\
&x_1 - 0.05x_3 \leq 50 \\
&x_1(1 - x_2/100) - 0.05x_3x_5 = 0 \\
&5 \leq x_1 \leq 250 \\
&x_2 = 92 \\
&60 \leq x_3 \leq 180 \\
&10000 \leq x_4 \leq 100000 \\
&10 \leq x_5 \leq 500 \\
&40 \leq x_6 \leq 90
\end{aligned} \tag{2.41}$$

2.3 Solutions

We've already discussed solutions throughout the explanations, but we haven't necessarily dedicated a moment to talk solely about this. The main point of any Optimization problem is to find solutions. We create variables, constraints, and formulate objective functions solely for this purpose.

Here, let's return to your common sense. Suppose you need to choose something—be it a new house, new sneakers, a romantic partner, or an ice cream flavor. Regardless of the context, there are good choices, bad choices, and choices that simply don't make sense. Maybe you know that buying strawberry ice cream is a good choice, regardless of the place. And maybe you know that buying a house in a dangerous part of the city might not be the best idea. Or, even if you wanted to, you couldn't buy Adidas sneakers in a Nike store.

2.3.1 *Feasible and Infeasible Solutions*

Now, let's translate these concepts to our Optimization scope. First, solutions can be feasible or infeasible. A solution is said to be **feasible** when a vector $\mathbf{x}$ satisfies all problem constraints, without exceptions.

And, on the other hand, an **infeasible solution** is one in which the vector $\mathbf{x}$ violates one or more constraints of our Optimization problem. Here, it matters little whether we're violating one, two, or all constraints: they would be infeasible just the same.

2.3.2 *Global and Local Minima (or Optima)*

In problems with only one objective, the definition of optimality is relatively straightforward. A solution $\mathbf{x}^*$ is considered globally optimal if, for every $\mathbf{x}$ that satisfies the problem's constraints:

$$f(\mathbf{x}^*) \leq f(\mathbf{x}) \quad \text{(for minimization problems)} \tag{2.42}$$

or, alternatively:

$$f(\mathbf{x}^*) \geq f(\mathbf{x}) \quad \text{(for maximization problems)} \tag{2.43}$$

In other words, there is no other feasible solution that can provide a better value for the objective function compared to the optimal solution.

Besides the global optimum, it's also important to mention the notion of a local optimum. A solution $\mathbf{x}^*$ is a local optimum if it's the best solution within a certain neighborhood. Mathematically, $\mathbf{x}^*$ is a local minimum if there exists a neighborhood $\mathcal{N}$ around $\mathbf{x}^*$ such that:

$$f(\mathbf{x}^*) \leq f(\mathbf{x}) \quad \text{for all } \mathbf{x} \in \mathcal{N} \text{ that satisfy the constraints} \tag{2.44}$$

In other words, if the problem in question is non-convex, there are multiple minima. In other places within the n-dimensional hypercube,[9] other values offer the best feasible solution within *that region*, but not globally. These solutions are known as the *local minima*.

The difference is significant from an algorithmic standpoint: some algorithms are quicker or consume fewer computer resources than others but are more prone to reach local minima. On the other extreme, other algorithms can have a lower risk of missing the global minimum at the expense of being slower. However, in convex problems, we have an easier time: they possess a special property that any local optimum is also a global optimum, which makes them more computationally tractable.

[9] That is, if we have two variables, we would have two dimensions of analysis. With three variables, we would have a cube—three dimensions of analysis, and so on.

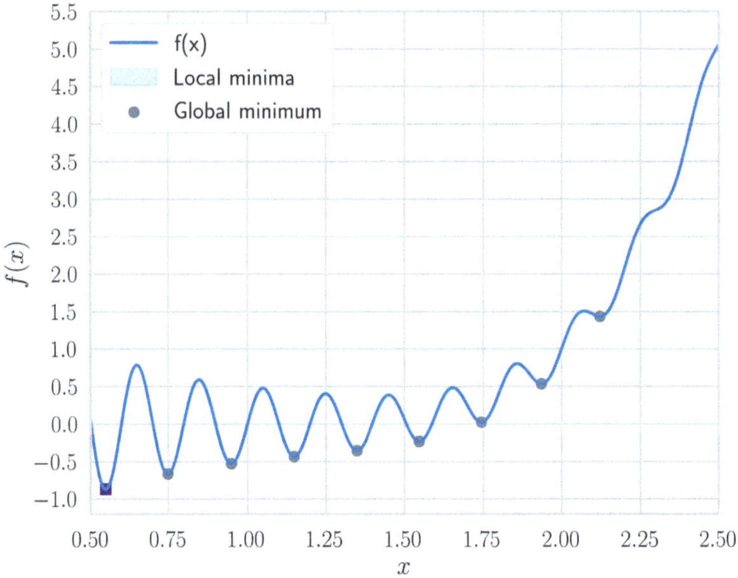

Fig. 2.4 Example of global and local minima based on the Gramacy and Lee function. *Source* The author, 2025

Figure 2.4 exemplifies this. It is based on the Gramacy and Lee function defined in Eq. 2.45. It is a one-dimensional test function containing multiple local minima, as evidenced by the gray dots. One of these local minima does not have a better result concerning $f(x)$. Therefore, this specific local minimum point is also the global minimum, as evidenced by the purple square.

$$f(x) = \frac{\sin(10\pi x)}{2x} + (x-1)^4 \tag{2.45}$$

Attention

We will have problems where a local minimum is *not* necessarily a global minimum in non-convex optimization problems. Given the complexity of real-life Optimization problems, it's very likely that most problems you'll work on will be non-convex.

Furthermore, for standardization purposes, we always refer to "local and global minima" and don't talk much about "local and global maxima." The reason for this, as we've mentioned in other sections, is that Optimization software commonly standardizes problems for minimization. So, if we want to maximize something, we just invert the sign to transform it into a minimization problem.

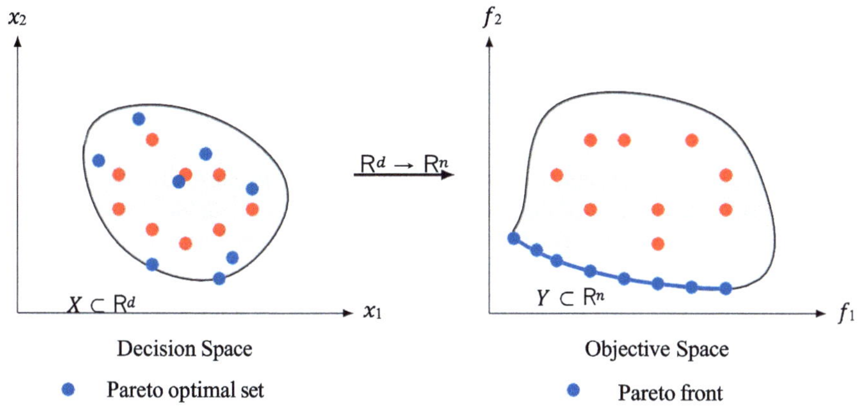

Fig. 2.5 Illustration of the Pareto-optimal set and its corresponding Pareto front. *Source* The author, 2025

Tips
Every global and local minimum is a feasible solution, but not every feasible solution is a local and global minimum. As an analogy, think of the following: you can have several t-shirts (feasible solutions), and among them, you might have your favorite t-shirt (global minimum). However, just because you have a favorite t-shirt doesn't mean all other t-shirts are also your favorites.

2.3.3 *Pareto Dominance*

We would like to invoke Fig. 2.4 again to address the correspondence between variables and objectives. In that image, the correspondence is clear: we have only one variable (x-axis) that corresponds to only one objective (y-axis). Now, what if we had more than one variable and one objective?

Let's look at Fig. 2.5. Here, we have a MOO problem containing two variables (x_1 and x_2, on the left) and two objectives (f_1 and f_2, on the right). In this figure, we can see two connected spaces: the decision space (on the left) and the objective space (on the right). In the decision space, each circle represents a possible solution or choice you can make, with coordinates corresponding to the decision variables (like the characteristics of a cellphone or an engine, as we saw earlier).

When we apply an evaluation function to these choices, they are mapped to the objective space, where each coordinate represents the performance of that solution concerning a specific objective (like price, performance, or efficiency).

Now, notice the feasible region in the decision space. This region is shown as a shaded blob with a black border with several solutions inside it. The feasible region of an optimization problem is often defined by constraints. It comprises multiple (infinite, perhaps) feasible solutions. Each optimization algorithm can find a different subset of feasible solutions, one slightly different from the other. This feasible region translates into *another* feasible region in the objective space, with a different shape.[10]

With this, we'd like to introduce an important concept to you: Pareto dominance. Some solutions are clearly better than others in all aspects—these dominate the others. But there's a special set of solutions, represented by the blue dots, that form the so-called "**Pareto-optimal set.**" These solutions are special because, to improve in one objective, they necessarily worsen in another.

When we map this to the objective space, this set forms the "**Pareto front**": a curve that represents the best possible compromise between conflicting objectives.

Important
In practice, when solving MOO problems, we seek to find this Pareto front. Thus, we offer the decision-maker a set of the best solutions from which they can choose according to their personal preferences regarding the relative importance of each objective.

2.3.3.1 Zooming Into the Objective Space

Looking at the objective space in Fig. 2.5, you might have noticed that we had several feasible solutions, and some of them formed the Pareto front approximation (also called the *Pareto front*).

Perhaps this is already clear to you, but it doesn't hurt to reinforce it with another example: even if we have two perfectly feasible solutions, one of them might be *better* than the other. It could be that in your house, you have coffee and beer available. However, it would generally be more appropriate to drink coffee than beer before leaving for work.

Thus, within feasible solutions, there are two other types of solutions: approximate and optimal. A solution is said to be optimal when it's the best possible. So, we could also call local and global minima *local and global optimal solutions*, or *local and global optima*. To help explain these differentiations, we really like to use Fig. 2.6. It's another example of an objective space, but now with a different Optimization scenario for illustrative purposes.

[10] As an analogy, it's as if a feasible region of ingredients (variables) corresponded to another feasible region of dishes that would be constructed from these ingredients (objectives).

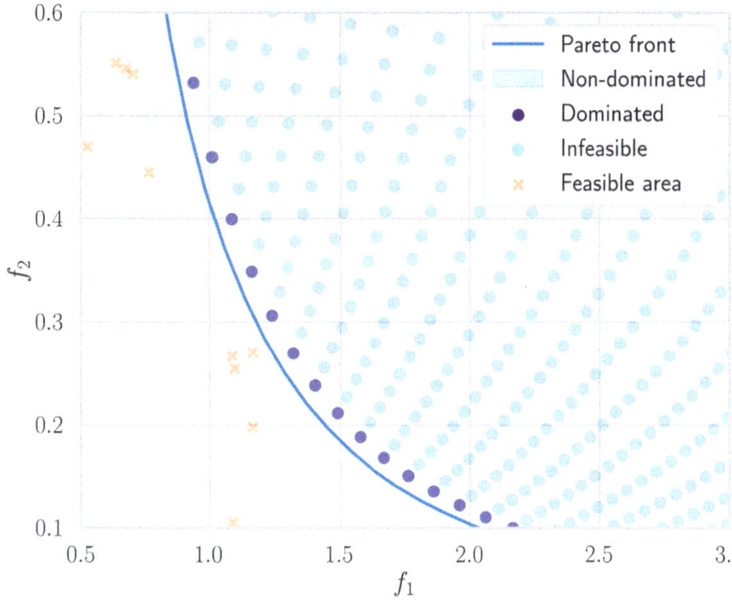

Fig. 2.6 Example of Pareto optimality for a two-objective minimization problem. It shows both infeasible, dominated, and non-dominated solutions. The feasible area and the actual Pareto front are also displayed. *Source* The author, 2025

Attention

Notice that in Fig. 2.6, we don't specify if the objectives are for minimization or maximization. Always assume that objectives are for minimization by default, except when explicitly stated otherwise. Therefore, the best solutions would be those located further to the left (considering f_1) and lower (considering f_2).

We also wanted to show you some examples of infeasible solutions, displayed by the crosses in the figure. Although they seem to have better values for f_1 or f_2 when compared to the solutions in the circles, they are outside the feasible region. It means, in practice, that these solutions do not respect at least one constraint of that objective function.

Having that said, let us now pay special attention to the solutions within the feasible region. We have both non-dominated and dominated solutions in that area, and all of them are feasible solutions. The non-dominated solutions (also referred to as the Pareto-optimal set) form a Pareto frontier or a Pareto-optimal front.

However, this curve is often unknown in practical applications. That is, the true set of Pareto-optimal solutions is unknown. Take as an example a route from your home to the nearest airport: what is the absolutely shortest time you could take to the nearest airport provided that (a) you are leaving immediately, and (b) you are respecting all traffic codes? You might open your favorite routing app in your phone and estimate you could get in there in 40 min, but then we could extend our thinking on it: why not 38 min, if you take a few shortcuts? Or 36 min, if you get lucky on all green lights? Or 35 min, if you perfect the usage of your throttle and brakes of the car? But why not 34 min and 40 s? Or 34 min and 20 s? The point is: finding the absolute minimum can be a rather exhausting or impossible activity, or simply not too rewarding—maybe that 40 min original estimate was alright for us from the start.

With that in mind, what we usually have when working with MOO problems is a set of Pareto-optimal solutions *with regard to* the final set of solutions found by the algorithm, which, again, does not necessarily represent the true Pareto-optimal front. Therefore, the non-dominated solutions found by an MOO algorithm form a Pareto-optimal curve *approximation* (or Pareto front approximation) instead, since they are as close as possible to the front but are not necessarily representative of the true Pareto-optimal front. This front *approximation* is, therefore, composed of approximated optimal solutions and not necessarily true optimal solutions.

2.4 Evaluating Solutions

To evaluate a solution, we calculate the value of the objective function for that specific point. Let's return to the multi-objective cellphone problem example we saw earlier. Suppose we have a cellphone with the following configurations:

$$\mathbf{x} = \begin{bmatrix} x_1 = 15 \text{ (battery life, in hours)} \\ x_2 = 70 \text{ (processor benchmark score)} \\ x_3 = 6 \text{ (RAM, in GB)} \\ x_4 = 128 \text{ (storage capacity, in GB)} \\ x_5 = 4000 \text{ (battery capacity, in mAh)} \\ x_6 = 8 \text{ (camera quality score)} \end{bmatrix} \tag{2.46}$$

And, recalling what we've seen in previous sections, the objective function for that problem was:

$$\min \mathbf{f}(\mathbf{x}) = \begin{bmatrix} f_1(\mathbf{x}) = 5x_1 + 2x_2 + 10x_3 + 0.500x_4 + 0.010x_5 + 2.000x_6 \\ -f_2(\mathbf{x}) = -(0.100x_1 + 0.500x_2 + 2x_3 + 0.050x_4 + 0.001x_5 + 0.500x_6) \end{bmatrix}$$

subject to

$$\begin{aligned} & 5x_1 + 2x_2 + 10x_3 + 0.500x_4 + 0.010x_5 + 2.000x_6 \leq 1000 \\ & 0.05x_2 + 0.5x_3 - 0.001x_5 = 10 \\ & 20 \leq x_1 \leq 72 \\ & 0 \leq x_2 \leq 100 \\ & x_3 \in \{2, 3, 4, 6, 8, 12, 16, 24, 32\} \\ & x_4 \in \{32, 64, 128, 256, 512, 1024, 2048\} \\ & 3000 \leq x_5 \leq 8000 \\ & x_6 = 10 \end{aligned} \tag{2.47}$$

With this, let's replace the values of each variable x_n. We thus have:

$$\min \mathbf{f}(\mathbf{x}) = \begin{bmatrix} f_1(\mathbf{x}) = 5(15) + 2(70) + 10(6) + 0.5(128) + 0.01(4000) + 2(8) = 395 \\ -f_2(\mathbf{x}) = -(0.1(15) + 0.5(70) + 2(6) + 0.05(128) + 0.001(4000) + 0.5(8)) = -62.9 \end{bmatrix}$$

subject to

$$\begin{aligned} & \checkmark 395 \leq 1000 \\ & \times 0.05(70) + 0.5(6) - 0.001(4000) = 3.5 + 3 - 4 = 2.5 \neq 10 \\ & \times 20 \leq 15 \leq 72 \\ & \checkmark 0 \leq 70 \leq 100 \\ & \checkmark 6 \in \{2, 3, 4, 6, 8, 12, 16, 24, 32\} \\ & \checkmark 128 \in \{32, 64, 128, 256, 512, 1024, 2048\} \\ & \checkmark 3000 \leq 4000 \leq 8000 \\ & \times 8 \neq 10 \end{aligned} \tag{2.48}$$

Notice that we ended up with a cellphone costing \$395 (f_1) and a performance of 62.9 (f_2)[11] on our scale. However, note that not all constraints were respected (marked with an $\times$ to the left of the constraints in that equation). Therefore, we would have an infeasible solution here.

For the solution to be feasible, we would have to change the variable values such that *all* constraints are met.

[11] In Eq. 2.48, we have a negative result for f_2 because this objective was initially for maximization. Therefore, to standardize the problem as minimization, we inverted the sign.

Tips

We could change the values of some variables until we find a combination that results in a feasible solution. However, this is slow work and sometimes difficult for a human to do. That's why we use Optimization algorithms once we've defined the objective function, its constraints, and its variables. We'll see more about this in the upcoming chapters.

2.4.1 Solutions in Multi-objective Problems

We've just seen an example of a single cellphone that would represent a hypothetical cellphone. And, also, we just mentioned that an Optimization algorithm could find one or more feasible solutions. In problems with multiple objectives, like our cellphone example, a single solution cannot be considered "the best" in all aspects. Instead, we obtain a set of non-dominated solutions in an approximation of the Pareto front, each representing a different compromise between the objectives.

That said, we'd like to bring Table 2.1 back for your analysis, but now with some minor adjustments. Let's assume that all the cellphones below are respecting all our constraints.[12] Each of these solutions represents a point on the Pareto front, offering a different compromise between price and performance. The choice between them will depend on the specific preferences of the decision-maker. The decision-maker (a consumer, for example) might want the cheapest possible cellphone or, if they can spend more, the most expensive one. Or, depending on the case, a cellphone that's a bit cheaper but still fast enough to open most applications. Do you see how each person can make a different decision *based on their own context*?

2.5 Conclusion

In this chapter, we discussed the mathematical foundations upon which practical optimization techniques are built. Specifically, we presented and discussed the formal definitions of variables, functions, objectives, and constraints. Understanding these mathematical definitions is a prerequisite for anyone wishing to study and work with optimization after intuitively grasping its concepts in Chap. 1. Indeed, mathematical language is essential for unambiguously defining problems, correctly interpreting algorithmic solutions, and selecting or developing appropriate optimization algorithms.

---

[12] In reality, they aren't, as we've been modifying some constraints since the last time we saw the examples in this table. However, let's focus only on the values of f_1 and f_2 for now.

This chapter also discussed properties like linearity and convexity, as well as different types of variables and constraints. These factors heavily influence the choice of optimization algorithms and the characteristics of the expected solutions. For instance, whether one can anticipate a single global optimum or a set of trade-off solutions on a Pareto front.

We also used examples in this chapter to illustrate translating real-world problems into this structured mathematical framework. Translating business and stakeholder requirements into an optimization problem is an important skill for both engineers and data scientists. Moreover, the quality and efficiency of the optimization process are directly linked to the understanding of these mathematical concepts. Furthermore, comprehending these concepts helps diagnose issues when an optimization model behaves unexpectedly, such as a solver reporting infeasibility or an algorithm prematurely stopping at a suboptimal point.

Chapter 3
Single-Objective Optimization Techniques

Abstract This chapter provides a detailed exploration of algorithms for Single-Objective Optimization (SOO), where the goal is to find the best solution for a single, well-defined objective function. It first addresses unconstrained optimization, where variables are not explicitly restricted, and constrained optimization, where the algorithm must respect specific equality or inequality conditions that define a feasible region. For unconstrained techniques, this chapter details gradient-based methods, including Gradient Descent (GD), Stochastic Gradient Descent (SGD), Newton's method, and quasi-Newton methods, alongside derivative-free approaches like Nelder-Mead, pattern search, and evolutionary algorithms. Regarding constrained optimization, it covers foundational concepts such as Lagrange multipliers and KKT conditions, as well as techniques including penalty and barrier approaches (interior-point methods), Sequential Quadratic Programming (SQP), the simplex method for Linear Programming (LP), and Branch and Bound/Cut techniques for Integer Programming (IP/MIP). We also highlight the real-world relevance of selecting suitable single-objective optimization algorithms based on problem features like differentiability, convexity, scale, and variable types.

Hopefully, if you've read Chaps. 1 and 2, or if you already have some prior knowledge about optimization, we now share a common language. We expect you to have a good idea of what terms like $f(\mathbf{x})$, "objective function," "variables," and "constraints" mean whenever we mention them from now on. You might not understand the different types of techniques that exist, but that's perfectly fine: that's exactly what we'll discuss next.

That said, perhaps the next question is: how would we effectively find the optimal solution, or a solution as close to optimal as possible, for a given optimization problem? Of course, we could list various types and, perhaps a bit dryly, explain how each one works. We could just comment on the theoretical definition of these techniques, and that would be the end of the story. However, that wouldn't necessarily explain to you how or where you could apply these techniques.

Therefore, starting with this chapter, we'll show you how optimization techniques (or algorithms) work. The prerequisite for all the techniques we'll see is the same:

W. R. Monteiro and G. R. Meza, *A Practical Guide to Optimization in Engineering and Data Science*, https://doi.org/10.1007/978-3-032-04633-8_3

you must have already defined your optimization problem with its objective function, variables, and constraints.

Tips

People entering Information Technology (IT) courses and working in IT frequently work with "algorithms." The term *algorithm* is sometimes used more broadly in non-technical contexts. For instance, it might be used by some to describe trending Artificial Intelligence (AI) solutions, such as prominent large language models and generative AI systems available at the time of writing, such as Gemini, ChatGPT, and Grok. However, that's not what we mean when we talk about algorithms in this text. In fact, understand an algorithm as a set of instructions that modify inputs to perform some operation. A cake recipe is an algorithm: we have inputs (ingredients) and outputs (the finished cake). Between the input and output, we have to follow a series of steps in a predetermined structure. The same applies to any other work and activity you do in your daily life, like driving a car to a place, tying your shoelaces, or making coffee to continue reading this book. That said, in our case, understand that we're talking about algorithms applied to a computational context and an optimization context. In other words, we will present various algorithms that, from an input (objective function, variables, and constraints), produce an output (the optimal solution(s)). Just as there are many carrot cake recipes and many ways to tie shoelaces, there are also many optimization techniques (algorithms). Some are more creative, others more direct. The choice of the correct algorithm will fundamentally depend on the nature of the problem at hand.

That said, in this chapter, we'll focus on understanding Single-Objective Optimization (SOO) techniques. These are problems where we seek to minimize or maximize a single-objective function, as discussed in Sect. 1.3.1. In Chap. 5, we will study multi-objective optimization problems. These problems will have an additional layer of complexity, as you may have already noticed in Sect. 2.2.5.

Now, the first division we'll make among SOO algorithms will be based on the presence or absence of constraints. Thus, we'll have constrained optimization and unconstrained optimization problems. We'll start with unconstrained optimization, as it's simpler to understand, test, and, as a bonus, many of its concepts can be reapplied to constrained optimization techniques.

3.1 Unconstrained Optimization

As the name suggests, unconstrained optimization problems deal with finding the minimum (or maximum) of a function $f(\mathbf{x})$ without any restrictions on the values that the variables $\mathbf{x}$ can assume. Mathematically, we represent this simply as[1] follows:

$$\min_{\mathbf{x}} f(\mathbf{x}) \tag{3.1}$$

Here, $\mathbf{x}$ can theoretically belong to the entire space $\mathbb{R}^n$ (for continuous variables) or $\mathbb{Z}^n$ (for discrete variables), although in practice, variables may have implicit limits or be defined in specific domains due to the nature of the problem.

Important
This might be a controversial point between us and your professors and managers or, perhaps, with other authors. Consider this: although formally we don't have explicit constraints (like $g(\mathbf{x}) = 0$ or $h(\mathbf{x}) \leq 0$), real-world problems modeled as unconstrained might have variables that, by definition, are limited (for example, variables like height, length, and mass couldn't be negative). We frequently handle this through the formulation of the function $f(\mathbf{x})$ or in the problem setup within the software. For instance, some software allows us to configure, separately, the optimization problem's constraints and the minimum and maximum limits (or bounds) for each variable. Furthermore, many constrained optimization problems are solved by transforming them into a sequence of unconstrained problems, as we'll see later.

There is a difference between the formal academic definition of an unconstrained optimization problem and how real-world scenarios involving variables with inherent bounds (like $g(\mathbf{x}) = 0$ or $h(\mathbf{x}) \leq 0$) are handled. Consider this: although a problem is academically termed unconstrained when its mathematical formulation lacks explicit constraint functions like $g_i(\mathbf{x}) = 0$ or $h_j(\mathbf{x}) \leq 0$, many unconstrained *implementations* involve variables that are naturally restricted (for example, physical quantities like height, length, or mass cannot be negative). We usually handle these limitations either by defining the domain of the objective function $f(\mathbf{x})$ appropriately (e.g., ensuring $f(\mathbf{x})$ is only evaluated for non-negative inputs for certain variables) or through specific configurations in the problem setup within the software (also known as *solver*). For instance, some software allows us to separately configure the optimization problem's formal constraints and the minimum and maximum limits (or bounds) for each variable. It's important for you to understand that applying such bounds does not alter the mathematical classification of the problem as unconstrained if no explicit g_i or h_j functions are defined. Furthermore, this

[1] It's worth remembering once more: Optimization software commonly standardizes all problems as minimization problems. If your problem is a maximization one, you would just invert the sign—that is, $-f(\mathbf{x})$.

precise understanding of what is an "unconstrained" problem is fundamental because many constrained optimization problems are solved by transforming them into a sequence of unconstrained subproblems, as we'll see later.

Also, there's an important difference in how software often handles these variable bounds compared to explicit constraints, which relates to the logic of solution validation. Variable bounds are often managed by the solver by restricting its search to the valid domain or by projecting solutions onto it, meaning the solver is typically designed to always satisfy these bounds. In contrast, explicit constraints ($g_i(\mathbf{x}) = 0$ or $h_j(\mathbf{x}) \leq 0$) are actively processed by the optimization algorithm, and satisfying them is a primary objective where infeasibilities might occur and be reported during the solution process. Note, however, that this practical distinction, often between what might be called "implicit" restrictions like bounds and "explicit" formal constraints, reflects choices in problem formulation for software and solver design, rather than altering the abstract mathematical definition of whether an optimization problem is constrained or unconstrained in the first place.

To help illustrate how a system might enforce different types of rules imagine a football[2] game. The rule that each team must have exactly 11 players on the field is a fundamental setup condition, similar to defining a variable's inherent domain (e.g., $length \geq 0$). The game proceeds with this condition intrinsically met. Now, some constraints or rules, like an offside or a foul, are dynamic and can be violated during play, requiring specific actions. These are more analogous to explicit constraints $g_i(\mathbf{x}) = 0$ or $h_j(\mathbf{x}) \leq 0$ that the optimization algorithm actively works to satisfy, where violations guide the search or result in penalties.

3.1.1 Why Should You Care?

The real world is full of limitations, rules, and restrictions, isn't it? Consequently, you might also conclude that many real-life optimization problems have constraints (in fact, it's quite likely we've said something similar throughout this and other texts). Therefore, the question remains: if most problems have constraints, why should we waste time on these unconstrained problems? Wouldn't it be better to just talk about constrained optimization problems?

Indeed, real-world problems do use unconstrained optimization. But before that, it's also important to talk about your learning process. Working with unconstrained problems allows us to focus on the heart of an optimization process, which is finding the lowest (or highest) point of a function. It's much easier to understand the logic of

[2] Also known as soccer. But, for us, it is football.

algorithms like the ones we will see in the next sections when we don't have to worry about constraints. And it does not mean you will have to relearn concepts entirely when addressing constraints—quite the contrary. We'd even say it would possibly be a simple adaptation process.

Well, that's the first point. The second point we deem equally relevant is applicability. In fact, we have fewer unconstrained optimization problems in real life compared to problems with constraints, but *fewer* is different from *zero*. Let's start with the world of statistics and machine learning, for example. We don't know if you're aware, but when we create a model to predict something (what we call *training*), to validate if that model is actually good or not, we use a quantitative metric, like error metrics. And making mistakes is bad—so, we want to minimize prediction errors. Data scientists might very well train three very similar models, but it's highly likely they'll choose the one with the lowest possible error. When we fit a linear regression, for example, we look for the coefficients β that would minimize the mean squared error (MSE) between the predictions and the actual values from some examples in our history.

Consider this: if in our validation we notice that the model makes few errors, there would be, in this case, a lower risk of also making errors with new examples in the future. That said, this would be an unconstrained and convex optimization problem: we seek the optimal solution of the error function (or, translating: the best combination of the best coefficients β that would result in the smallest possible error).

We could also think of an estimation problem in statistics. In various data-driven situations within organizations, we might need to find the parameters of a statistical model, such as the mean and standard deviation of a normal distribution in our data, or the rate λ of an exponential distribution. In these cases, we would have to find these parameters from a maximum likelihood estimation (MLE). This task is also an unconstrained optimization problem.[3]

Fun fact

The training of many machine learning (ML) models from a simple logistic regression to neural networks (and prominent large language models like a Generative Pre-Transformer (GPT)) is also based on minimizing loss functions using algorithms we'll see in this book. However, it's common not to think much about the differences between these loss functions or even how they work. In fact, this was one of the motivators for creating this book.

That said, it's also common to work with unconstrained optimization problems in the exploratory stages of an engineering project in organizations. Think with us: imagine you work in the aerospace industry and need to produce an airfoil. Indeed, it would be interesting to have an airfoil profile that minimizes drag under certain conditions. Of course, a real-life problem like this isn't so simple: there are

[3] In the case of distributions, we actually have some constraints. For example, the standard deviation and the rate λ would need to be above zero. However, as we mentioned earlier, there's the possibility of implicitly restricting this *before* the problem is optimized.

various conditions we'd need to consider, such as temperature, humidity, atmospheric pressure, and material properties. In fact, we could make this problem as complex as necessary. On the other hand, we'd lose a lot of time producing an initial version. So, the idea here would be to use an unconstrained optimization problem over the parameters that would define its geometry to enable initial exploration. The same logic would apply if we were calibrating sensors, or working with optics problems: the intention is to experiment, explore, and get an initial idea of the type of solution we might expect. After that, we'd start introducing constraints to refine our problem. Notice how this resembles a continuous improvement cycle (or, as IT professionals know it, the agile methodologies).

Furthermore, in various engineering applications, we face the challenge of curve fitting. This involves finding the coefficients of a polynomial, exponential, logarithmic, or other types of equations. As an example, we use curves to predict the growth rate of livestock, and this, as you might suppose, is very useful for planning in food industries. Therefore, minimizing the errors of these curves by finding the correct coefficients means saving (a lot of) money. And, guess what? It's a problem frequently addressed by unconstrained optimization techniques.

3.1.2 Gradient-Based Methods

A significant portion of techniques for unconstrained optimization is applicable to continuous and differentiable functions.[4] First, we'll look at gradient-based methods.

To explain this method, we'd like to use the analogy of a mountain, a range, or a hill. In fact, we'd like you to imagine you're driving a car with your friends, and you're at the top of this mountain range. You need to reach the valley (the lowest point) of the range quickly for two reasons. The first is convenience: the nearest town is in the valley, and that's where the next hotel you'll be staying at is located. The second is urgency: you have a terrible headache and need to buy medicine at the pharmacy next to the hotel.

The path you must follow is illustrated in Fig. 3.1. However, you can't see anything in front of you because of heavy fog and, to make matters worse, the GPS isn't working. Let's recap: your goal is to descend to the valley, but limited visibility prevents you from seeing the final destination. Nevertheless, by holding the steering wheel and seeing a few meters ahead, you can feel the slope of the road—whether it's going up, down, or leveling out. You can also see if there's a curve just ahead, even if you can't see the entire path.

Following this logic, we expect you to head in the direction where the road descends most steeply, trusting the slope to guide you. As you advance, curve after curve, you adjust your trajectory with each new stretch of road, always seeking the

---

[4] That is, functions for which we can calculate derivatives.

Fig. 3.1 Illustrative analogy of a winding mountain road, representing a search space with potential paths for an optimization algorithm. *Source* Pexels

path that leads further down. Eventually, you reach a point where the road becomes flat or you start seeing signs indicating you've finally arrived at the town in the valley.[5]

3.1.2.1 Gradients

We told this short story because it's an analogy for how **gradient-based methods** work. They use information from the derivative (the gradient of the function) to determine the direction of steepest descent (for minimization) and take iterative steps in that direction.

This "sensitivity" to the local slope of the road in our mountain analogy has a mathematical name when dealing with functions of several variables (like altitude f, which depends on our position $\mathbf{x}$ with its coordinates): the gradient.

Let's return to our analogy: to know where you are on that mountain path, you need (at least) two coordinates, right? Like latitude and longitude, or coordinates x_1 and x_2[6] on a map. Your position is a set of numbers, or a vector $\mathbf{x} = [x_1, x_2]$. If we

[5] This analogy is also useful as it introduces the concept of step size or learning rate, akin to driving speed. Consider this: if you drive too fast on a dangerous road under these conditions, there's a real risk of losing control and not ending up in the valley, but rather in a ditch, right? Similarly, if you drive too slowly, there's a chance you'll take so long that everyone will lose patience and give up on riding with you midway.

[6] Or *variables* x_1 and x_2.

thought in more dimensions and variables, this same vector could be represented by $\mathbf{x} = [x_1, x_2, \ldots, x_n]$.

What we're trying to predict (that is, whether we're in the mountain valley or not; in other words, the altitude) would be a function f that depends on your entire position $\mathbf{x}$. That is, f receives a vector $\mathbf{x}$ as input and returns a value (altitude) at that point.

That said, a gradient-based method is iterative. As we descend the mountain, we occupy different coordinates, correct? First we're at the first position (which we could call $\mathbf{x}_1$), then a second position ($\mathbf{x}_2$), and so on until we reach the k-th iteration ($\mathbf{x}_k$). Throughout these iterations, we can determine the direction and intensity for where we should go. In other words, we mathematically define a gradient as follows:

$$\nabla f(\mathbf{x}_k) \tag{3.2}$$

This is a vector: an arrow with direction and intensity, calculated at that vector point $\mathbf{x}_k$. This vector reveals the direction in which we would have the greatest increase in altitude from that point, and also how steep this change would be ($\|\nabla f(\mathbf{x}_k)\|$).

Now, notice that we mentioned the gradient vector $\nabla f(\mathbf{x}_k)$ points toward the steepest ascent, but our problem is one of *descent*. Therefore, we just need to follow the opposite vector: the negative gradient $-\nabla f(\mathbf{x}_k)$ points us to the vector direction of the steepest descent from $\mathbf{x}_k$ or, translating to less technical language, to the quickest way down.

In fact, gradient-based methods use the result of $-\nabla f(\mathbf{x}_k)$ to determine the next step (i.e., $\mathbf{x}_{k+1}$), just as you would determine where you should go in the next few seconds with your car on that road we mentioned initially.

3.1.2.2 Gradient Descent (GD)

The most fundamental gradient-based method is gradient descent. We've already discussed the main idea: starting from an initial point $\mathbf{x}_0$, we calculate the gradient $\nabla f(\mathbf{x}_k)$ and update our position by taking a step in the direction opposite to the gradient. Mathematically, this would be

$$\mathbf{x}_{k+1} = \mathbf{x}_k - \alpha \nabla f(\mathbf{x}_k) \tag{3.3}$$

Now, how does this method work step by step? Look at Algorithm 1. This description follows all the notation we've seen so far, which includes, for example, $\mathbf{x}_k$ and the gradient $\nabla f(\mathbf{x}_k)$. Additionally, g_k is a variable used to store the gradient calculated in iteration k.[7] Figure 3.2 also illustrates the result of an example execution of this algorithm.

Additionally, α is the learning rate. We could understand this as the size of the step we take at each iteration. That is, how far we move in the direction $-\nabla f(\mathbf{x}_k)$.

[7] It's common to store values in variables for reuse at other points in the algorithm.

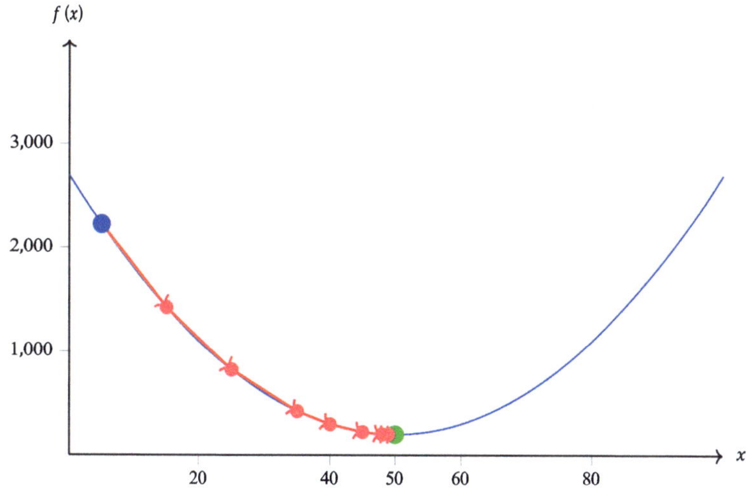

Fig. 3.2 Illustration of the iterative steps of the gradient descent. *Source* The author, 2025

Algorithm 1 Gradient Descent (Batch GD version)

Require: Objective function $f(\mathbf{x})$ to minimize
Require: Initial parameters $\mathbf{x}_0$
Require: Learning rate α
Require: Number of iterations K (or other convergence criteria)
Ensure: Optimized parameters $\mathbf{x}^*$ that approximately minimize $f(\mathbf{x})$
1: Initialize parameters: $\mathbf{x} \leftarrow \mathbf{x}_0$
2: **for** $k = 1$ to K **do**
3: Compute gradient: $g_k \leftarrow \nabla f(\mathbf{x})$ ▷ Gradient computed over the entire dataset
4: Update parameters: $\mathbf{x} \leftarrow \mathbf{x} - \alpha \cdot g_k$
5: **end for**
6: **return** $\mathbf{x}$ as $\mathbf{x}^*$

We described this algorithm with a fixed α for simplicity, but in practice, we can use a learning rate α_k that decreases with each iteration k. These effects are also illustrated in Fig. 3.3.

Tip about α
The choice of the learning rate α is one of the most critical and, at the same time, artisanal aspects of this method. If α is too small, the algorithm will converge (that is, it will reach the optimal solution), but it will take a very long time for this to happen. In extreme cases, it might even seem like the algorithm has "stopped" due to numerical precision limitations before actually reaching the optimal point. It's as if, in our mountain analogy, we were driving very slowly, making minimal progress, akin to moving at a snail's pace.

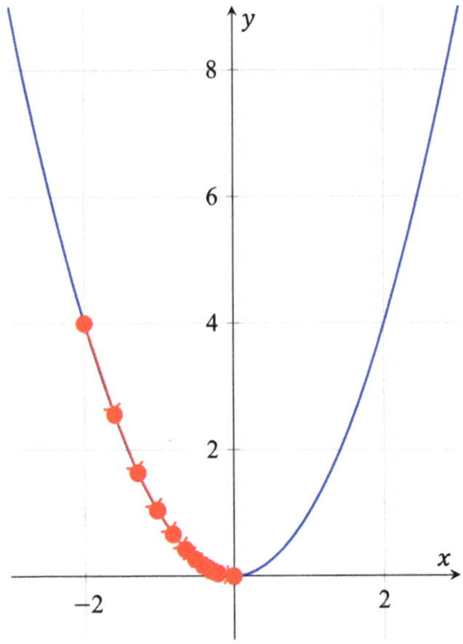

(a) Gradient descent with low α.

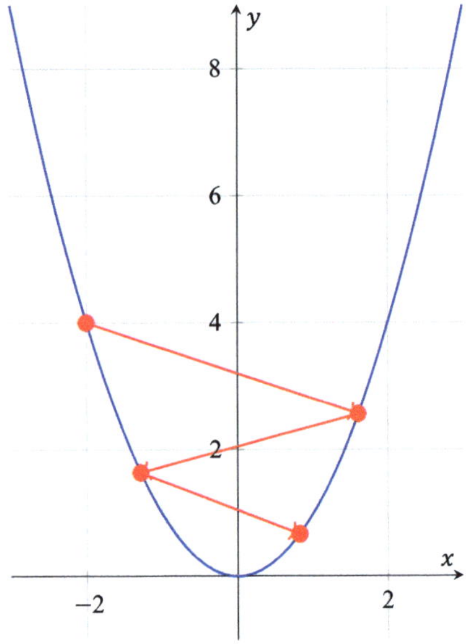

(b) Gradient descent with high α.

Fig. 3.3 Effect of the learning rate (α) on gradient descent convergence. *Source* The author, 2025

Now, if the value of α is too large,[8] the algorithm might pass right by the minimum value, failing to find it. With each step, it might overshoot the minimum, causing oscillations or divergence, much like a pinball machine. Returning to our mountain analogy, it's like driving so fast that we miss all the entrances and road changes, passing through all places except the most important one: the final destination.

Therefore, it's common to test values on a logarithmic scale, starting with small values. It's common to start at 0.0001, and then test with 0.001, 0.01, and 0.1. Or, perhaps, 0.003, 0.03, and 0.3. There are also some cases where we delegate to the computer to test these different values. We have, for example, brute-force test cases to find the best hyperparameters in machine learning models that use GD or, even, adaptive methods of deep learning algorithms, like Adam, AdaGrad, and RMSProp.

You might also notice that we mentioned in Algorithm 1 a "stopping criterion." Optimization algorithms can run for a long time, like a few hours or days. And, sometimes, we don't have that much time on our hands and want to interrupt the execution at some point. As an example, we can establish a predetermined maximum

[8] Here, the value of α is always on a scale between 0.0 and 1.0. It's very rare to have cases where the value of α is greater than 1.0.

value of iterations K, or a very low value to establish whether we are still improving or not. The latter example would be $||\nabla f(\mathbf{x}_k)|| > \epsilon$, where ϵ is a very small positive number (for example, 10^{-5} or 10^{-6}). This means, in practice, something like "while the slope is still significantly different from zero, keep descending."

Simple Example: Minimizing $f(x) = x^2$
Let's minimize $f(x) = x^2$. The gradient (derivative in 1D) is $f'(x) = 2x$. The update rule is $x_{k+1} = x_k - \alpha(2x_k)$. If we start at $x_0 = 4$ and use $\alpha = 0.1$:

- $k = 0$: $x_1 = 4 - 0.1 \times (2 \times 4) = 4 - 0.8 = 3.2$,
- $k = 1$: $x_2 = 3.2 - 0.1 \times (2 \times 3.2) = 3.2 - 0.64 = 2.56$,
- $k = 2$: $x_3 = 2.56 - 0.1 \times (2 \times 2.56) = 2.56 - 0.512 = 2.048$,
- ... and so on, converging to $x = 0$, which is the global minimum.

See what happens if $\alpha = 0.9$ or $\alpha = 1.1$.

Now, this whole explanation seems quite promising, right? Indeed, we started explaining this method because it's conceptually simple, and also easy to implement in the programming language you're most comfortable with. It also works well for various functions, especially convex ones. And, furthermore, this logic is the basis for other more advanced algorithms, like the SGD we'll see next.

That said, it's not a solution we can use everywhere, or a type of algorithm where we don't need to worry about adapting it for each use case. On the contrary: as we've just seen, it's very sensitive to the value of the learning rate α. Depending on the value we choose and the behavior of the function, there are two extra risks: the algorithm might never find an optimal solution, or it might get stuck in local minima in non-convex functions, as we saw in Sect. 2.3.2. Therefore, while effective, it has certain limitations.

3.1.2.3 Stochastic Gradient Descent (SGD)

In Algorithm 1, we saw that GD requires calculating the gradient using the entire dataset at each step. This isn't a problem when testing with a few cases on a sheet of paper, but it becomes a major performance issue if there are millions of calculations to be made.

Important
As you well know, in optimization problems, we typically want to minimize a function $f(\mathbf{x})$ that directly depends on the decision variables in a vector $\mathbf{x}$. That said, many complex real-world problems look like this:

$$f(\mathbf{x}) = \sum_{i=1}^{N} f_i(\mathbf{x}) \tag{3.4}$$

Or, perhaps, like this:

$$f(\mathbf{x}) = \frac{1}{N} \sum_{i=1}^{N} f_i(\mathbf{x}) \tag{3.5}$$

That is: many optimization problems are, in fact, the sum, the average, or another mathematical function that depends on many smaller components. This is relatively easy to calculate when we have a few variables, but becomes very difficult to calculate when we have thousands or millions of variables.

And, to be honest, it's common to have problems with that many variables. Consider this: when we design an airplane wing, we need to minimize drag, correct? However, this drag can vary depending on various conditions, such as speed, altitude, angle of attack, and even atmospheric conditions. Therefore, the wing we design will behave differently under different conditions. In other words: we want to minimize not one, but several drags (that is, each condition would have its own drag). We could even adapt Eq. 3.5 for this case, and we would have N different conditions.

Or, perhaps, we could have an ML problem. Let's say we want to train an AI model to predict how much bank customers would spend on their credit cards next month based on their characteristics like age, spending history, and socioeconomic information. We don't know if you're aware, but the process of training an ML model involves finding the best parameters in order to reduce the errors of each prediction based on a history. Imagine we have a linear function, like Eq. 3.6.

$$f(\mathbf{x}) = a_1x_1 + a_2x_2 + \cdots + a_nx_n \tag{3.6}$$

In it, each variable x_i is a customer feature. The objective, in this case, would be to find the best combination of coefficients a_i that result in the correct credit card spending considering the entire history we have so far for all customers. That is, we could say I spent \$1000 last month, and you spent \$2000. The idea would be to find the coefficients that would make the model predict a value as close as possible to what I, you, and all other customers have spent so far. These coefficients cannot change for each customer, and we will use these values to predict what would happen in the future.

This task can be relatively simple when we have few customers, but it becomes a very complex job when we have to calculate errors for millions of customers in a historical database, right?

We could even use GD to solve the problems we've exemplified here, but it would be an immense task. Consider this: GD would calculate the gradient for all N components before moving on to the next step. This means the problem would be computationally prohibitive to calculate at each iteration or, in simple terms, it would take a lifetime to compute.

Thus, there's a variation of GD called SGD or stochastic gradient descent. The idea is simple, yet intelligent: we introduce the concept of a mini-batch. Instead of calculating the complete sum of gradients, it estimates the true gradient $\nabla f(\mathbf{x})$ using only a very small subset of these components at each step.

Now, how do we define this mini-batch? Let's assume we have 50,000 components ($N = 50000$). With that, we define a size that each mini-batch should have, and we'll call this b. That said:

- The closer b is to N, the more stable the gradient will be (and the closer its operation to GD). With this, each step is computationally more expensive, and we can use a larger α_k. It might converge in fewer epochs,[9] but that doesn't mean it will converge in less clock time. Additionally, we'll need more computer memory.
- The closer b is to 1,[10] the greater the noise. Each step will be faster and might require a smaller α_k. On the other hand, it can become very unstable.

Tip

It's common to use a value of b that is a multiple of 2 to maximize the use of computer memory and processing. In practice, it's common to test numbers like [32, 64, 128, 256, 512, 1024, 2048, ...].

That being said, what would the SGD algorithm look like? Let's look at the representation in Algorithm 2. Although it's a modification of what we saw earlier for GD in Algorithm 1, there are some changes worth considering. Let's see:

- We need a dataset D. In the airplane wing example we mentioned earlier, D would be the set of N flight conditions. In the credit card example for training an ML model, D would be a dataset containing numerous variables for N customers.
- The objective function is now called $L(\mathbf{x}; D)$, indicating that it's usually a loss function calculated *over* the dataset D.

  This is important because SGD is mostly used to solve ML problems. The loss function L measures how "bad" our model's predictions are compared to the actual training data (in D). Examples of loss functions are MSE, cross-entropy, and Mean Absolute Error (MAE), among others. In fact, it's common to see the term "loss function" when researching SGD—hence this renaming.
- Furthermore, we denote that the learning rate is explicitly α_k—that is, it should vary at each step k.

---

[9] We'll talk about this shortly, later in this section.

[10] $b = 1$ means we would be calculating one sample at a time.

- We changed the stopping criterion to consider a limit of epochs E. Each epoch will process all B mini-batches. At the beginning of each epoch, we shuffle[11] the dataset D and divide it into mini-batches of size b sequentially. This, in practice, implies that we're calculating mini-batches with data grouped differently at each epoch.

Algorithm 2 Stochastic Gradient Descent (SGD) with Mini-Batches

Require: Dataset $D = \{(\text{sample}^{(j)}, \text{label}^{(j)})\}_{j=1}^{\text{NumSamples}}$
Require: Objective/Loss function $L(\mathbf{x}; \text{sample}, \text{label})$
Require: Gradient estimator $\nabla \tilde{L}(\mathbf{x}; B)$ for mini-batch B
Require: Initial parameters $\mathbf{x}_0$
Require: Learning rate schedule or value(s) for α_k
Require: Number of epochs E
Require: Mini-batch size b
Ensure: Optimized parameters $\mathbf{x}^*$
Initialization:
Initialize parameters: $\mathbf{x} \leftarrow \mathbf{x}_0$
Initialize global step counter: $k \leftarrow 0$
for $e = 1$ to E **do** ▷ Loop over epochs
  Shuffle the dataset D randomly (optional, but common)
  Partition D into mini-batches $B_1, B_2, \ldots, B_{\lfloor \text{NumSamples}/b \rfloor}$
  for each mini-batch B_i in the partitioned D **do**
    Determine current learning rate α_k (can depend on k, e)
    Compute gradient for mini-batch: $\mathbf{g} \leftarrow \nabla \tilde{L}(\mathbf{x}; B_i)$
    ▷ (e.g., $\frac{1}{b} \sum_{(\text{sample},\text{label}) \in B_i} \nabla_{\mathbf{x}} L(\mathbf{x}; \text{sample}, \text{label})$)
    Update parameters: $\mathbf{x} \leftarrow \mathbf{x} - \alpha_k \cdot \mathbf{g}$
    $k \leftarrow k + 1$ ▷ Increment global step counter
  end for
end for
$\mathbf{x}^* \leftarrow \mathbf{x}$
return $\mathbf{x}^*$

Additionally, the update rule has been updated. The gradient $\nabla f(\mathbf{x}_k)$ is now an estimate based on a mini-batch:

$$\mathbf{x}_{k+1} = \mathbf{x}_k - \alpha_k \nabla \tilde{L}(\mathbf{x}_k; B_i) \tag{3.7}$$

Here, the average of the gradient over b samples is a much more stable estimate than with just one sample, and also with a much lower computational cost per step than if we were to calculate it over the entire dataset.[12]

[11] Imagine your dataset D is ordered in some way (for example, by some date criterion, or some identifier code). If we always calculated the mini-batches in the same order and distribution (i.e., the first b samples, then the next b, etc.), the algorithm could be biased by that order.

[12] There's also another "hidden" advantage: operations with mini-batches can be highly optimized on modern hardware (GPUs) using vectorized computation.

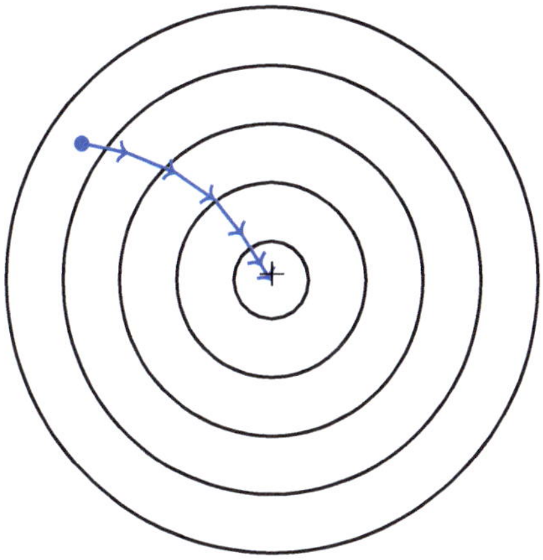

(a) Gradient descent.

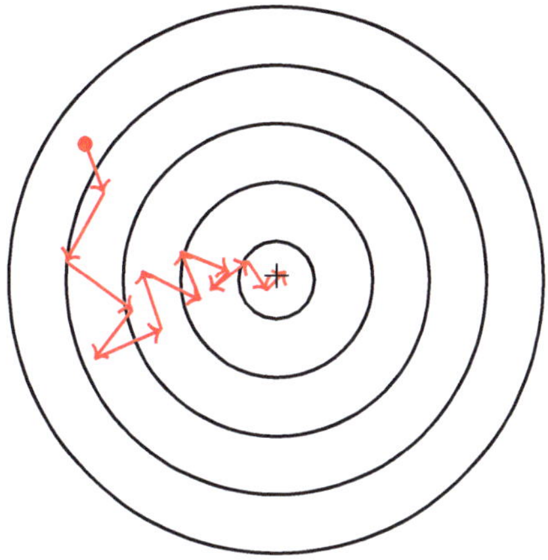

(b) Stochastic gradient descent.

Fig. 3.4 Comparison of typical trajectories of Gradient Descent (GD) and Stochastic Gradient Descent (SGD). *Source* The author, 2025

Naturally, this estimate $\tilde{L}(\mathbf{x}; B)$ will not be exactly equal to the true gradient $\nabla L(\mathbf{x}_k)$. It will have a random fluctuation (also called "noise") around the true direction, depending on the specific samples that fell into mini-batch B_i. With this, the path to be taken by SGD will be more erratic than that taken by GD, as we can see in the example in Fig. 3.4.

It might even seem that this noise is a bad thing, but in real problems, it ends up being quite useful. Firstly, SGD has a higher chance of escaping local minima in non-convex problems. Secondly, in high-dimensional problems, we can have a flat function in some directions and curved in others.[13] At these points, GD could slow down, just as we might think we're in the valley just because we have a flat road in our initial analogy of descending the mountain. On the other hand, the noise of SGD helps the algorithm escape these flat points and continue its descent in other directions.

That said, SGD is better than GD for iterating through large datasets and also for helping us escape local minima and saddle points. On the other hand, this same concept of noise implies that we can find a solution very close to the global optimum, but not necessarily it. And, furthermore, you might need to test a few other sizes of B for your problem.

[13] We call this a *saddle point*.

You only have ChatGPT thanks to SGD
Defining the ideal size of each mini-batch is part of the daily routine for all data scientists looking to train deep learning neural network algorithms, like ChatGPT itself. Consider this: algorithms like Adam, RMSProp, and their variations are, in fact, more sophisticated and adaptive versions of the SGD we just discussed.

These algorithms are used to optimize hundreds of billions, or even trillions, of parameters (the weights of a neural network). And they are trained on volumes of text that correspond to a good portion of the Internet. Calculating the true gradient of the loss function over all this data to take a single optimization step (as GD would do) would be impractical, even with the largest supercomputers. It would take too long and require an absurd amount of memory. That's why we use mini-batches to solve the problem. Furthermore, optimizers like Adam intelligently adjust the learning rate α_k for each of the trillions of parameters individually, helping to address the complex and non-convex optimization landscape of these giant models.

3.1.2.4 Newton's Method

If GD and SGD use only first-order information (the gradient), Newton's method uses second-order information for the optimization process. But, before we examine this method, it's important to discuss the Hessian.

Hessian
Let's return to our analogy of descending a mountain. But now, allow us to make a simplification. The concept of the gradient $\nabla f(\mathbf{x})$ we saw earlier would be like a combination of a compass (i.e., it shows in which cardinal direction the ascent is steepest) and an accelerometer (i.e., it shows how inclined this ascent is). On the other hand, we can't know what lies ahead. What if after taking a small step in the right direction, you encounter a large barrier, like a river you can't cross? Wouldn't it be easier to make a detour before encountering it?

Don't you think we'd get more details if we had a drone to see what's around us[14]?

In this sense, there's a concept in calculus called the Hessian matrix (also symbolized by $H(\mathbf{x})$). If $\nabla f(\mathbf{x})$ gives us the "slope" of the function, $H(\mathbf{x})$ (also known as $\nabla^2 f(\mathbf{x})$) informs us about the local "curvature" of our objective function.

It's a square matrix that organizes all the second partial derivatives of the function $f(\mathbf{x})$. Therefore, if we have n variables, the Hessian would be an $n \times n$ matrix, where the element in row i and column j is the second partial derivative of f, first with respect to x_i and then with respect to x_j:

$$H_{ij}(\mathbf{x}) = \frac{\partial^2 f}{\partial x_i \partial x_j}$$

By analyzing the properties of this matrix, such as its eigenvalues, we can determine if a point where the gradient is zero is a local minimum (Hessian "positive definite": curves upward in all directions), a local maximum (Hessian "negative definite": curves downward in all directions), or a saddle point (Hessian "indefinite": curves upward in some directions and downward in others).

That said, let's look at how this algorithm works in Algorithm 3. This is the most direct and basic implementation of the method. However, the implementations we find ready-made in software have additional modifications to make it more robust, such as the introduction of a line search or adjustments to the Hessian.

That being said, the central idea is to locally approximate the function $f(\mathbf{x})$ by a quadratic function[15] and then jump directly to the minimum of this quadratic approximation. The update rule involves the inverse of the Hessian matrix $\mathbf{H}(\mathbf{x}_k)$ (the matrix of second derivatives):

$$\mathbf{x}_{k+1} = \mathbf{x}_k - [\mathbf{H}(\mathbf{x}_k)]^{-1} \nabla f(\mathbf{x}_k) \tag{3.8}$$

The term $\mathbf{p}_k = -[\mathbf{H}(\mathbf{x}_k)]^{-1} \nabla f(\mathbf{x}_k)$ is called Newton's step.

In this more direct implementation of the method, we always have a step size $\alpha = 1$ for all iterations. That said, also notice that we are calculating the Hessian $\mathbf{H}_k$ (with n^2 second derivatives) and solving the linear system at each iteration. This can be very expensive[16] to solve in problems with many variables (i.e., n is large).

That said, let's look at Fig. 3.5. Notice that Newton's method reaches the minimum faster than GD. Fair enough after all the praise we gave the Hessian, right? In

---

[14] Considering the local curvature, and not seeing what is "ahead" of us.

[15] Using Taylor series up to the second order.

[16] Here, "expensive" refers to high computational cost. That is, we use many computational resources to calculate this. In practice, the more processor cycles and memory we use, the more clock time and/or more expensive and powerful computers we need.

Algorithm 3 Newton's Method for Optimization

Require: Objective function $f(\mathbf{x})$ to minimize, $\mathbf{x} \in \mathbb{R}^n$
Require: Gradient $\nabla f(\mathbf{x})$, Hessian $\mathbf{H}(\mathbf{x}) = \nabla^2 f(\mathbf{x})$
Require: Initial point $\mathbf{x}_0$
Require: Tolerance $\epsilon > 0$, ϵ_{step}
Require: Maximum number of iterations K
Ensure: Optimized point $\mathbf{x}^*$ that approximately minimizes $f(\mathbf{x})$

Initialization:
$k \leftarrow 0$
$\mathbf{x}_k \leftarrow \mathbf{x}_0$
Main Loop:
while $k < K$ **do**
    Calculate gradient: $\mathbf{g}_k \leftarrow \nabla f(\mathbf{x}_k)$
    if $||\mathbf{g}_k|| < \epsilon$ **then** ▷ Check for convergence based on gradient norm
        break ▷ Convergence reached (stationarity)
    end if
    Calculate Hessian: $\mathbf{H}_k \leftarrow \mathbf{H}(\mathbf{x}_k)$
    if $\mathbf{H}_k$ is not positive definite or is singular **then**
        ▷ Hessian modification needed (e.g., Levenberg-Marquardt) or switch strategy
        break ▷ Or apply recovery/modification strategy
    end if
    Solve the linear system $\mathbf{H}_k \mathbf{p}_k = -\mathbf{g}_k$ for search direction $\mathbf{p}_k$.
    if $||\mathbf{p}_k|| < \epsilon_{step}$ **then** ▷ Optional: Check for convergence based on step size
        break ▷ Convergence reached (small step)
    end if
    Set step size $\alpha_k \leftarrow 1$ ▷ For standard Newton's method
    Update point: $\mathbf{x}_{next} \leftarrow \mathbf{x}_k + \alpha_k \mathbf{p}_k$
    ▷ Optional: Line search could be performed here to find a better α_k
    ▷ Optional: Check for sufficient decrease if α_k is not 1 or if H_k was modified
    $\mathbf{x}_k \leftarrow \mathbf{x}_{next}$ ▷ Update current point for the next iteration
    $k \leftarrow k + 1$
end while
$\mathbf{x}^* \leftarrow \mathbf{x}_k$
return $\mathbf{x}^*$

fact, if the function is well behaved (i.e., differentiable, convex, and without extreme local curvatures), Newton's method will reach an optimal solution much faster. Furthermore, since the step size is determined by the Hessian, we wouldn't have an adjustment of the learning rate in this version.

Well-Behaved Functions

Of course, this description of a well-behaved function might make a lot of sense in a book, but it's not necessarily something easy to understand in a real context. Meaning: in a professional setting, an objective function is unlikely to be explicitly labeled as "well-behaved" by a supervisor.

That said, perhaps an analysis of the objective function is worthwhile to determine this. Firstly, if the function is convex, that would already be a good sign. With that, we could move on to the second point: look, if there are points in

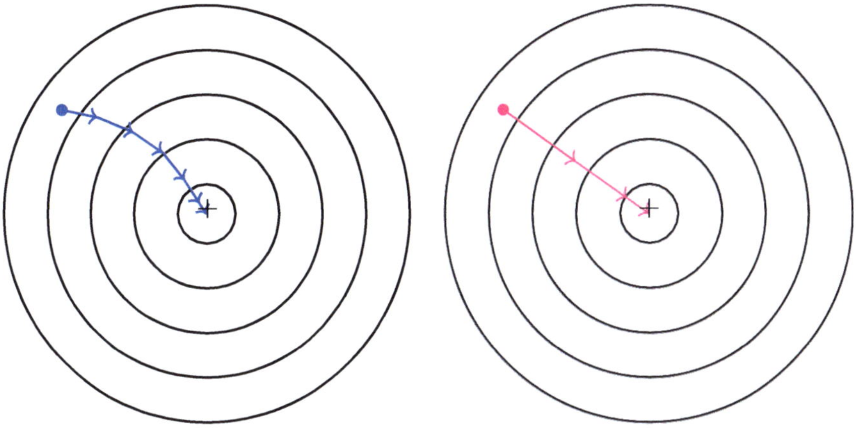

(a) Gradient descent. (b) Newton's Method.

Fig. 3.5 Comparison between the trajectories of gradient descent and Newton's method. *Source* The author, 2025

your objective function where the gradient and the Hessian become undefined (division by zero, log of a negative number, etc.), we wouldn't have a very smooth function to solve with this method.

However, these analyses can be relatively easy to calculate in simpler problems, but not in situations where we have many variables or an objective function whose equation we cannot directly access, as is the case with some machine learning problems. In that case, one of the things we usually do at work is to think practically: for that specific problem, is there a smooth relationship between $\mathbf{x}$ and $f(\mathbf{x})$? Or, depending on the rules and constraints, would there be some non-linearities in the problem? This can be an indicator that Newton's method might be a good option (or not).

Finally, if the Hessian brings advantages, it also brings some additional headaches for this method. First, calculating the Hessian and its inverse at each iteration, as we mentioned earlier, is computationally expensive when we have many variables involved. Second, it can be sensitive to the choice of the initial point—the farther we are from the optimum, the greater the risks of not finding it.[17]

[17] There are algorithms where we define a random point as $\mathbf{x}_0$. In others, we define a value that we judge to be close to the optimum based on our experience. In others, we start with a default value, like an average of the limits of each variable.

Quasi-Newton Methods and AI
To mitigate the computational cost of Newton's method, quasi-Newton methods were developed. The main idea is to approximate the Hessian matrix (or its inverse) using only information from gradients calculated in successive iterations. Popular algorithms like BFGS (Broyden-Fletcher-Goldfarb-Shanno) and L-BFGS (limited-memory BFGS) belong to this family.

These methods avoid the explicit calculation and inversion of the Hessian matrix at each step, operations that are computationally very expensive. L-BFGS, in particular, stores only a few of the most recent gradient and update vectors, making it viable even for problems with millions of variables, where it wouldn't even be possible to store the approximate Hessian of BFGS.

They offer a good balance between convergence speed (faster than GD) and computational cost (less costly than Newton's method). Therefore, we use them to train machine learning models like support vector machines (SVMs), logistic regression, and some types of neural networks. Furthermore, some software uses these types of optimization methods as the default for single-objective non-linear unconstrained optimization problems.

3.1.3 Derivative-Free Optimization Methods

We've discussed some advantages and disadvantages of certain methods in terms of convergence speed, computational cost per iteration, and robustness, but all those we saw in Sect. 3.1.2 share a fundamental premise: they depend on the ability to calculate, variably, the derivatives of $f(\mathbf{x})$.

So, what would we do if the function $f(\mathbf{x})$ is very complex, non-differentiable, or if we couldn't calculate the gradient for any reason? In such cases, we need methods that evaluate the function only by its values, without using derivatives. Let's look at some of them now.

Tip
Derivative-free methods are an alternative when gradients are not available, but they are typically less efficient than gradient-based methods in situations where the latter can be used.

3.1.3.1 Nelder-Mead or Downhill Simplex

One of the most classic and ingenious algorithms that operates without using derivatives is the Nelder-Mead method, also known as the Downhill Simplex method.[18] It's purely geometric, which radically differs from the techniques we've seen so far. In this case, Nelder-Mead works with a geometric shape called a simplex.

Tip
As a reminder: a simplex is a geometric figure with $n + 1$ vertices in an n-dimensional space. Thus, a simplex would be

- 1D space: a simplex would have two points (a segment);
- 2D space: a simplex would have three points (a triangle);
- 3D space: a simplex would have four points (a tetrahedron);
- n-D space: a simplex would have $n + 1$ points (a tetrahedron).

And, in our optimization context, each vertex of the simplex represents a candidate solution **x** being evaluated.

The algorithm's objective is to make this simplex "slide" and "adapt" through the space of function f(**x**), moving and changing shape iteratively, always trying to "flow" toward regions where the values of f are lower, until it eventually stabilizes and contracts around a minimum point. We know this geometric description might initially seem abstract, particularly in a text with numerous mathematical concepts. Therefore, we'd like to invoke another analogy: this time, of water droplets trickling down a surface, like the example in Fig. 3.6.

With that, suppose we have a small cluster of $n + 1$ droplets very close to each other, resting on the irregular surface of a leaf (and we'd like you to imagine this surface as the landscape of the function $f(\mathbf{x})$). So, the droplets can only travel this predetermined space of the surface. Due to gravity and the micro-undulations of the leaf, each droplet $\mathbf{x}_i$ is at a location with an associated function value $f(\mathbf{x}_i)$. Here, the objective of this cluster is to find the position of lowest possible energy, perhaps a small depression or concavity on the leaf, where it can stabilize. How do they do this without a sensor or GPS to locate themselves? Let's think of a simple step-by-step process.

Firstly, we check the "locality" $f(\mathbf{x})$ of each of the $n + 1$ droplets and order them from the best solution to the worst solution. In our context, the best solution would be the lowest droplet, which we'll call $\mathbf{x}_{best}$, and the worst solution would be the highest droplet, which we'll call $\mathbf{x}_{worst}$. The second worst solution would be the droplet in the second highest position. We'll call this $\mathbf{x}_{secondworst}$.

[18] This is different from the Simplex for Linear Programming, which we will see in Sect. 3.2.6.

Fig. 3.6 Analogy for the Nelder-Mead method: water droplets seeking the lowest point on a leaf surface, representing the function landscape. *Source* Pexels

After that, our mission is to discard this worst position ($\mathbf{x}_{worst}$) and replace it with something better. To do this, we would calculate the centroid (the vector average) $\mathbf{x}_{centroid}$ among all our solutions except for $\mathbf{x}_{worst}$.

Now, we know that the point $\mathbf{x}_{worst}$ of the leaf is bad. So, the idea is to test a new point symmetrically opposite in relation to the center of the other good points (i.e., symmetrically opposite in relation to $\mathbf{x}_{centroid}$). This phase is called *reflection*, and the name we give to this new point is $\mathbf{x}_{reflection}$.

At this moment, we have a few choices:

1. Alternative 1: $\mathbf{x}_{reflection}$ is better than $\mathbf{x}_s$, but still hasn't outperformed the best we had previously (i.e., $f(\mathbf{x}_{best}) \leq f(\mathbf{x}_{reflection}) < f(\mathbf{x}_s)$).
   In this case, we replace $\mathbf{x}_{worst}$ with $\mathbf{x}_{reflection}$ and return to the first step.
2. Alternative 2: It worked out very well. That is, $\mathbf{x}_{reflection}$ is better than all other points, including $\mathbf{x}_{best}$.
   In this case, we continue with the *expansion* step. Here, we'll try to move $\mathbf{x}_{reflection}$ even further from $\mathbf{x}_{centroid}$ and $\mathbf{x}_{worst}$. We'll call this new point $\mathbf{x}_{expanded}$. With that, we have two more alternatives:
   Alternative 2A: $\mathbf{x}_{expanded}$ is even better than $\mathbf{x}_{reflection}$. That is:

$$f(\mathbf{x}_{expanded}) < f(\mathbf{x}_{reflection})$$

With this, we replace $\mathbf{x}_{worst}$ with $\mathbf{x}_{expanded}$ and return to the first step.
Alternative 2B: $f(\mathbf{x}_{expanded}) \geq f(\mathbf{x}_{reflection})$. With this, we replace $\mathbf{x}_{worst}$ with $\mathbf{x}_{reflection}$ and return to the first step.

3. Alternative 3: It didn't work. That is, $\mathbf{x}_{reflected}$ is worse than $\mathbf{x}_s$. Here, we use an idea similar to the expansion step, but now it's called *contraction*. Now, we'll try to bring $\mathbf{x}_{reflection}$ closer to $\mathbf{x}_{centroid}$. This new point is called $\mathbf{x}_{contraction}$.
Alternative 3A: If $f(\mathbf{x}_{contraction}) < f(\mathbf{x}_{worst})$, we swap $\mathbf{x}_{worst}$ for $\mathbf{x}_{contraction}$ and return to the first step.
Alternative 3B: $f(\mathbf{x}_{contraction}) \geq f(\mathbf{x}_{worst})$. So, we're in a bad region and we shrink the simplex. With that, we move all $\mathbf{x}$, except for $\mathbf{x}_{best}$, closer to $\mathbf{x}_{best}$.

Thus, we keep repeating the cycle of ordering the solutions (meaning: the droplets in our example), trying to reflect the highest one, and deciding whether to expand, accept, contract, or shrink. The cluster of droplets (the simplex) moves and deforms on the leaf's surface, tending to "flow" to lower points.

This algorithm will generally terminate in two situations: either the "cluster" becomes too small (the distance between droplets is less than a tolerance value we specify ourselves), or when the positions of all droplets are very close (indicating a flat region, probably a minimum).

That said, let's see an example implementation in Algorithm 4. Unlike gradient-based methods, Nelder-Mead introduces a different set of concepts and parameters to control its behavior. Besides the notation we've seen so far, notice that we also have

- α is the reflection coefficient. It defines how far $\mathbf{x}_{reflected}$ would be projected through $\mathbf{x}_{centroid}$. This needs to be a value above zero. We usually use $\alpha = 1$, which means the reflected point is exactly the same distance from the centroid as the original point $\mathbf{x}_{worst}$, but on the opposite side.
- γ is the expansion coefficient. It's a value greater than 1, and we usually use a value greater than α. It determines how much we would "stretch" beyond $\mathbf{x}_{reflected}$. It's common to use $\gamma = 2$, which indicates we would try to advance twice the distance from $\mathbf{x}_{centroid}$ in the direction of reflection.
- ρ: This coefficient defines how much we would contract if the reflection doesn't work out. We normally use a value $0 < \rho \leq 0.5$, but the most common value is 0.5. That is, we would try to insert a point halfway between $\mathbf{x}_{centroid}$ and $\mathbf{x}_{reflected}$ or $\mathbf{x}_{worst}$.
- σ: This coefficient defines how much we would move all points to get closer to $\mathbf{x}_{best}$, except for itself. It would be a value $0 < \sigma < 1$, but the most common value is 0.5. That is, we would move all points to half the distance that originally separated them from $\mathbf{x}_{best}$.

That said, the great advantage of Nelder-Mead lies in its ability to operate without any derivative information. It's conceptually simpler than higher order methods, which is why it's one of the first optimization algorithms we learn in universities. Visual descriptions, like Fig. 3.7, illustrate this. Furthermore, it often finds decent

Algorithm 4 Nelder-Mead Simplex Algorithm

Require: Objective function $f(\mathbf{x})$ to be minimized, $\mathbf{x} \in \mathbb{R}^n$
Require: Initial simplex $S = \{\mathbf{x}_1, \ldots, \mathbf{x}_{n+1}\}$ ($n + 1$ points)
Require: Reflection $\alpha > 0$ (typ. 1); Expansion $\gamma > 1$ (typ. 2)
Require: Contraction $0 < \rho \leq 0.5$ (typ. 0.5); Shrink $0 < \sigma < 1$ (typ. 0.5)
Require: Termination: max iterations K, tolerance ϵ
Ensure: Best point found $\mathbf{x}^*$

$k \leftarrow 0$
while $k < K$ **do** ▷ Main iteration loop
    Evaluate $f_i \leftarrow f(\mathbf{x}_i)$ for all $\mathbf{x}_i \in S$.
    Order vertices in S such that $f_1 \leq f_2 \leq \cdots \leq f_{n+1}$.
    Let $\mathbf{x}_{best} \leftarrow \mathbf{x}_1$, $f_{best} \leftarrow f_1$.
    Let $\mathbf{x}_s \leftarrow \mathbf{x}_n$ (second worst), $f_s \leftarrow f_n$.
    Let $\mathbf{x}_{worst} \leftarrow \mathbf{x}_{n+1}$ (worst), $f_{worst} \leftarrow f_{n+1}$.
    if stopping criterion met (e.g., stddev$(f_1, \ldots, f_{n+1}) < \epsilon$ or simplex diameter small) **then**
        break
    end if
    $\mathbf{x}_{centroid} \leftarrow \frac{1}{n} \sum_{i=1}^{n} \mathbf{x}_i$ ▷ Centroid of the n best points (all except $\mathbf{x}_{worst}$)
    // 1. Reflection
    $\mathbf{x}_{reflection} \leftarrow \mathbf{x}_{centroid} + \alpha(\mathbf{x}_{centroid} - \mathbf{x}_{worst})$
    $f_{reflection} \leftarrow f(\mathbf{x}_{reflection})$
    if $f_{best} \leq f_{reflection} < f_s$ **then** ▷ Reflection is not worse than second worst, not better than best
        Replace $\mathbf{x}_{worst}$ in S with $\mathbf{x}_{reflection}$.
    else if $f_{reflection} < f_{best}$ **then** ▷ Reflection is better than current best, attempt Expansion
        $\mathbf{x}_{expanded} \leftarrow \mathbf{x}_{centroid} + \gamma(\mathbf{x}_{reflection} - \mathbf{x}_{centroid})$
        $f_{expanded} \leftarrow f(\mathbf{x}_{expanded})$
        if $f_{expanded} < f_{reflection}$ **then**
            Replace $\mathbf{x}_{worst}$ in S with $\mathbf{x}_{expanded}$. ▷ Accept expansion
        else
            Replace $\mathbf{x}_{worst}$ in S with $\mathbf{x}_{reflection}$. ▷ Accept reflection (expansion failed)
        end if
    else ▷ $f_{reflection} \geq f_s$. Reflection is poor, perform Contraction
        $\mathbf{x}_{contraction} \leftarrow \mathbf{x}_{centroid} + \rho(\mathbf{x}_{worst} - \mathbf{x}_{centroid})$ ▷ Contracts worst point towards centroid
        $f_{contraction} \leftarrow f(\mathbf{x}_{contraction})$
        if $f_{contraction} < f_{worst}$ **then**
            Replace $\mathbf{x}_{worst}$ in S with $\mathbf{x}_{contraction}$. ▷ Accept contraction
        else
            // Shrink Simplex
            for $i \leftarrow 2$ to $n + 1$ **do** ▷ For all points in S except current $\mathbf{x}_{best}$
                $\mathbf{x}_i \leftarrow \mathbf{x}_{best} + \sigma(\mathbf{x}_i - \mathbf{x}_{best})$ ▷ Move $\mathbf{x}_i$ closer to $\mathbf{x}_{best}$
            end for ▷ Shrunk points will be re-evaluated at the start of the next iteration
        end if
    end if
    $k \leftarrow k + 1$
end while
$\mathbf{x}^* \leftarrow \mathbf{x}_1$ (the best point in the final simplex after ordering)
return $\mathbf{x}^*$

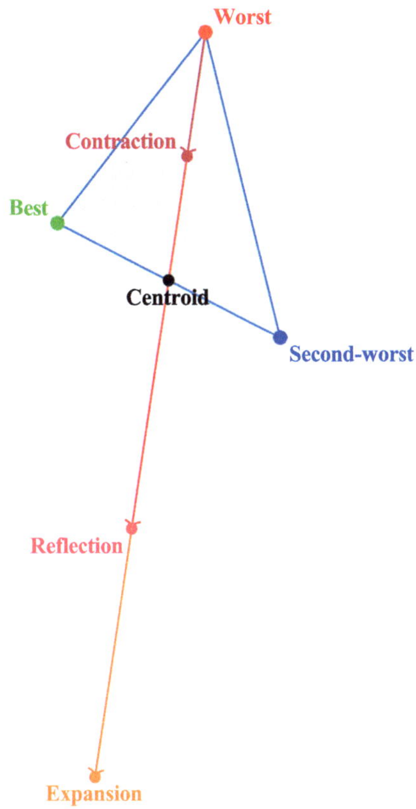

Fig. 3.7 An example of point manipulation in one iteration of a Nelder-Mead algorithm. *Source* The author, 2025

solutions with relative ease for low-dimensional problems, so it ends up being useful for some problems.

On the other hand, its disadvantages become more pronounced as the complexity or dimensionality of the problem increases. That's why we rarely see this algorithm being used within companies for solving real problems.

3.1.3.2 Pattern Search

This family of methods operates by exploring points on a "mesh" or pattern defined around the current solution, $\mathbf{x}_k$, without using derivatives. Thus, there isn't "the" pattern search algorithm, but rather various types of algorithms with this nature, such as

- Hooke-Jeeves.
- Coordinate search.
- Axis-aligned search.
- Generalized pattern search.
- Mesh adaptive direct search.

The central idea is to define a set of search directions $\mathcal{D} = \{\mathbf{d}_1, \ldots, \mathbf{d}_p\}$ (the pattern) and a step size (or mesh size) $\Delta_k > 0$. The simplest pattern uses positive and negative coordinate directions (e.g., in 2D: $\mathcal{D} = \{(1, 0), (-1, 0), (0, 1), (0, -1)\}$).

To understand this better, let's suppose you're in a river at point $\mathbf{x}_k$, and you need to find the deepest point. To make things difficult, imagine the river is murky, and you can't see anything underwater. To help you navigate, you have a staff in your hands, and this staff has a length Δ_k, but it can increase and decrease in size because it's foldable. You probe the terrain at a distance Δ_k in a fixed pattern of directions (e.g., north, south, east, and west). If you find a lower point in any of these probes, you move there (success) and perhaps even use a larger staff next time (increase Δ). If none of the probes reveal a lower point (failure), you stay where you are, but fold the staff (decrease Δ) to make closer probes in the next attempt. You repeat the process until the step size (staff length) becomes minimal.

That said, let's look at Algorithm 5 for a generic implementation of this family of algorithms. Figure 3.8 also illustrates how this type of algorithm works. At each iteration k, the algorithm evaluates the objective function $f(\mathbf{x})$ at the mesh points: $\mathbf{x}_k + \Delta_k \mathbf{d}_i$ for each direction $\mathbf{d}_i \in \mathcal{D}$. This step is called polling or mesh search. The goal is to find a point on the mesh that is better than the current point $\mathbf{x}_k$.

With this, we basically have two possible outcomes. First, let's consider a scenario where we were successful: in situations where some point on the mesh, say $\mathbf{x}_{trial} = \mathbf{x}_k + \Delta_k \mathbf{d}_j$, presented a better function value than the current one ($f(\mathbf{x}_{trial}) < f(\mathbf{x}_k)$), the iteration is considered successful. The new solution becomes $\mathbf{x}_{k+1} = \mathbf{x}_{trial}$. In this case, the step size for the next iteration, Δ_{k+1}, can be maintained or even increased (e.g., $\Delta_{k+1} = \gamma_{increase} \Delta_k$, with an expansion factor $\gamma_{increase} \geq 1$, and usually 2) to try to accelerate the search in the promising direction.

The second scenario would be failure. That is, if no point on the mesh is better than $\mathbf{x}_k$ (i.e., $f(\mathbf{x}_k + \Delta_k \mathbf{d}_i) \geq f(\mathbf{x}_k)$ for all directions $\mathbf{d}_i \in \mathcal{D}$), the iteration fails. The current solution is maintained ($\mathbf{x}_{k+1} = \mathbf{x}_k$), but the step size is necessarily reduced (e.g., $\Delta_{k+1} = \gamma_{decrease} \Delta_k$, with a contraction factor $0 < \gamma_{decrease} < 1$, typically 0.5). This allows the algorithm to refine the search in a smaller neighborhood around $\mathbf{x}_k$.

This algorithm also has a stopping criterion. In this case, it will terminate when the step size Δ_k becomes smaller than a predefined tolerance (Δ_{tol}), indicating that we cannot find significant improvements with smaller steps, or after a maximum number of iterations or function evaluations.

That being said, algorithms from this family can be used in problems where derivatives are not available, are difficult to calculate, or the function is non-smooth or very noisy. Additionally, important variants, such as Generalized Pattern Search (GPS), are more robust when compared to gradient-based methods—so much so that they are implemented in commercial software like MATLAB, for example.

On the other hand, their convergence can be extremely slow in problems with many variables or in problems whose variables have very different numerical scales. Furthermore, it can converge to local minima in non-convex functions, without necessarily finding the global optimal solution.

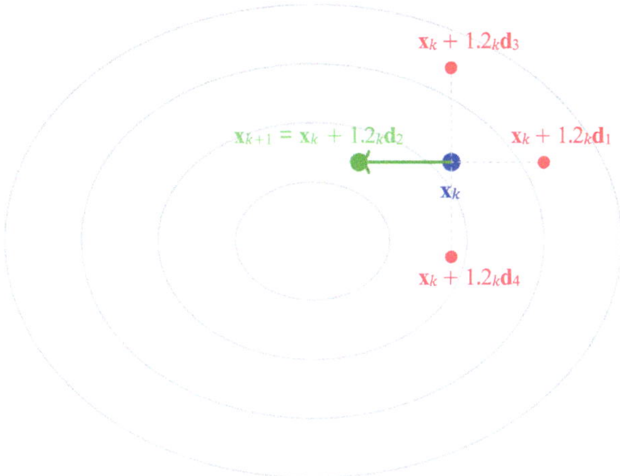

(a) Success case. Better point found: $f(\mathbf{x}_{k+1}) < f(\mathbf{x}_k)$. Mesh size unchanged.

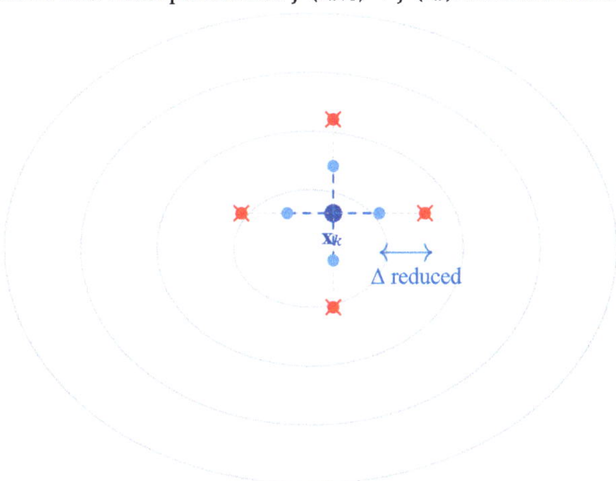

(b) Failure case. No improvement found: $f(\mathbf{x}_k + \Delta_k\mathbf{d}_i) \geq f(\mathbf{x}_k)$. Mesh size reduced.

Fig. 3.8 Illustration of the polling step in pattern search. The search evaluates points on a mesh around the current solution $\mathbf{x}_k$. In blue, we have the current point $\mathbf{x}_k$. In red, we have the mesh points $\mathbf{x}_k + \Delta_k\mathbf{d}_i$. In green, we have the next iterate $\mathbf{x}_{k+1}$. The gray dashed line represents the mesh grid. *Source* The author, 2025

3.1.3.3 Evolutionary Algorithms

This is one of our favorites. Consider the following analogy: suppose there's a species of moth that lives in a forest near you. Since it's a forest, there are various other species coexisting, from plants to animals and, naturally, predators. For moths, the main survival goal is to minimize the chance of being eaten by predators. Some species use different techniques to protect themselves. In the case of moths, they

Algorithm 5 Pattern Search Algorithm

Require: Objective function $f(\mathbf{x})$ to minimize
Require: Initial point $\mathbf{x}_0$ (n-dimensional vector)
Require: Set of pattern directions $\mathcal{D} = \{\mathbf{d}_1, \ldots, \mathbf{d}_p\}$ (n-dimensional vectors)
Require: Initial step size $\Delta_0 > 0$
Require: Step size tolerance $\Delta_{tol} > 0$ (stopping criterion)
Require: Step expansion factor $\gamma_{inc} \geq 1$
Require: Step contraction factor $0 < \gamma_{dec} < 1$
Require: (Optional: Maximum number of iterations k_{max})
Ensure: Final point found $\mathbf{x}^*$

Initialization:
$k \leftarrow 0$
$\mathbf{x}_k \leftarrow \mathbf{x}_0$
$\Delta_k \leftarrow \Delta_0$
$f_k \leftarrow f(\mathbf{x}_k)$
Main Loop:
while $\Delta_k \geq \Delta_{tol}$ (**and** $k < k_{max}$ if k_{max} is used) **do**
  // Exploration Step (Polling)
  $found\_better \leftarrow$ **false**
  $\mathbf{x}_{best\_poll} \leftarrow \mathbf{x}_k$ ▷ Best point found in this polling step
  $f_{best\_poll} \leftarrow f_k$
  for all $\mathbf{d} \in \mathcal{D}$ **do** ▷ Iterate through each pattern direction
    $\mathbf{x}_{trial} \leftarrow \mathbf{x}_k + \Delta_k \mathbf{d}$
    $f_{trial} \leftarrow f(\mathbf{x}_{trial})$
    if $f_{trial} < f_{best\_poll}$ **then**
      $\mathbf{x}_{best\_poll} \leftarrow \mathbf{x}_{trial}$
      $f_{best\_poll} \leftarrow f_{trial}$
      $found\_better \leftarrow$ **true** ▷ Optional: break here if "first improvement" strategy is used
    end if
  end for
  // Update Step
  if $found\_better$ **then** ▷ Successful poll: move to the best point found
    $\mathbf{x}_{k+1} \leftarrow \mathbf{x}_{best\_poll}$
    $f_{k+1} \leftarrow f_{best\_poll}$
    $\Delta_{k+1} \leftarrow \gamma_{inc} \cdot \Delta_k$ ▷ Maintain or increase step size
  else ▷ Unsuccessful poll: no improvement found
    $\mathbf{x}_{k+1} \leftarrow \mathbf{x}_k$ ▷ Stay at the current point
    $f_{k+1} \leftarrow f_k$
    $\Delta_{k+1} \leftarrow \gamma_{dec} \cdot \Delta_k$ ▷ Reduce step size
  end if
  // Prepare for Next Iteration
  $\mathbf{x}_k \leftarrow \mathbf{x}_{k+1}$
  $f_k \leftarrow f_{k+1}$
  $\Delta_k \leftarrow \Delta_{k+1}$
  $k \leftarrow k + 1$
end while
$\mathbf{x}^* \leftarrow \mathbf{x}_k$
return $\mathbf{x}^*$

use camouflage: the better the color and pattern of their wings blend with the tree trunks, the lower the chance of being seen. Therefore, the objective here would be to minimize the species' visibility (or maximize camouflage), and not that of a single individual. After all, we want to perpetuate the species in this forest.

However, the initial moth population in this forest is diverse. This means we have, in this forest, moths with wings of different colors and shades. Some have spots, and others don't. Some have more conspicuous colors, and others have less conspicuous colors. And each moth is an indiv.idual, and its physical characteristics are defined by its genes. We'll call this initial population $P(0)$.

Fig. 3.9 A tree bark with lichen. *Source* Pexels

Now, an important piece of information about the forest: it has several trees with lichen, like the example in Fig. 3.9. Moths land on these trunks while predators walk and fly over the forest. Consequently, the moths from $P(0)$ that stand out most against the lichen and tree bark will be more easily seen by predators. While a few conspicuous individuals might survive by chance, the majority are likely to be predated. Here, let's call this survival aptitude of moths *fitness*.

As you probably know from your school days, moths that survive are those that manage to reproduce and pass on their genes. In our example, these are the ones that camouflage best. Thus, nature chooses individuals with better fitness. Perhaps we could call this *natural selection*, but for generalization purposes, let's just call this step *selection*.

That said, the surviving moths reproduce among themselves. We don't necessarily define which individual will reproduce with which, but we know we'll have offspring moths from the survivors. These offsprings will inherit a combination of their parents' genes, which we call *crossover*. Additionally, sometimes we can have small random changes in the offsprings' genes, such as a slightly different color than their parents had, or perhaps some other change in their physical aspects—that is, a *mutation*. Some mutations can help, and others can worsen each individual's survival.

Thus, eventually, we have a new generation of moths, which we could call $P(1)$. Then, we could have $P(2)$ for a next generation, and so on. For each generation, we repeat the same steps of selection and reproduction with crossover and mutation. In this way, each generation will be slightly better adapted to the tree trunks we saw in Fig. 3.9.

The example we provided is inspired by biological evolution, and these same concepts guide the family of evolutionary algorithms (EAs). This type of algorithm is excellent for avoiding local minima, but they don't guarantee optimality.

They maintain a population of candidate solutions and apply operators inspired by biological evolution (selection, crossover, mutation) to generate new solutions and converge toward promising regions. These methods are very good at exploring the search space and quickly finding the general region where the best solution (or one of the best) is located, but slow to "polish" that solution and achieve high numerical precision at the exact optimal point. Thus, they end up being good for finding very good solutions close to the ideal, but not necessarily finding the best possible.[19]

As we will see in Chaps. 4 and 5, there are several types of EAs. However, we wouldn't want to keep you waiting until then. Therefore, we describe in Algorithm 6 the general and iterative structure of a generic EA, designed for unconstrained SOO.

The process begins with the initialization phase, in which an initial population $P(0)$ containing N individuals (candidate solutions) is created, and the fitness of each individual $\mathbf{x}$ is calculated based on the objective function f. Then, the main loop is executed for a maximum number of generations G_{max} or until another stopping criterion is met.

Notice that much of the logic occurs within the loop—that is, within each generation t. The first step may involve the selection of pairs of individuals that would reproduce from the current population $P(t)$, depending on the variation strategy. Then, the key step of variation generates a set of offspring $Offspring(t)$ containing a number $N_{offspring}$. Here, we apply generic evolutionary operators (which may include recombination, mutation, or others) to the selected parents or to the total population.

Afterward, we evaluate the fitness of these new offspring, then selecting the N fittest survivors from the combination of $Offspring(t)$ (and, depending on the technique, also considering $P(t)$) to form the next generation $P(t+1)$. At this point, you'll notice we have two parameters: μ, which indicates the size of the parent population, and λ, which indicates the number of offspring generated. Typically, these two parameters are equal to N.

At this point, we have reached the end of the evaluations for one generation and have chosen which new individuals should proceed. Thus, we proceed to the next generation to repeat the same operations. That said, at the end of the algorithm, we choose the best individual $\mathbf{x}_{best}$ from the final population.

[19] The aphorism "perfect is the enemy of good" fits well in this type of situation.

Algorithm 6 Generic Evolutionary Algorithm (EA)

```
Require: Population size N (i.e., μ)
Require: Number of offspring to generate N_offspring (i.e., λ)
Require: Fitness function f(x) (conventionally, higher is better; adapt if minimizing)
Require: Parent selection method SelectParents(P, num_needed)      ▷ We often use tournament selection here
Require: Number of parents to select num_needed (usually 2)
Require: Recombination method Recombine(p1, p2) and probability p_c
Require: Mutation method Mutate(ind) and probability p_m
Require: Survivor selection method SelectSurvivors(P_current, O_offspring, N_target)
Require: Termination condition (e.g., max generations G_max)
Ensure: Best individual found x_best
Initialization:
P(0) ← InitializePopulation(N)
EvaluateFitness(P(0), f)
t ← 0
x_best ← FindBestIndividual(P(0))                                  ▷ Overall best individual found so far
f_best_so_far ← f(x_best)
Main Evolutionary Loop:
while Termination condition not met (e.g., t < G_max) do
                                                    ▷ 1. Variation: Generate N_offspring Offspring —
    Offspring(t) ← ∅
    for k ← 1 to ⌈N_offspring/2⌉ do                 ▷ Assuming recombination produces two offspring
        (parent1, parent2) ← SelectParents(P(t), num_needed)
        if RandomReal(0, 1) < p_c then
            (child1, child2) ← Recombine(parent1, parent2)
        else
            child1 ← parent1; child2 ← parent2                    ▷ Clones if no crossover
        end if
        child1 ← Mutate(child1, p_m)                              ▷ Mutation applied based on p_m
        child2 ← Mutate(child2, p_m)
        Add child1 to Offspring(t)
        if |Offspring(t)| < N_offspring then                      ▷ Ensure N_offspring are added
            Add child2 to Offspring(t)
        end if
    end for
    if |Offspring(t)| > N_offspring then
        Trim Offspring(t) to size N_offspring
    end if
                                                                  ▷ 2. Evaluate New Offspring
    EvaluateFitness(Offspring(t), f)
                           ▷ 3. Survivor Selection (Form Next Generation Parent Population)
    P(t + 1) ← SelectSurvivors(P(t), Offspring(t), N)
                                ▷ e.g., (μ + λ) selects N best from P(t) ∪ Offspring(t)
                   ▷ e.g., (μ, λ) selects N best from Offspring(t) (requires N_offspring ≥ N)
                                                            ▷ 4. Update Overall Best Solution
    x_best_in_gen ← FindBestIndividual(P(t + 1))
    if f(x_best_in_gen) > f_best_so_far then                      ▷ Assumes f is maximized
        x_best ← x_best_in_gen
        f_best_so_far ← f(x_best)
    end if
    t ← t + 1
end while
return x_best
```

3.1.4 *Choosing the Best Technique at Work*

That said, which of these techniques could you use to solve a problem (unconstrained, remember) in a company? The main question would be regarding access: for a specific optimization problem, do you have access to the derivatives (i.e., the gradient and Hessian)?

If we have access to the derivatives and can calculate them, then we can test some of the gradient-based methods. In ML problems, the answer will almost always be SGD and its variants. For problems with a large number of variables, we can start with one of the quasi-Newtonian methods, like L-BFGS, which we discussed in Sect. 3.1.2.4. Or, if we have a smaller number of variables (something like up to 1000 variables, depending on the problem), we could test BFGS. Newton's method, in turn, could be tested in situations where we have a smaller number of variables, the Hessian is computable, and if its inversion is also feasible.

Now, let's suppose you don't have access to the derivatives. Or, perhaps, these derivatives are very expensive to calculate. Or, maybe, you realize that the objective function isn't smooth. In these cases, we would need to opt for derivative-free methods. And then it would depend on the number of variables: if we have up to about 10 variables and if the objective function is relatively smooth,[20] we could opt for Nelder-Mead and pattern search.

Thinking about the real world
Try starting with the easiest and simplest method available in the software you have access to. Practically all optimization software has L-BFGS (when you have gradients) and Nelder-Mead (when you don't), at least. We usually test Nelder-Mead with up to 10 variables, for example. Start with one of them to save time, and then experiment with one or two more techniques. After all, it might be that for your specific problem, one algorithm could work better than another.

On the other hand, if we have more than 10 variables, or if the objective function isn't smooth, if it has many local optima, or even if the objective function is so complex that we can't read and understand its logic (i.e., a "black box"), it might be more interesting to test evolutionary algorithms.

3.2 Constrained Optimization

Most real-world optimization problems involve constraints. As we saw in Chap. 2, such constraints can be equality constraints ($g_i(\mathbf{x}) = 0$) or inequality constraints

[20] Stochastic simulations, data from sensors, samplings, and estimates introduce noise into the objective function.

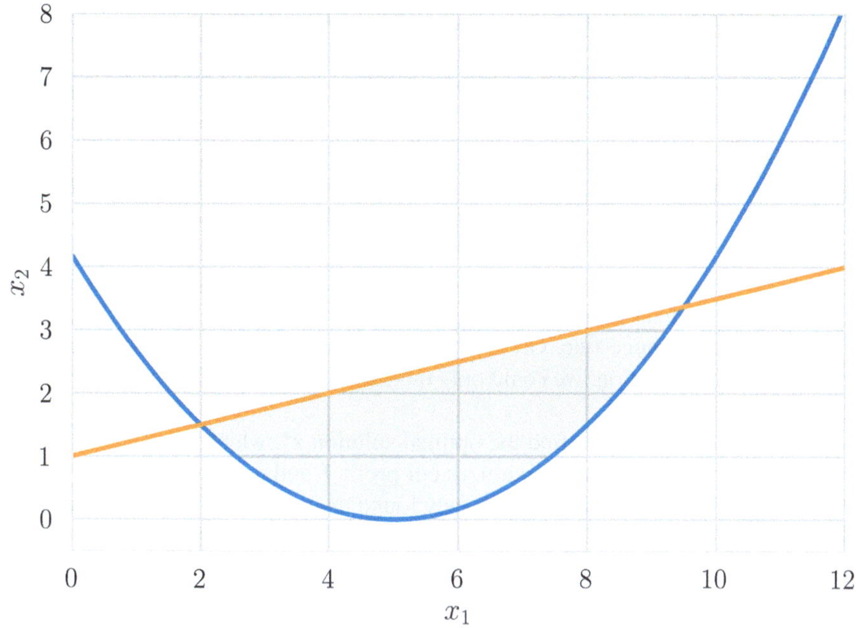

Fig. 3.10 Illustration of a feasible region and the difference between the unconstrained minimum (outside the region) and the constrained minimum (on the boundary or within the region). Each line represents a constraint. *Source* The author, 2025

($h_j(\mathbf{x}) \leq 0$). They define the *feasible region*, which is the set of all solutions $\mathbf{x}$ that satisfy all constraints. To recall the concept of the feasible region seen in Fig. 2.1, see Fig. 3.10.

Thus, our objective now is

$$\begin{aligned} \min_{\mathbf{x}} \quad & f(\mathbf{x}) \\ \text{subject to} \quad & g_i(\mathbf{x}) = 0, \quad i = 1, \ldots, m \\ & h_j(\mathbf{x}) \leq 0, \quad j = 1, \ldots, p \end{aligned} \tag{3.9}$$

These constraints translate into some problems for us. First, we can no longer simply follow the gradient of the objective function, as this could lead us out of the feasible region. Therefore, we'll need to adapt the techniques we learned in Sect. 3.1 or, perhaps, think of new techniques.

But, before proceeding, it's important that we understand the conditions that an optimal solution (which we'll call $\mathbf{x}^*$ here) must satisfy in a constrained problem, assuming our objective function is differentiable.

3.2.1 Lagrange Multipliers (Equality Constraints)

We'd like to draw your attention to another analogy to discuss Lagrange multipliers. In this case, let's return to the donut example from Chap. 1. Now, we have a small factory that produces only two flavors, $V1$ and $V2$. To produce one unit of $V1$, you spend a certain amount of electrical energy e_1 (after all, we have to think about the oven and all the appliances used in production), and to produce one unit of $V2$, we spend e_2. And, knowing this, we'd like to maximize the total profit f of the factory.

However, we have a rule: the electric company has a contract with our factory, and we'll need to use a fixed amount of energy per day e_{day}—no more, no less. Otherwise, we'd have to pay a huge fine. Or, thinking mathematically, $e_1 \times V1 + e_2 \times V2 = e_{day}$. Now, in this equation, we could only modify $V1$ and $V2$, right? So, they would be our variables.

With this, we're trying to find the optimal solution $\mathbf{x}^*$, which is the combination of producing $V1$ and $V2$ that maximizes our profit f and uses exactly e_{day}.

And, to help, let's assume we've already found the optimal solution $\mathbf{x}^*$. But, what if at this point we could still use a little more energy? For example, an extra 1 kWh? At this point, we could test what our additional profit would be if we allocated this extra energy to produce $V1$ and what the additional profit would be if, instead, we allocated it to $V2$. If there's a difference between these two profits, we would know that, in practice, we could further improve this solution by reducing the production of what has the lower profit. Meaning: if at that moment we know that if we made a little more of $V2$ we could further improve our profit, why not reduce $V1$ a bit to make more $V2$, right?

In fact, we could do this indefinitely until we reached a point of equilibrium—that is, when any attempt to produce more of one product to increase profit would require more energy, violating the constraint, and the only way to free up energy would be to reduce the other product, but the gain on one side would be equal to (or less, at the optimum) the loss on the other in terms of marginal profit per energy. Or, in other words, we wouldn't be able to improve f anymore.

This is the concept of Lagrange multipliers: they are the mathematical tool that helps us find this equilibrium point.

We formalize this by introducing the Lagrangian function $\mathcal{L}(\mathbf{x}, \boldsymbol{\lambda})$:

$$\mathcal{L}(\mathbf{x}, \boldsymbol{\lambda}) = f(\mathbf{x}) + \sum_{i=1}^{m} \lambda_i g_i(\mathbf{x}) \tag{3.10}$$

where λ_i are these Lagrange multipliers.[21]

The necessary condition for $\mathbf{x}^*$ to be a local optimum is that there exists a vector $\boldsymbol{\lambda}^*$ such that the gradient of the Lagrangian with respect to $\mathbf{x}$ is zero, and the original constraints are satisfied:

---

[21] Some books, instead of adopting λ_i, use an alphabetical sequence starting from λ: $[\lambda, \mu, \nu, \xi, ...]$. However, we adopt the λ_i notation because we can reuse the other letters shortly.

$$\nabla_{\mathbf{x}} \mathcal{L}(\mathbf{x}^*, \boldsymbol{\lambda}^*) = \nabla f(\mathbf{x}^*) + \sum_{i=1}^{m} \lambda_i^* \nabla g_i(\mathbf{x}^*) = \mathbf{0} \tag{3.11a}$$

$$g_i(\mathbf{x}^*) = 0, \quad i = 1, \ldots, m \tag{3.11b}$$

Solving these $n + m$ equations for the $n + m$ unknowns ($\mathbf{x}^*$ and $\boldsymbol{\lambda}^*$) gives us the candidate points for local optima.

3.2.2 KKT Conditions (Equality and Inequality Constraints)

Now, what if we have inequality constraints ($h_j(\mathbf{x}) \leq 0$)? In these cases, the situation is a bit more complex. An inequality constraint can be active at the optimum ($\mathbf{x}^*$), meaning $h_j(\mathbf{x}^*) = 0$, or inactive, where $h_j(\mathbf{x}^*) < 0$. That said, we'd like to talk a bit about the Karush-Kuhn-Tucker (KKT) conditions, which generalize Lagrange multipliers to include inequalities. They are necessary conditions for optimality, under certain regularity conditions. Let's see: for a point $\mathbf{x}^*$ to be a local optimum, there must exist multipliers $\boldsymbol{\lambda}^*$ (for equalities) and $\boldsymbol{\mu}^*$ (for inequalities) respecting the criteria below.

Stationarity

This means that the gradient of the Lagrangian (now including the inequality terms) with respect to $\mathbf{x}$ is zero:

$$\nabla f(\mathbf{x}^*) + \sum_{i=1}^{m} \lambda_i^* \nabla g_i(\mathbf{x}^*) + \sum_{j=1}^{p} \mu_j^* \nabla h_j(\mathbf{x}^*) = \mathbf{0} \tag{3.12}$$

Primal feasibility

This means that we are satisfying the original constraints:

$$g_i(\mathbf{x}^*) = 0, \quad i = 1, \ldots, m \tag{3.13}$$

$$h_j(\mathbf{x}^*) \leq 0, \quad j = 1, \ldots, p \tag{3.14}$$

Dual feasibility

This means that the multipliers of the inequality constraints must be non-negative to ensure that following the gradient of an active inequality constraint moves us out of the feasible region, which is necessary so that f cannot decrease further in that direction. Or, mathematically speaking:

$$\mu_j^* \geq 0, \quad j = 1, \ldots, p \tag{3.15}$$

Complementary slackness

This is: if a constraint is inactive, it does not influence the solution locally. Therefore, its multiplier is zero. Or, in other words: for each inequality constraint, either it is active ($h_j(\mathbf{x}^*) = 0$) or its multiplier is zero ($\mu_j^* = 0$), or both:

$$\mu_j^* h_j(\mathbf{x}^*) = 0, \quad j = 1, \ldots, p \tag{3.16}$$

Important
Although we don't see many mentions of KKT conditions in work meetings or Optimization software, it's important that you know they exist. After all, they are fundamental in constrained optimization theory because, among other reasons:

- They provide a criterion to check if a point found by an algorithm is (potentially) a local optimum.
- They are the theoretical basis for the development of many constrained optimization algorithms (such as interior-point methods and sequential quadratic programming).
- For convex problems, the KKT conditions are also sufficient for global optimality.

That said, solving the KKT conditions directly is generally not how algorithms work in practice (except in very specific cases). Instead, we use iterative algorithms that try to find points satisfying these conditions. We'll look at some of these algorithms now.

3.2.3 *Penalty Methods*

A smart idea for solving constrained problems is to transform them into a sequence of unconstrained problems that are easier to solve. In this regard, we can use methods that add a term to the objective function that penalizes constraint violations.

Important
Penalty methods are not an optimization algorithm per se, but rather a transformation technique. That is: we can convert a problem with constraints into a problem without constraints and, with that, use an unconstrained optimization algorithm, like those presented in Sect. 3.1.

For this, we'd like you to think about the points system on a driver's license adopted by some countries. A driver's main objective is to reach their destination as quickly as possible (minimize travel time, or $f(\mathbf{x})$). To do this, the driver needs to respect various rules, such as speed limits, traffic signs, and traffic lights, among others. In this case, imagine these are constraints like $h_j(\mathbf{x}) \leq 0$.

That said, what happens when the driver violates one of these rules, for example, by exceeding the speed limit $h_j(\mathbf{x}) > 0$? This driver will receive a penalty ρ in the form of a fine and some points on their license. The more severe the infraction or the more infractions committed, the greater the accumulated penalty will be. And, the more points on the license, the greater the chances of its suspension.

Therefore, notice that the driver is not just trying to minimize travel time $f(\mathbf{x})$, but rather optimizing a combination that considers travel time and the total cost of penalties. The lower the penalty ρ, the greater this driver's appetite for risk might be—meaning, they might decide it's worth violating the limits a bit to significantly reduce travel time, since the consequences would be low. On the other hand, if ρ is high, the cost of violating the rules becomes prohibitive. Thus, the driver would be encouraged to drive within the limits to reach their destination, even if it increases their travel time.

Penalty methods in optimization work analogously: we add an artificial "cost" (the penalty term, which depends on ρ) to the original objective function $f(\mathbf{x})$. This cost increases the more the constraints are violated. By minimizing the modified function, the algorithm is guided to find solutions that violate the constraints less, especially when we increase ρ.

That said, we could adapt the constrained problem from (3.9) to become an unconstrained function:

$$P(\mathbf{x}; \rho) = f(\mathbf{x}) + \rho \sum_{i=1}^{m} [g_i(\mathbf{x})]^2 + \rho \sum_{j=1}^{p} [\max(0, h_j(\mathbf{x}))]^2 \tag{3.17}$$

where, as we mentioned earlier, $\rho > 0$ is a penalty parameter. Here, we minimize $P(\mathbf{x}; \rho)$ for an increasing sequence of ρ. As $\rho \to \infty$, the solution of the unconstrained problem tends to the solution of the original constrained problem.

We have only one small problem here: if we want high precision, the value of ρ will need to be very large.[22] In fact, there's no single "ideal" value of ρ. It's common

[22] Or, technically speaking, the practical drawback of penalty methods is **ill-conditioning**. As ρ becomes very large, the Hessian matrix of the penalized function $P(\mathbf{x}; \rho)$ can become ill-

to test sequentially (for example, starting with 1, 10, or 100). Eventually, we'll find an optimal solution, even if it violates some constraints. In this case, we increase the value of ρ, multiplying it by a factor (for example, [1, 10, 100, 1000, ...]) until the optimized solution violates the constraints at an acceptable level. Naturally, the definition of what is acceptable is subjective and depends on each problem. There are scenarios and people who will try to adjust the value of ρ until no constraint is violated. In other cases, until only a small percentage of constraints are violated.

In the industry It's common to find cases where we use these methods in conjunction with algorithms that don't originally handle constraints. For example, some of the techniques we'll see in Chap. 4 are great for complex problems, but are normally formulated for unconstrained problems. Thus, we can use penalties to adapt our constrained problem for these algorithms.

3.2.4 Barrier Methods (or Interior-Point Methods)

We've just seen how penalty methods deal with constraints in a "reactive" way: they allow the solution to violate constraints, but add an increasing cost or penalty for this violation. That's why we used the driver's license analogy. After all, we only receive a fine after committing an infraction, right? Now, let's explore a different approach: barrier methods.

The philosophy here is preventive. Instead of penalizing those who "cross the line," these methods create a kind of invisible barrier within the feasible region. This barrier term, added to the original objective function, has a low or moderate value well within the interior of the allowed region, but grows abruptly and tends to infinity as the solution $\mathbf{x}$ approaches the boundary (where $h_j(\mathbf{x}) = 0$).

For this, we'd like to introduce you to Kora: a Shiba Inu puppy. She, as you can see in Fig. 3.11, is a good girl. However, she's also very independent and stubborn. Consider this: she loved running around the lawn and the house, but there were some more dangerous places we didn't want her to go, like the kitchen, laundry room, bathrooms, and the garden flowers. On the other hand, it would be impractical to put a fence in all these places.

In this case, notice that our intention is for her to explore as much as she can (or, in our mathematical terms, maximize $f(\mathbf{x})$) as long as she doesn't enter these areas we mentioned (i.e., something like a constraint $h(\mathbf{x}) \leq 0$). In other words, we want to prevent her from reaching these forbidden spaces, and for her to know that she's approaching a forbidden place.

---

conditioned, leading to numerical difficulties and slow convergence for the unconstrained optimization algorithm used to minimize it.

Fig. 3.11 Kora, a good girl. *Source* The author, 2025

Now, how did we do this during her training? Every time she got close to these restricted areas, we could give her a command like a "*No!*". If she ignores us and continues approaching these restricted areas, our command gets louder, something like a "*Nooo!!!*", and then "***Noooooo!!!!!!***", and it keeps increasing in intensity as her stubbornness also increases. If she still disobeys us and continues advancing, the moment she's about to put her paw in one of these forbidden areas, we hold her and make her go back somewhere else. Over time, she learned that this command means something bad—after all, she wants to play without being interrupted, right?

The key here is that the intensity of our command (the "barrier") increases to the point of becoming so prohibitive right at the boundary that, for Kora, it's as if there's an invisible glass wall there. The consequence of trying to touch or cross it is perceived as infinitely bad (in her case, the consequence could be a temporary restriction from the desired area), and so she learns to never even touch the boundary line. And thus, she would always operate within the interior of the allowed region.

Similarly, the algorithm will try to find solutions that respect the constraints, since violating them becomes prohibitively expensive for the total value of the function we're trying to optimize.[23] Here, we add a term that "explodes" when **x** approaches

[23] Likewise, Kora will try to stay in a space where no one is constantly saying "*No!*" in her ear.

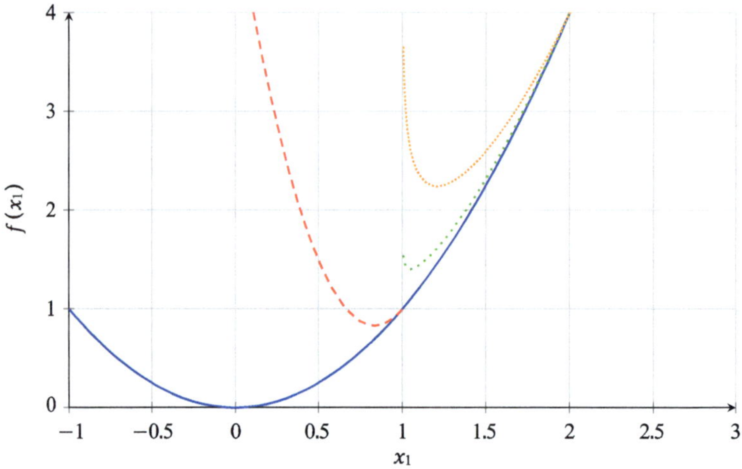

Fig. 3.12 Conceptual illustration of penalty functions (increases outside the feasible region) and barrier functions (increases near the feasible boundary). The blue line represents the unconstrained objective. The dashed red line represents a penalty function, while the black and orange dotted lines represent two different barrier functions. Here, the feasible boundary is set at $x_1 = 1$. *Source* The author, 2025

the boundary of the feasible region defined by the inequalities $h_j(\mathbf{x}) \leq 0$. This forces the solutions to remain strictly within the feasible region. A common barrier is the logarithmic one:

$$B(\mathbf{x}; \tau) = f(\mathbf{x}) - \tau \sum_{j=1}^{p} \ln(-h_j(\mathbf{x})) \tag{3.18}$$

where $\tau > 0$ is the barrier parameter. We minimize $B(\mathbf{x}; \tau)$ for a decreasing sequence of $\tau \to 0$.

Let's return to Kora's example: the barrier parameter τ would be like how "close" to that invisible barrier she can get before our strongest command (which would basically be a loud "*no!*" and, furthermore, we'd take her somewhere else). A larger τ means we would start restricting earlier, keeping her further from the edge. And, the more we reduce the value of τ (i.e., $\tau \to 0^+$), the more she could explore the spaces near that barrier to access the forbidden areas. In fact, we have an interesting example of its operation in Fig. 3.12.

Furthermore, adjusting the value of τ is similar to how we did it for penalty methods. That is: it's iterative and depends on each problem. It's common to start with a higher value of τ, like 1 or 0.1. In practice, this would mean we're a bit further from the boundaries. Then, we test smaller values, usually dividing by 10 (i.e., $[10^0, 10^{-1}, 10^{-2}, 10^{-3}, \ldots, 10^{-8}]$) until we're comfortable with the optimal solution we've found.

Penalty or Barrier?
There are situations where one or the other type of method is better. For example, if we don't care about the precision of a solution's feasibility, but rather an approximation of it, we would generally lean toward penalty methods. Also, if we have more inequality constraints than equality constraints, we go in the direction of barrier methods.

Problems like linear programming, convex quadratic programming (as is the case with financial problems), and smooth non-linear problems are excellent application cases for a barrier method. Penalty methods, on the other hand, would be used to solve problems with derivative-free techniques or those evolutionary algorithms we saw earlier, for example.

Also, note that barrier methods are typically applied to inequality constraints $h_j(\mathbf{x}) \leq 0$ (or $h_j(\mathbf{x}) \geq 0$ by transformation) and require an initial strictly feasible point.

In the industry That said, our explanation was conceptual. In practice, it's common to have methods that intelligently combine functions to solve problems, as is the case with Interior-Point Methods (IPMs). In fact, the implementation of IPMs combines barrier functions and some concepts from Newton's method to solve subproblems. Thus, applications used by companies, like CPLEX and Gurobi, use an algorithm that, behind the scenes, employs this type of logic.

3.2.5 Sequential Quadratic Programming (SQP)

The methods we've discussed previously aren't necessarily algorithms themselves, but rather techniques we use in conjunction with existing algorithms. This is not the case with SQP.

This is an iterative method, which means it refines the solution with each iteration. To explain this iteration process, we'll use, guess what? An analogy.

Now, let us consider an analogy from strategy video games. Among those games, we have complex real-time strategy games, like StarCraft, Age of Empires, and Kerbal Space Program. In these games, we need to move a unit (like a hero, a spaceship, or a warrior) toward some objective. Since there's a chance you've never played a game like this, allow us to generalize as well.

Thus, imagine a generic space game where you need to take your spaceship from an orbit around Earth to an orbit around the Moon efficiently (i.e., minimize fuel consumption $f(\mathbf{x})$) while respecting the laws of orbital physics (non-linear) and the limits of your engine and fuel (constraints $g_i = 0$, $h_j \leq 0$). However, as you well know, the distance between Earth and the Moon is enormous—even more so considering the size of a spaceship. Therefore, it's very difficult to calculate the entire trajectory of this spaceship with precision in advance.

Because of this, we wouldn't calculate the entire trip in advance, but focus only on the next step we need to take, like firing the engines and burning more fuel. And, for this planning, we don't need to focus on all the laws of Physics and the Universe, but only on what influences us most at that specific point in space. With this, we can simplify and estimate many calculations. In other words, we divide a large objective (reaching the Moon) into small steps, or QP subproblems.

For each QP, we can also estimate when and where we should make the next course adjustment. Every time we burn more fuel, we can also make some course adjustments. After all, we wouldn't necessarily execute what we planned—sometimes we realize we can burn a little more or a little less fuel, mainly so we don't deviate too much from the planned route and don't spend all the fuel before time. Consider this: even though we're planning the next step along these orbits, we shouldn't forget the final objective, which is to land on the Moon.

Now, how does this translate to SQP? The central idea is, at each iteration $\mathbf{x}_k$ (in our example, at each position or state of the spaceship between Earth and the Moon where we plan a maneuver):

1. Approximate the Lagrangian function by a quadratic function. In our example, this would be like estimating the "cost" of the next maneuver, both from a fuel perspective and in terms of how it affects reaching the final objective. In this case, we do this using a simplified quadratic formula instead of the complete and non-linear orbital physics.
2. Linearize the constraint functions. In our example, this would be assuming, in a simplified way, that the laws of physics, like gravity or thrust, act linearly or constantly during the short duration of the maneuver being planned.
3. Solve this *quadratic programming subproblem* (QP) to find a search direction $\mathbf{p}_k$. A QP subproblem is given by Eq. 3.19, where $\mathbf{B}_k$ is an approximation of the Hessian of the Lagrangian (often using quasi-Newton updates like BFGS). Returning to our example, this would be solving a "simplified maneuver plan" that tells us the ideal direction and intensity (p_k) for the next engine burn, according to the simplifications and approximations we've made so far.
4. Perform a line search along the direction p_k to find $\mathbf{x}_{k+1} = \mathbf{x}_k + \alpha_k p_k$. In practice, we're ensuring an improvement in a **merit function**, which combines the objective and constraint violation. That is, it would be like executing the calculated burn while monitoring the actual fuel spent and the resulting trajectory in relation to the final objective to decide the exact moment to cut the engine, avoiding waste or excessive deviations.

$$\begin{aligned} \min_{\mathbf{p}} \quad & \frac{1}{2}\mathbf{p}^T \mathbf{B}_k \mathbf{p} + \nabla f(\mathbf{x}_k)^T \mathbf{p} \\ \text{subject to} \quad & \nabla g_i(\mathbf{x}_k)^T \mathbf{p} + g_i(\mathbf{x}_k) = 0, \quad i = 1, \ldots, m \\ & \nabla h_j(\mathbf{x}_k)^T \mathbf{p} + h_j(\mathbf{x}_k) \leq 0, \quad j = 1, \ldots, p \end{aligned} \tag{3.19}$$

Merit function The merit function is an auxiliary function constructed to balance two conflicting objectives: minimizing the objective function f and satisfying the constraints g_i and h_j. Here, we generally use a penalty term to measure how infeasible $\mathbf{x}$ would be.

That being said, let's consider the implementation of a generic SQP algorithm, as in the example in Algorithm 7. In it, we have some new variables, such as:

- B_k: An $n \times n$ matrix that approximates the Hessian of the Lagrangian function at iteration k. It captures the curvature information of the problem in the vicinity of $\mathbf{x}_k$. Without it, we can't converge quickly in SQP. It's common to update it at each iteration k using methods like BFGS.
- p_k: An n-dimensional vector representing the search direction calculated at iteration k. It's the solution of the QP subproblem and indicates the step that the quadratic approximation suggests taking to improve the solution.
- α_k: A scalar we call "step length." It determines how far to move in direction p_k. It's a value usually in the range $(0, 1]$.
- $\phi(\mathbf{x})$: The merit function is used to evaluate the quality of a candidate point, balancing improvement in the objective f with reduction in constraint violation.
- λ_{k+1}: An estimate of the Lagrange multipliers at the new iteration.

That said, we use this type of algorithm in small- to medium-sized non-linear problems that have derivatives. As an example, we have model predictive control problems[24] and resource allocation problems with non-linear returns/costs and budgetary or market constraints that are solved with SQP.

3.2.6 Simplex Method (for Linear Programming)

Now, we shift focus to a widely used class of problems for constrained optimization: Linear Programming (LP). As we saw in Sect. 1.3.5, a problem is linear if both the objective function and all constraints are linear. To solve this specific type of problem, there's a classic and very powerful algorithm called the Simplex method.

Be careful with the name "Simplex"!
Don't confuse the Simplex method for LP, which we'll see now, with the Nelder-Mead method (or Downhill Simplex) presented in Sect. 3.1.3.1. Although they share names, they are fundamentally different algorithms:

---

[24] Where we optimize a sequence of future control actions subject to system dynamics (non-linear) and actuator/state limits.

Algorithm 7 Sequential Quadratic Programming (SQP)

Require: Objective function $f(\mathbf{x})$; Equality constraints $g_i(\mathbf{x}) = 0, i = 1, \ldots, m$; Inequality constraints $h_j(\mathbf{x}) \leq 0, j = 1, \ldots, p$
Require: Initial point $\mathbf{x}_0$
Require: Initial positive definite Hessian approximation $\mathbf{B}_0$ (e.g., $\mathbf{I}$)
Require: Tolerance ϵ; Maximum iterations k_{max}
Ensure: Optimal solution $\mathbf{x}^*$ (and optionally, optimal Lagrange multipliers $\boldsymbol{\lambda}^*$, $\boldsymbol{\mu}^*$)
Initialization:
$k \leftarrow 0$
$\mathbf{x}_k \leftarrow \mathbf{x}_0$
$\mathbf{B}_k \leftarrow \mathbf{B}_0$
Main Loop:
while $k < k_{max}$ **do**
  // Solve QP Subproblem
  Calculate $f_k = f(\mathbf{x}_k)$, $\nabla f_k = \nabla f(\mathbf{x}_k)$, $g_{i,k} = g_i(\mathbf{x}_k)$, $\nabla g_{i,k} = \nabla g_i(\mathbf{x}_k)$, $h_{j,k} = h_j(\mathbf{x}_k)$, $\nabla h_{j,k} = \nabla h_j(\mathbf{x}_k)$.
  Solve for $\mathbf{p}_k$:

$$\min_{\mathbf{p}} \quad \frac{1}{2}\mathbf{p}^T \mathbf{B}_k \mathbf{p} + \nabla f_k^T \mathbf{p}$$
$$\text{s.t.} \quad \nabla g_{i,k}^T \mathbf{p} + g_{i,k} = 0, \quad i = 1, \ldots, m$$
$$\nabla h_{j,k}^T \mathbf{p} + h_{j,k} \leq 0, \quad j = 1, \ldots, p$$

  Let $\mathbf{p}_k$ be the solution, and $\boldsymbol{\lambda}_{QP,k}$, $\boldsymbol{\mu}_{QP,k}$ be the QP Lagrange multipliers.
  if QP subproblem is infeasible or $\mathbf{p}_k$ is not a suitable descent direction **then**
    ▷ Handle: e.g., modify $\mathbf{B}_k$, use trust region, filter, or other recovery
    break ▷ Or attempt recovery strategy if implemented
  end if
  // Check Convergence
  if $||\mathbf{p}_k|| < \epsilon$ **or** KKT conditions satisfied by $(\mathbf{x}_k, \boldsymbol{\lambda}_{QP,k}, \boldsymbol{\mu}_{QP,k})$ within ϵ **then**
    $\mathbf{x}^* \leftarrow \mathbf{x}_k$; $\boldsymbol{\lambda}^* \leftarrow \boldsymbol{\lambda}_{QP,k}$; $\boldsymbol{\mu}^* \leftarrow \boldsymbol{\mu}_{QP,k}$
    break ▷ Convergence reached
  end if
  // Line Search
  Find step length $\alpha_k \in (0, 1]$ (e.g., using a merit function $\Phi(\mathbf{x}, \nu)$ and Armijo/Wolfe conditions) such that $\Phi(\mathbf{x}_k + \alpha_k \mathbf{p}_k, \nu_k) < \Phi(\mathbf{x}_k, \nu_k)$ (sufficient decrease).
  // Update Solution
  $\mathbf{x}_{k+1} \leftarrow \mathbf{x}_k + \alpha_k \mathbf{p}_k$
  // Update Hessian Approximation (e.g., BFGS)
  $\mathbf{s}_k \leftarrow \mathbf{x}_{k+1} - \mathbf{x}_k$
  $\nabla_x L_k \leftarrow \nabla f(\mathbf{x}_k) + \sum_i (\boldsymbol{\lambda}_{QP,k})_i \nabla g_i(\mathbf{x}_k) + \sum_j (\boldsymbol{\mu}_{QP,k})_j \nabla h_j(\mathbf{x}_k)$
  $\nabla_x L_{k+1} \leftarrow \nabla f(\mathbf{x}_{k+1}) + \sum_i (\boldsymbol{\lambda}_{QP,k})_i \nabla g_i(\mathbf{x}_{k+1}) + \sum_j (\boldsymbol{\mu}_{QP,k})_j \nabla h_j(\mathbf{x}_{k+1})$
  $\mathbf{y}_k \leftarrow \nabla_x L_{k+1} - \nabla_x L_k$
  if $\mathbf{s}_k^T \mathbf{y}_k > 0$ **then** ▷ Ensure curvature condition for BFGS

$$\mathbf{B}_{k+1} \leftarrow \mathbf{B}_k + \frac{\mathbf{y}_k \mathbf{y}_k^T}{\mathbf{y}_k^T \mathbf{s}_k} - \frac{\mathbf{B}_k \mathbf{s}_k \mathbf{s}_k^T \mathbf{B}_k}{\mathbf{s}_k^T \mathbf{B}_k \mathbf{s}_k}$$

  else
    $\mathbf{B}_{k+1} \leftarrow \mathbf{B}_k$ ▷ Skip update or use a modified update
  end if
  // Prepare Next Iteration
  $\mathbf{x}_k \leftarrow \mathbf{x}_{k+1}$
  $\mathbf{B}_k \leftarrow \mathbf{B}_{k+1}$
  $k \leftarrow k + 1$
end while
if $k = k_{max}$ and $||\mathbf{p}_k|| \geq \epsilon$ **then** ▷ Max iterations reached without satisfying tolerance
  $\mathbf{x}^* \leftarrow \mathbf{x}_k$ ▷ Return current best if loop finished by k_{max}
end if
return $\mathbf{x}^*$ (and optionally $\boldsymbol{\lambda}^*$, $\boldsymbol{\mu}^*$ if available from last step)

- Nelder-Mead: Uses a geometric shape that moves and deforms in the search space for unconstrained problems, without using derivatives.
- Simplex for LP (here): Is an algorithm for constrained linear problems that works by "jumping" between the vertices of the feasible region defined by the constraints, seeking the optimal vertex.

If the optimization problem is linear (linear objective function and all linear constraints), we have an LP problem, as discussed in Sect. 1.3.5. That said, any LP problem can be converted to the standard form established by Eq. 3.20.

$$\begin{aligned} \min_{\mathbf{x}} \quad & \mathbf{c}^T\mathbf{x} \\ \text{subject to} \quad & \mathbf{A}\mathbf{x} = \mathbf{b} \\ & \mathbf{x} \geq \mathbf{0} \end{aligned} \tag{3.20}$$

This equation means the following: we want to minimize a linear objective function $\mathbf{c}^T\mathbf{x}$ considering a vector of decision variables $\mathbf{x} = [x_1, x_2, \ldots, x_n]^T$, and a vector of cost coefficients $\mathbf{c} = [c_1, c_2, \ldots, c_n]^T$. To calculate the dot product $c_1x_1 + c_2x_2 + \cdots + c_nx_n$, we would need to transpose one of them—hence $\mathbf{c}^T$.

This problem needs to respect the set of equality constraints $\mathbf{A}\mathbf{x} = \mathbf{b}$, where $\mathbf{A}$ is the matrix of constraint coefficients. If there are m constraints and n variables, $\mathbf{A}$ is an $m \times n$ matrix:

$$\mathbf{A} = \begin{bmatrix} a_{11} & a_{12} & \ldots & a_{1n} \\ a_{21} & a_{22} & \ldots & a_{2n} \\ \vdots & \vdots & \ddots & \vdots \\ a_{m1} & a_{m2} & \ldots & a_{mn} \end{bmatrix} \tag{3.21}$$

where element a_{ij} is the coefficient of variable x_j in the i-th constraint equation. $\mathbf{b}$ is a column vector with m elements.

$$\mathbf{b} = \begin{bmatrix} b_1 \\ b_2 \\ \vdots \\ b_m \end{bmatrix} \tag{3.22}$$

In practice, we'll have a complete system of m linear equations:

$$\begin{cases} a_{11}x_1 + a_{12}x_2 + \cdots + a_{1n}x_n = b_1 \\ a_{21}x_1 + a_{22}x_2 + \cdots + a_{2n}x_n = b_2 \\ \vdots \\ a_{m1}x_1 + a_{m2}x_2 + \cdots + a_{mn}x_n = b_m \end{cases} \tag{3.23}$$

Furthermore, we have the non-negativity constraints $\mathbf{x} \geq \mathbf{0}$. This is nothing more than a compact way of saying that each decision variable x_i must be greater than or equal to zero (i.e., $x_i \geq 0$ for all $i = 1, \ldots, n$). In fact, this is a common requirement for many LP algorithms.

Of course, this formalization is useful, but how does the Simplex method use it to find the optimal solution? The key lies in a fundamental geometric property of linear programming.

Geometry of LP and its Vertices

A fundamental property of LP is that the region defined by linear constraints ($\mathbf{A}\mathbf{x}^T = \mathbf{b}, \mathbf{x} \geq \mathbf{0}$) forms a geometric figure called a *polytope*. Think of a well-behaved polygon in 2D or a crystal with many faces and corners (or vertices) in 3D.

Each vertex corresponds to a specific solution that satisfies all constraints. Here, the important information for us is: if an LP problem has a finite optimal solution (i.e., it's not unbounded), then at least one of these vertices of the feasible region will be an optimal solution. With this, our search is simplified: instead of searching infinite points within or on the faces of the polytope, we focus only on the vertices, which are a finite number.

To help better understand this technique, we'd like you to think of an ant lost in a tangle of branches on the lawn. The ant's goal is simple: escape the branches and return to the anthill. Since it can't jump or fly, this means it needs to go to the lowest point of this tangle, which is where one of the branches touches the lawn so it can get out.

In this case, some branches touch at specific points (a vertex). The ant would start at any vertex and analyze all the branches that are part of that vertex. Its next step is to choose the branch that offers the steepest incline. After all, it can get closer to the ground this way. Then, it will walk along this branch until it finds the next vertex, and repeat the analysis—that is, see which would be the next branch with the steepest incline. As you can imagine, the ant would stop these checks when it reached a vertex where no branch would lead to an improvement in the objective or, in other words, when it reached the lawn.

What do you think about seeing a pseudocode based on this analogy in Algorithm 8?

Notice that we mentioned the need to start with a feasible vertex (BFS). "Phase I" is a process whose sole objective is to find any basic feasible solution for the original

Algorithm 8 Simplex Method for Linear Programming

Inputs:
LP Problem in Standard Form: min $\mathbf{c}^T\mathbf{x}$ subject to $\mathbf{A}\mathbf{x}^T = \mathbf{b}, \mathbf{x} \geq \mathbf{0}$
Output:
Solution status (Optimal, Unbounded, or Infeasible)
If Optimal: Optimal solution $\mathbf{x}^*$ and optimal value $f(\mathbf{x}^*)$
Initialization:
Find an initial Basic Feasible Solution (BFS) $\mathbf{x}_0$ (a vertex of the feasible region).
▷ This may require a Phase I of Simplex if the origin is not feasible.
if no initial BFS is found **then**
  Return Status = Infeasible ▷ Problem has no feasible solution.
end if
$\mathbf{x}_{current} \leftarrow \mathbf{x}_0$
Main Loop (Simplex Iterations):
loop
  ▷ 1. Check Optimality Condition
  Calculate reduced costs ($\bar{c}_j$) for all variables x_j not in the current basis.
  ▷ The reduced cost indicates the rate of change in the objective if x_j increases.
  if all $\bar{c}_j \geq 0$ (for a minimization problem) **then**
    $\mathbf{x}^* \leftarrow \mathbf{x}_{current}$ ▷ Current vertex is optimal.
    $f(\mathbf{x}^*) \leftarrow \mathbf{c}^T\mathbf{x}^*$
    Return Status = Optimal, $\mathbf{x}^*$, $f(\mathbf{x}^*)$
    break ▷ End of algorithm.
  end if
  ▷ 2. Choose Direction (Variable to Enter Basis)
  Select a non-basic variable x_j with reduced cost $\bar{c}_j < 0$.
  ▷ This variable will enter the basis, defining the edge to follow.
  ▷ (Common rule: choose j with the most negative $\bar{c}_j$).
  ▷ 3. Choose Maximum Step (Variable to Leave Basis)
  Determine which current basic variable, x_{B_i}, will limit the increase of x_j (Ratio Test).
  ▷ Finds how far we can go in the chosen direction before violating $\mathbf{x} \geq \mathbf{0}$.
  if no basic variable limits the step (x_j can increase indefinitely) **then**
    Return Status = Unbounded ▷ The objective can go to $-\infty$.
    break
  end if
  Let x_{B_r} be the basic variable chosen to leave the basis.
  ▷ 4. Move to Next Vertex (Pivot)
  Perform the pivoting operation: swap x_j (enters) for x_{B_r} (leaves) in the basis.
  Update the basic feasible solution $\mathbf{x}_{current}$ to the new vertex.
  ▷ (Algebraically, this updates the solution representation and the basis).
end loop

constraints. We completely ignore the original objective function, as the idea is not yet to minimize, but rather to find any valid starting point.

After finding this solution, we'll move from vertex to vertex. In this case, we check if any movement to a neighboring vertex can improve the objective. We do this through *reduced costs* (a key concept in Simplex implementation). If no reduced cost is favorable (negative for minimization), the current point is optimal. Now, if it's not optimal, we choose the "best" direction to follow, selecting the non-basic variable x_j that will give us the greatest rate of improvement (i.e., the most favorable reduced cost).

After that, we choose how far to the neighboring vertex we can go in that direction before violating the non-negativity of some variable. This is done by the "Ratio Test," which identifies which current basic variable (x_{B_r}) will leave the basis.

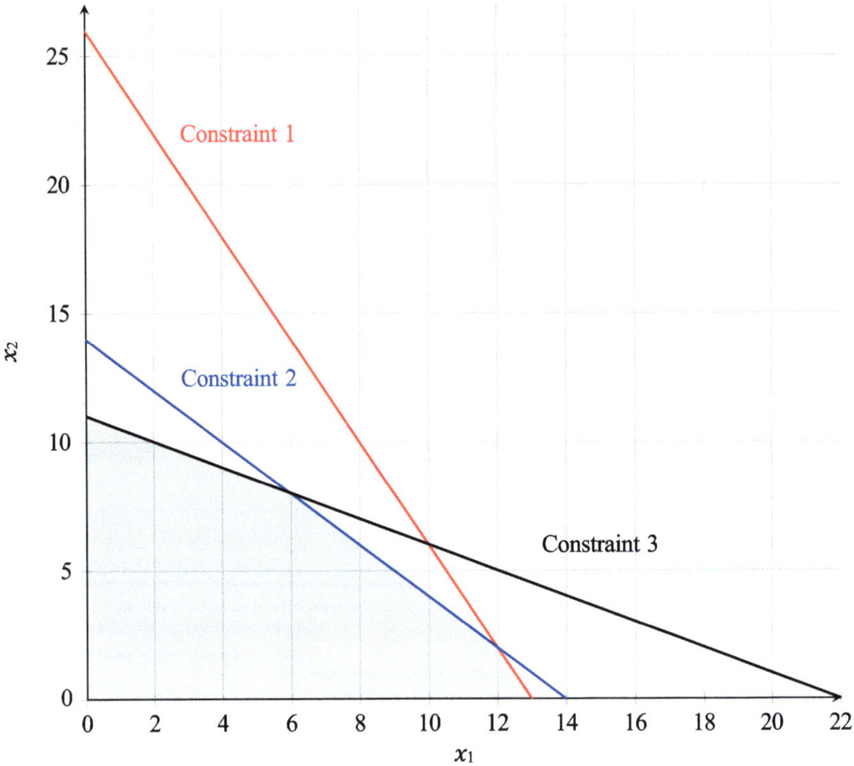

Fig. 3.13 Illustration of the Simplex method moving between vertices of the feasible region of a linear programming problem. *Source* The author, 2025

Then, we swap the variables in the basis and update the solution to the coordinates of the new vertex to repeat the same procedure.

Basic and Non-Basic Variables
Remember that, in our ant analogy and in the pseudocode, Simplex jumps between vertices of the feasible region. Algebraically, this means that each vertex corresponds to a Basic Solution of the system of equations $\mathbf{Ax} = \mathbf{b}$ respecting the non-negativity condition $\mathbf{x} \geq \mathbf{0}$.

Now, consider a solution composed of the vector $\mathbf{x}$. Its n variables are divided between basic variables: a quantity m of variables corresponding to the m columns of matrix $\mathbf{A}$ that are linearly independent, and non-basic variables, which are the remaining variables.

Figure 3.13 illustrates this movement between vertices in a simple two-dimensional problem.

Besides finding the optimal solution, the Simplex method is also capable of detecting if an LP problem is unbounded. This happens when the algorithm identifies an edge along which the objective function can improve indefinitely without ever finding an "end" (another vertex to limit movement in that direction).

That being said, a large portion of real-life LP problems can be efficiently solved by Simplex—it's no coincidence that there's a vast amount of excellent books and resources dealing with LP with a strong emphasis on Simplex. However, don't consider it the best algorithm in the world for solving any LP problem. In fact, some researchers and analysts prefer using other techniques, like barrier methods, for solving large-scale LP problems.

Optimization Solvers

Fortunately, we rarely need to implement these algorithms from scratch (unless for academic or very specific research purposes). There are many optimization libraries and software (solvers), both commercial (CPLEX, Gurobi, MATLAB Optimization Toolbox) and open-source (SciPy Optimize, PuLP, OR-Tools, IPOPT), that implement these methods robustly and efficiently. Our job as engineers and data scientists usually involves

1. Correctly modeling the problem.
2. Choosing the appropriate solver/algorithm for the problem's structure.
   And that's why this book exists. After all, how can we choose the adequate technique if we don't even understand them conceptually?
3. Interpreting the results.

3.2.7 Evolutionary Algorithms

We've already talked about evolutionary algorithms in Sect. 3.1.3.3. However, we focused on unconstrained problems, where the only measure of quality ("fitness") was the value of the objective function $f(\mathbf{x})$. However, we have a problem when working with constraints.

The challenge arises because the variation operators of EAs (crossover and mutation) can easily generate new individuals that violate these constraints, falling outside the feasible region. Discarding these infeasible solutions wouldn't be very effective, because we could potentially reuse some useful genetic information. Therefore, we would need Constraint Handling Techniques (CHTs).

Among these, we'd like to comment on the most common approaches:

Penalty functions

This is probably the most intuitive and common. This is exactly the technique we discussed in Sect. 3.2.3. There are several ways to apply penalties, including

- *Static penalties*: ρ is fixed throughout the entire run. It's simple, but hard to calibrate.
- *Dynamic penalties*: ρ increases over generations, making the penalty more severe over time.
- *Adaptive penalties*: ρ adjusts automatically based on the population's state (e.g., proportion of feasible individuals).

The main advantage is the simplicity of implementation, but the disadvantage is the need to adjust penalty parameters to balance the optimization of $f(\mathbf{x})$ with constraint satisfaction. Meaning, it's not just a matter of pressing an "optimize" button and waiting for the result.

Feasibility rules

This is the second most common approach. Instead of modifying fitness, this approach modifies the selection process. When comparing two individuals, feasibility takes precedence over the objective function value. Deb's rules are a classic and widely used example, especially in tournament selection:

- If one individual is feasible and the other infeasible, the feasible one is always preferred.
- If both are feasible, the preferred one is that with the better $f(\mathbf{x})$ value.
- If both are infeasible, the preferred one is that with the smallest amount of constraint violation.

Notice that, with these comparisons, we don't use any additional parameters like ρ. However, depending on the quantity and complexity of the constraints, it can take a long time to find a feasible solution.

Repair mechanisms

This is another interesting strategy. The idea is to insert an extra "repair" step before evaluating and selecting individuals for the next generation. This extra step evaluates all infeasible individuals and adjusts the values of the variables that don't respect the constraints to make them feasible.

The challenge here is complexity: first, we need to thoroughly understand the problem to insert these repair rules. Second, complex constraints can be difficult to implement or greatly affect the algorithm's performance.

Multi-objective approaches

Another possibility would be to transform a single-objective problem into a multi-objective one: minimize $f(\mathbf{x})$ and minimize the constraint violation, which we could call $V(\mathbf{x})$. We can then use a Multi-Objective Evolutionary Algorithm (MOEA, which we'll see in Sect. 5.4) to find the Pareto front between these two objectives. Although it's an interesting approach, it's more complex to implement.

3.2.8 Bundle Methods

There are situations where gradient-based methods don't apply directly, such as cases where functions are not smooth. An option that can be useful for non-smooth convex and large-scale problems is bundle method.

To explain this method, we'd like to revisit the mountain analogy we used for gradient-based methods, but with some changes. Now, imagine we're guides whose job is to lead others climbing a mountain. And, as you know, a mountain is anything but smooth. Naturally, we wouldn't memorize every irregularity of a mountain, but we have a good sense of the general slope of the mountain region where we are now (a subgradient $\mathbf{g}$). Our objective remains the same: reach the valley (minimize $f(\mathbf{x})$).

Other people (meaning, optimization techniques) might analyze the slope of the current point only, but that wouldn't be very smart given the mountain's complexity. However, we (meaning, this bundle method) would keep a history of all slopes ($\mathbf{g}_i$) and altitudes ($f(\mathbf{x}_i)$) from various locations $\mathbf{x}_i$ where we've recently passed.

Putting all this information together, we would have a simplified model that would serve as an approximation of f and, with that, we could try to navigate to a new point $\mathbf{x}_{trial}$. Think of it this way: if in the last few minutes we've been descending a promising path, the trend is that the next few minutes will continue to be promising, right? Now, if in the last few minutes we've been walking through a place that continually becomes more difficult, we also understand that the trend is for things to worsen if we continue in the same direction. With $\mathbf{x}_{trial}$, we would evaluate $f(\mathbf{x}_{trial})$ and $\mathbf{g}_{trial}$. We compare the actual value with what we were estimating would happen. Now, we have two main possibilities:

- Option 1: We had a good improvement compared to the current point $\mathbf{x}_k$. So, we move to the new point ($\mathbf{x}_{k+1} = \mathbf{x}_{trial}$) and update our information bundle.
- Option 2: We didn't have a substantial improvement. Even if we don't move (i.e., $\mathbf{x}_{k+1} = \mathbf{x}_k$), we don't discard the information from $\mathbf{x}_{trial}$. It is, in fact, updated into the algorithm and we use this information to our advantage to look for another point in the next iteration.

These methods are widely used in large-scale optimization problems, especially in situations where we need to decompose problems or, even, in stochastic optimization. However, they are more complex to implement when compared to other techniques. Furthermore, they are predominantly used in convex problems. Thus, although they can be adapted for non-convex problems, it might be interesting to test other techniques for these situations.

3.3 Integer and Mixed-Integer Optimization

So far, our discussion has primarily focused on problems where the decision variables **x** can take any real value within an interval or, in other words, we've focused on continuous optimization problems. However, in many practical situations, some or all variables need to be integers. Just imagine:

- We can't produce 45.2 donuts: we either produce 45 or 46 donuts.
- We can't allocate 1.5 employees to a shift: we either allocate 1 or 2 employees.
- We can't send 0.4 trucks on a route: we either send a whole truck (and waste space and money because it's not full), or we don't send any.

Well, we believe you've got the point, right? Problems where all decision variables must be integers are called **Integer Programming** (IP). Now, if some variables are integers and others are continuous, we have **Mixed-Integer Programming** (MIP). Frequently, when the objective functions and constraints are linear, we talk about **Integer Linear Programming** (ILP) and **Mixed-Integer Linear Programming** (MILP), which are very common and important classes of problems.

And why is it a challenge?
The requirement for variables to be integers makes optimization problems much harder to solve than if they were continuous. Meaning: if we could solve an LP problem in polynomial time, a similar ILP or MILP problem could grow exponentially with the size of the problem.

You might then suggest just rounding the continuous values of an LP solution. However, this wouldn't mean our solution would remain feasible or even that it would remain an optimal solution.

That said, we'd like to comment on two main IP/MIP techniques which, incidentally, we use together in most specialized solvers for these problems: Branch and Bound and Cutting Planes.[25]

[25] We call the combination of these two techniques *Branch and Cut*.

3.3.1 Branch and Bound (B&B)

Enumerating all possibilities in ILP or MILP problems is generally infeasible. Thus, one strategy we can use is to branch the original problem into smaller subproblems, and then bound the search by removing those subproblems that wouldn't have a solution outperforming the best solution found so far.

Let's think, for example, of the classic Traveling Salesperson Problem (TSP). In this problem, we have a salesperson who needs to visit a set of n cities, starting and ending in the same city, and passing through each of the other cities exactly once. The objective is to find the route (sequence of cities) that minimizes the total distance traveled. The decisions here are integers: which city to visit after which[26]?

To solve a problem like this, we could even try all possible routes. This might work when we have few cities, but the number of routes grows factorially (i.e., $\frac{(n-1)!}{2}$), quickly making brute force impossible.

For cases like these, we would use branch and bound. Instead of testing all possible integer combinations (which would be infeasible), it explores the solution space more strategically. The general strategy involves a decision tree in which each node represents a subproblem, as illustrated in Fig. 3.14.

Furthermore, the general idea of the algorithm is available in Algorithm 9. Let's see:

1. First, we initialize our algorithm. We start with the original problem P_0 which has some integer variables I. We also create a complementary list called L. This list will be useful for storing all active subproblems in the next steps. For now, this list will only contain P_0.
   P_0 is the complete problem—that is, the objective function f, $\mathbf{x}$, and the constraints. This will be useful for creating subproblems later.
2. We also need to store what the best solution has been so far in a variable called $\mathbf{x}_{best}$, as well as $f(\mathbf{x}_{best})$ in another variable we'll call U. At the beginning of the algorithm, we don't yet know $\mathbf{x}_{best}$. Thus, it will be null and U will receive ∞ because it's a minimization problem.
3. Then, we enter the main loop. It keeps running as long as we have promising subproblems in list L. At each iteration, we select a subproblem P_k from L to analyze (the choice strategy can vary, like picking the branch with the most promising lower bound).
4. Next, we solve an easier version of the subproblem (usually the continuous relaxation P_k^{relax})[27] to obtain a lower bound (L_k) on the best possible integer value

[26] The variables are integers because each of them is represented by an integer. In this case, a possible solution would be something like "*Start at City 1. Then, go to City 3. Then, to City 2. Then, return to City 1.*"

[27] Solving the "continuous relaxation" P_k^{relax} means temporarily ignoring the constraints that some variables x_i in subproblem P_k must be integers (for example, treating a binary variable $x_i \in \{0, 1\}$ as if it could take any continuous value $0 \leq x_i \leq 1$, or removing the requirement $x_i \in \mathbb{Z}$). This transforms the difficult integer/mixed problem P_k into a continuous problem (often an LP problem,

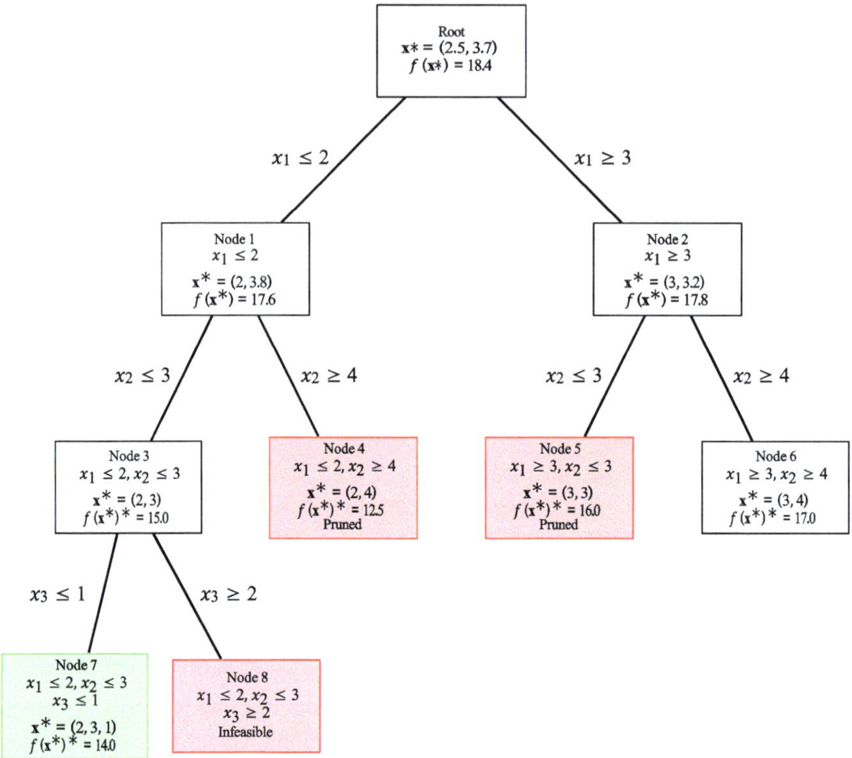

Fig. 3.14 Illustration of the decision tree explored by the branch and bound algorithm. The optimal solution is marked in green, while the pruned nodes are in red. *Source* The author, 2025

at that moment and a relaxed solution x_k^*, which should be treated only as an estimate—after all, we relaxed the constraints to produce this value.

5. Then, we compare our estimate L_k with the value U of the best integer solution (incumbent) found so far.
   If $L_k \geq U$ or if the solution is infeasible, it means there's no point in dedicating more effort to expanding the current subproblem into smaller subproblems, as it wouldn't be possible to improve upon what we already have in $\mathbf{x}_{best}$. This would save us a good amount of search time, as we would discard several other combinations generated from the current subproblem.
   Now, if $L_k < U$, we've just found a new best known integer solution. We update $\mathbf{x}_{best}$ with the value of $\mathbf{x}_k^*$ and U with L_k. Since there would be nothing to improve

---

if the original was MILP), which is much easier and faster to solve using algorithms like Simplex. The optimal solution L_k of this relaxation serves as a lower bound, in the case of minimization, for the optimal integer value of P_k, being essential for the pruning step (bound), in addition to providing the solution $\mathbf{x}_k^*$ that we use to check for integrality and guide branching.

here, it wouldn't be worth dividing this subproblem into smaller subproblems, and we would also save search time.

6. With this, we can proceed with one more step: If the relaxed solution $\mathbf{x}_k^*$ is feasible, but if it's not integer, we can then choose a fractional variable[28] and divide problem P_k into at least two new subproblems. The most common way to do this is by generating one subproblem where we force a fractional variable x_i down and another where we force it up.
   Then, we add these new subproblems to list L to resume our loop.
7. We would end the search when L is empty. At this moment, $\mathbf{x}_{best}$ would already store the global optimum.

3.3.2 Cutting Planes

B&B is very powerful for integer problems, but it has an issue: its efficiency heavily depends on the "quality" of the L_k we obtain when solving the continuous relaxation at each node of the tree. If the feasible region of the relaxation is much larger than the actual set of integer solutions, the bound L_k can be too optimistic, and B&B might end up exploring an excessive number of nodes before finding the optimal solution.

Therefore, to improve the efficiency of B&B, we could make the bounds L_k tighter. In practice, we do this by adding cutting planes. As we briefly mentioned at the beginning of this section, most modern MILP solvers use this combination of B&B with cutting planes, which is also known as **Branch and Cut**.

Observe Fig. 3.15. Notice how the addition of the cutting plane makes the new relaxed region smaller which, in practice, means better performance for our algorithm.

Tip
There are many types of cutting planes algorithms, each exploring different mathematical structures of the problem. Thus, you'll see software using techniques like Gomory cuts, cover cuts, clique cuts, flow cover cuts, and lift-and-project cuts, among others. That said, our concern is for you to understand the concept of the technique first.

Mathematically, a cutting plane is a new linear inequality constraint that we add to the relaxation model P_k^{relax}. This new constraint is satisfied by all feasible integer solutions of the original problem but is violated by the current fractional solution $\mathbf{x}_k^*$.

That said, we generally integrate this logic within the nodes of the B&B tree, and usually when the relaxation solution $\mathbf{x}_k^*$ is fractional. By adding these cuts, we're effectively "tightening" the representation of the feasible region of the relaxation,

[28] Meaning, one of those variables that originally should have been integers, but which we relaxed to accept continuous values in our tests.

Algorithm 9 Branch and Bound

Require: Original optimization problem P_0 (with integer variables $x_i, i \in I$), objective function f to minimize.
Ensure: Optimal integer solution $\mathbf{x}_{best}$ and its objective value U. (If infeasible, $\mathbf{x}_{best}$ is null and $U = \infty$).
Initialization:
$L \leftarrow \{P_0\}$ ▷ List (or priority queue) of active subproblems (nodes)
$U \leftarrow \infty$ ▷ Global Upper Bound: best known integer objective value
$\mathbf{x}_{best} \leftarrow$ null ▷ Best known integer solution (incumbent)
Main Loop:
while L is not empty **do**
  Choose and remove a subproblem P_k from list L. ▷ E.g., using best-first (smallest L_k) or depth-first strategy
  ▷ Solve Relaxation (Bounding)
  Solve the continuous relaxation of P_k (denoted P_k^{relax}).
  Let $\mathbf{x}_k^*$ be its solution and $L_k = f(\mathbf{x}_k^*)$ its objective value (lower bound).
  if P_k^{relax} is infeasible **then**
    $L_k \leftarrow \infty$ ▷ Explicitly set lower bound to infinity
  end if
  ▷ Prune
  if $L_k \geq U$ **then**
    continue ▷ Prune by bound or infeasibility: P_k cannot yield a better solution
  end if
  ▷ Check Integrality and Update Incumbent
  if $\mathbf{x}_k^*$ is feasible for P_0 (i.e., $x_{i,k}^* \in \mathbb{Z}$ for all $i \in I$) **then**
    if $L_k < U$ **then** ▷ Found a better integer solution
      $U \leftarrow L_k$ ▷ Update Global Upper Bound
      $\mathbf{x}_{best} \leftarrow \mathbf{x}_k^*$ ▷ Update incumbent solution
    end if
    continue ▷ Prune by optimality: Node solved, no need to branch
  else ▷ $\mathbf{x}_k^*$ is feasible for P_k^{relax} but not integer for P_0
    ▷ Branch
    Select a variable x_j ($j \in I$) such that $x_{j,k}^*$ is fractional.
    Create two new subproblems:
      $P_{k,1} \leftarrow P_k \cup \{x_j \leq \lfloor x_{j,k}^* \rfloor\}$
      $P_{k,2} \leftarrow P_k \cup \{x_j \geq \lceil x_{j,k}^* \rceil\}$
    Add $P_{k,1}$ and $P_{k,2}$ to L.
  end if
end while
▷ End
if $\mathbf{x}_{best}$ = null **then**
  return "Original Problem Infeasible" ▷ Explicit check for infeasibility
else
  return $\mathbf{x}_{best}$ (with optimal value U)
end if

making it closer to the actual shape of the set of integer solutions. Stronger lower bounds (L'_k closer to U) allow pruning more nodes from the B&B tree, which means we'll find an optimal integer solution even faster. Analyze Algorithm 10. It's a modification of the B&B algorithm we saw in Algorithm 9 to consider these adjustments.

The biggest change we made was the addition of a cut loop. If cuts are generated and added, we force the algorithm to go back and re-solve the relaxation P_k^{relax}, which is now stronger due to the cuts. This can lead to a larger lower bound L_k. We'll end the loop if no useful cuts are generated (or if a limit on cut rounds is reached), allowing the algorithm to proceed to the next step.

Now, you might ask: "*Are there algorithms that are pure B&B being used in the industry, or is everything branch and cut?*" Look: when we use a solver from some commercial software to solve an IP or MILP problem, it's very likely we're

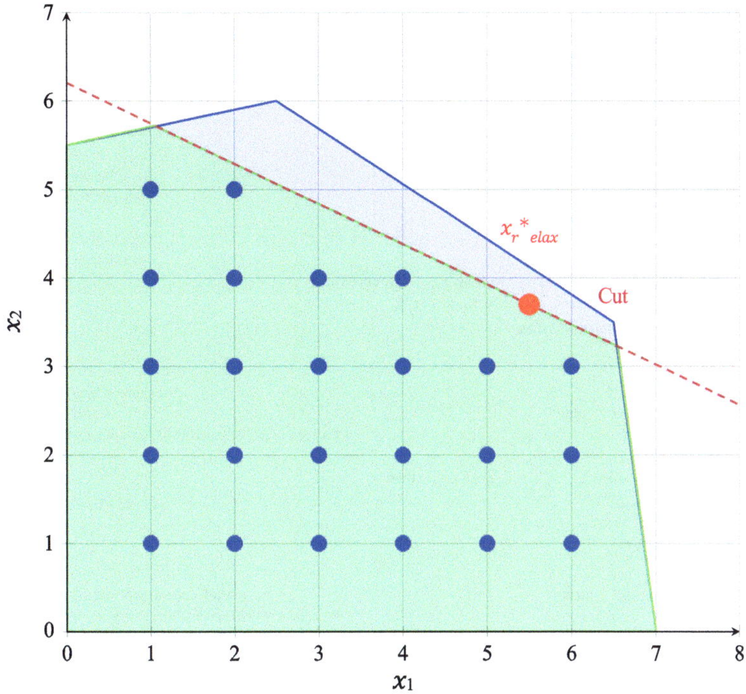

Fig. 3.15 Illustration of a cutting plane: The initial shaded region in blue is the LP relaxation. Blue dots are feasible integer solutions. The optimal solution of the relaxation (x^*_{relax}) is fractional. The dashed line is a cutting plane: it "cuts off" x^*_{relax} but keeps all integer solutions. The new relaxed region (intersection of the original with the half-plane defined by the cut) is smaller and provides a stronger lower bound. It is the green shaded area. *Source* The author, 2025

using branch and cut. Consider this: companies typically have complex problems that would take hours, days, or even be impossible to solve with pure B&B, but can be solved in minutes or seconds with a good branch and cut algorithm. That said, we can find use cases of pure B&B in simpler software and educational tools. Or, even to solve non-linear integer problems, since generating valid cutting planes becomes a much harder task for these cases.

3.4 Global Versus Local Optimization Strategies

We've already talked about the difference between local and global minima in Sect. 2.3.2. Furthermore, when discussing some techniques, we also mentioned the risk of getting stuck in local minima, especially in non-convex problems. Just to recap, this means the algorithm might converge to a solution that's the best in its immediate vicinity, but not necessarily the best possible solution in the entire feasible region—the global minimum.

Algorithm 10 Branch and Cut (Minimization)

Require: Original optimization problem P_0 (with integer variables $x_i, i \in I$).
Require: Objective function f to minimize and cut_limits.
Ensure: Optimal integer solution $\mathbf{x}_{best}$ and its objective value U (or indicates infeasibility).

Initialization:
$L \leftarrow \{P_0\}$ ▷ List of active subproblems (nodes)
$U \leftarrow \infty$ ▷ Global Upper Bound
$\mathbf{x}_{best} \leftarrow$ null ▷ Incumbent integer solution
while L is not empty **do** ▷ Main Branch and Bound loop
  Select and remove a subproblem P_k from L.
  $node\_processed \leftarrow$ **false**
  while not $node\_processed$ **do** ▷ Cut loop for current node P_k
    Solve the continuous relaxation of P_k (with any previously added cuts).
    Let solution be $\mathbf{x}_k^*$ and objective value be $L_k = f(\mathbf{x}_k^*)$.
    if P_k's relaxation is infeasible **then**
      $L_k \leftarrow \infty$
    end if
    if $L_k \geq U$ **then** ▷ Pruning by bound or infeasibility
      $node\_processed \leftarrow$ **true** ▷ Mark node as processed
      continue ▷ Proceed to end of inner while, then outer while takes next node
    end if
    if $\mathbf{x}_k^*$ is integer feasible (i.e., $x_{i,k}^* \in \mathbb{Z}$ for all $i \in I$) **then**
      if $L_k < U$ **then** ▷ Found a new best integer solution
        $U \leftarrow L_k$
        $\mathbf{x}_{best} \leftarrow \mathbf{x}_k^*$
      end if
      $node\_processed \leftarrow$ **true** ▷ Mark node as processed (prune by optimality)
      continue ▷ Proceed to end of inner while, then outer while takes next node
    else ▷ $\mathbf{x}_k^*$ is fractional, attempt to add cuts
      $C_{new} \leftarrow$ GENERATECUTS($P_k, \mathbf{x}_k^*$) ▷ Attempt to find violated valid cuts
      if C_{new} is not empty **and** *cut_limits not reached (optional)* **then**
        Add C_{new} to the formulation of P_k.
        ▷ Re-solve P_k with new cuts in the next iteration of this inner loop
      else ▷ No more (effective) cuts found or generation stopped
        ▷ Branching Step
        Select a variable x_i ($i \in I$) such that $x_{i,k}^*$ is fractional.
        Create $P_{k,1} \leftarrow P_k \cup \{x_i \leq \lfloor x_{i,k}^* \rfloor\}$
        Create $P_{k,2} \leftarrow P_k \cup \{x_i \geq \lceil x_{i,k}^* \rceil\}$
        Add $P_{k,1}$ and $P_{k,2}$ to L.
        $node\_processed \leftarrow$ **true** ▷ Node branched, processing for P_k finished
      end if
    end if
  end while ▷ End of cut loop for P_k
end while ▷ End of main B&B loop
if $\mathbf{x}_{best}$ = null **then**
  return "Original Problem Infeasible"
else
  return $\mathbf{x}_{best}$ (with optimal value U)
end if

Now, if we work on real problems, like those in engineering and design, this can be a risk. After all, finding the true global optimum can result in significant cost savings, much better performance, or a more functional design. Thus, the question we might have is: "*when is global optimality guaranteed?*"

Consider this: for convex optimization problems, the situation is simple. Any local minimum found is also a global minimum. Therefore, algorithms like GD, Newton's methods, and interior-point methods are guaranteed (in theory) to converge to the global optimum if your problem is convex. LP and convex QP are important examples of this.

Now, if the problem isn't convex, we have a much greater challenge. We don't have just one method that will always work for all problems, but some strategies can increase the chance of finding the global optimum or, at least, a high-quality solution. Let's look at some cases:

- Multi-start: It's simple, but very common. You remember that some techniques always start with an initial solution, right? In this case, we would run a local optimization algorithm like Newton's method or GD from several different starting points, spread across the feasible region. Thus, our chance of finding a global optimum increases, but it's still not a guarantee that the best solution we find with this strategy is, in fact, the global optimum.
- Heuristics and metaheuristics: We'll see more about this in Chap. 4. This is the second most common alternative. These are techniques generally inspired by natural processes, and typically handle continuous and integer variables. They also work well with complex search spaces, but still don't guarantee that the best solution we find is, in fact, the global optimum.
- Deterministic optimization: There are some algorithms that use the B&B we saw in Sect. 3.3.1 to guarantee finding the global optimum. The problem with this is performance: because they are very computationally heavy, they are usually applicable only to problems of lower complexity.

3.5 Thinking About Employment

Software and Comparisons
We will discuss the software that implements the techniques we saw in this chapter as well as metaheuristics in Chap. 4. Furthermore, we will compare all these techniques in that same chapter.

As you may have noticed, so far we've used various analogies and application examples to explain concepts and techniques, right? That being said, we'll now comment on some examples of applying techniques to solve problems in the Industry.

Always think about the job interview. Our objective is to think about the application—that is, to understand which techniques professionals use most in the industry and why. So, consider that each of the questions below was asked to you in a job interview. Knowing the correct answer could indicate experience or conceptual understanding of the application.

3.5.1 Mechanical Engineering: Spring Design

Following our job interview suggestion, imagine you were asked the following question: "*We need to design a helical spring to minimize its volume (and, therefore, weight and material cost), subject to constraints such as maximum stress, minimum deflection under load, and minimum wire diameter, among others. How would you solve this problem?*".

Consider this: This is a single-objective (only volume) and non-linear optimization problem, because when creating the objective function, we would realize we need to multiply some of these variables. Additionally, we would have some equally non-linear constraints, such as the shear stress in the springs.

That said, the Physics equations we generally use in real-life problems of this kind are differentiable. Therefore, we would also have a constrained non-linear programming problem.

Because of this, we could use a method like SQP or some barrier method, like IPMs, due to the non-linearity, the constraints, and the fact that it's a smooth problem.

We could even use B&B, but it wouldn't be suitable: a variable like diameter is continuous, and to use these techniques we would have to treat them as integers and, in turn, this would make the problem unnecessarily more difficult. We also couldn't use any unconstrained method we saw in Sect. 3.1, because our problem has constraints. And, furthermore, we can't use any LP method, because it's a non-linear problem.

3.5.2 Electrical Engineering: Simplified Economic Dispatch

Now, what if the problem was: "*We need to design the power output of two thermo-electric plants P_1 and P_2 to meet the total demand of a region, while minimizing the total generation cost.*"?

Consider this: power output in plants is a continuous value. Thus, we have continuous variables. The cost C_i of a plant P_i is generally calculated as $C_i(P_i) = a_i + b_i P_i + c_i P_i^2$, where a_i represents fixed costs, $b_i P_i$ represents linearly varying cost, and $c_i P_i^2$ represents the efficiency of converting fuel into electrical energy. Here, we would have a quadratic and convex function with linear constraints, such as demand and the generation limits of each plant.

This, in other words, would be a single-objective optimization problem. Since the objective is convex, it would also be a convex QP problem.

For this, we return to IPMs from Sect. 3.2.4. After all, we need to explore the specific structure of a problem, and they are more efficient than a non-linear programming technique. This is another case where we couldn't use an LP algorithm nor an unconstrained method. In fact, we could only use an LP technique like Simplex if the cost functions were linear.

3.5.3 *Data Science: Logistic Regression with L2 Regularization*

Now, let's imagine you're working on a Data Science team. The challenge is to build a logistic regression model (commonly used for binary classification problems). To avoid overfitting, we'll use L2 regularization (Ridge). Now, how would we find the model parameters β that minimize the Negative Log-Likelihood (NLL) loss function plus the L2 regularization term?

Consider this: this NLL function is one we can use to tell us if the model is good or not. It's smooth and convex, just like the regularization term. And, furthermore, we have no constraints. In other words, we have a single-objective, unconstrained, convex, and smooth optimization problem.

Here, we could have two answers. Since we want to evaluate this loss function for each instance (i.e., each row of a table) in our dataset, we would depend on the size of this dataset to make a decision. SGD and its variants are good for large datasets (remember the concept of SGD mini-batches we saw in Sect. 3.1.2.3?). For a smaller amount of data, we could test a technique that would converge faster, like quasi-Newtonian methods (e.g., L-BFGS).

Tip

There's a reason why one of the most popular Machine Learning libraries, `scikit-learn`, has methods like SGD and L-BFGS already implemented by default. There's also a reason why TensorFlow and Keras, libraries we use to train neural network models, have SGD-based techniques like Adam and RMSProp readily available. What about exploring the other techniques that exist in this and other libraries?

3.5.4 *Production and Logistics Engineering: Petroleum and Refineries*

Now, let's consider the following question: "*A refinery needs to produce a certain amount of gasoline respecting pre-approved octane and sulfur limits. It has received crude oil of different types and from different locations, with each shipment of crude oil having its own cost and chemical properties. How could we minimize the total cost of crude oil in this case?*".

Consider this: the objective is the crude oil, not the gasoline. Meaning, we have the produced gasoline as a linear equality constraint. Additionally, we can't use more oil than we have available for each type, which would be other linear constraints. Furthermore, the quantities of each type of oil would be continuous and non-negative variables (after all, we can't consume a negative amount of oil). And, still, the total cost is linear, since we could model it as the sum of the costs of each component c_i multiplied by the quantity of each type of oil x_i.

We could also have some non-linear constraints, such as the division of chemical properties. The average sulfur content of the gasoline we would produce, for example, is calculated from the properties of each x_i. The same would apply to octane and other characteristics. However, since the amount of gasoline to be produced is fixed, we could linearize these constraints by reusing it in these equations.

Thus, we would have a good LP problem here—single-objective (and a linear objective), continuous variables, and linear constraints, with smooth behavior. Since it's LP, we could use techniques like the Simplex method and, again, IPMs.

Tip

Incorrect or incomplete modeling will lead to solutions that are not useful in practice, even if the optimization algorithm works perfectly from a mathematical standpoint. That's why investing a good amount of time to understand and model the problem correctly can be even more important than choosing and applying the optimization algorithm.

3.6 Conclusion

This chapter has covered various SOO techniques, providing you with an essential toolkit for problems ranging from unconstrained continuous functions to constrained integer scenarios. It's also important to remember the nuanced characteristics of optimization problems: no single algorithm works for everything, and the best choice always depends on the problem's specific mathematical properties. Understanding problem characteristics such as linearity, convexity, and differentiability allows for the selection of techniques ranging from basic gradient-based approaches to evolutionary algorithms. Furthermore, the effectiveness of the solutions depends on the accuracy of the mathematical formulation representing the real-world problem.

Moreover, while a deep understanding of the concepts behind these algorithms is very helpful, actual implementation in professional settings often involves using specialized solvers and libraries. This means that in industry, one is not typically expected to implement any optimization technique from scratch but rather to use existing implementations from these tools. Since these tools handle the complexities of the methods, more time can be invested in modeling the problem and interpreting the results effectively.

That said, we invite you to consider which SOO techniques are most suitable as you think about the kind of optimization problems you might face in a professional setting. When an exact or derivative-based method is insufficient due to problem complexity, non-differentiability, or the challenge of local optima, consider using metaheuristics. These techniques, detailed in the next chapter, offer effective and flexible alternatives for such complex scenarios.

Chapter 4
Metaheuristics for Single-Objective Optimization

Abstract This chapter explores metaheuristics as effective strategies for single-objective optimization problems. These strategies are useful for complex, black-box, or non-convex problems where traditional exact methods may be impractical or become stuck in local optima. This chapter introduces and explains popular metaheuristics, starting with simulated annealing, Tabu search, and variable neighborhood search. It then explores population-based approaches, such as evolutionary algorithms. Techniques covered include genetic algorithms, evolution strategies, and differential evolution. Swarm intelligence techniques, such as particle swarm optimization and ant colony optimization, are also explored. The goal is to provide a clear understanding of the concepts behind each technique. Consequently, for each strategy, we explain its core principles, mechanisms, and key parameters. Given their complexity, analogies are often used to balance technical depth with a conceptual overview.

In Chap. 3, we explored several interesting methods for solving Single-Objective Optimization (SOO) problems, such as Gradient Descent (GD), Newton's method, and branch and bound. We also briefly touched upon a type of metaheuristic: Evolutionary Algorithms (EAs). However, as previously mentioned, this chapter will further explore EAs and other metaheuristics. Understanding the need for these techniques is essential. Well, to discuss heuristics and metaheuristics, we first are required to understand *why* it is necessary to learn about these techniques.

Consider this: the exact methods we've seen so far, like GD, Newton's methods, simplex, and branch and bound (B&B), are excellent at guaranteeing they find a global optimal solution (or an optimal solution close to the global one). However, they can also have some issues, such as guaranteeing optimality in large problems (as is the case with exact methods for integer programming (IP) and mixed-integer programming (MIP)), or getting stuck in local optima in non-convex problems (as with gradient-based methods). Or, as we discussed with EAs in Sects. 3.1 and 3.2, we might encounter cases where some problems don't have easily calculable gradients or have so much noise that we're prevented from using exact methods efficiently.

And, since many engineering and data science problems exhibit these complexities, it's important that we're familiar with additional methods that allow us to find acceptable solutions within a reasonable timeframe.

W. R. Monteiro and G. R. Meza, *A Practical Guide to Optimization in Engineering and Data Science*, https://doi.org/10.1007/978-3-032-04633-8_4

4.1 Heuristics Versus Metaheuristics

With that said, the first aspect is to understand the difference between heuristics and metaheuristics. Both terms can be intertwined, depending on the book or documentation. However, here we consider a heuristic to be a practical method designed for a specific problem with the aim of finding a good solution in an adequate amount of time. In other words, it's made to be fast, easy to implement, and specific to a particular problem. Stochastic Gradient Descent (SGD), for example, is a heuristic as it's a specific, rule-based iterative approach. The specific logics we saw in Sect. 3.3.1 for relaxing and rounding values to obtain an integer are also examples of heuristics.

On the other hand, metaheuristics are high-level strategies that use smaller heuristics to more intelligently explore the search space for solutions. The issue with this is complexity: they are harder to implement and use more computational resources than simpler methods. Conversely, they are also much more flexible and can find better solutions than simple heuristics or exact methods.

Important
In this chapter, our focus will be on **metaheuristics**, as they offer a more powerful and general framework for dealing with complex real-world optimization problems, where finding the exact global optimum is often impossible or impractical.

4.2 Simulated Annealing (SA)

The task of conceptualizing metaheuristics is easier than for the other methods we saw in Chap. 3. After all, instead of spending time thinking of an analogy you can understand, the techniques themselves are inspired by real-world processes.

Let's start with Simulated Annealing (SA). In metallurgy, there's a process called annealing. In this case, metallurgists heat a metal to high temperatures and then cool it slowly to alter its physical properties. At high temperatures, the atoms move freely. When we slowly lower the temperature, the atoms have enough time to settle into the desired arrangement.

SA mimics this physical process: we start with a parameter that simulates temperature (called T). This value is high at the beginning to allow for broad exploration and the ability to escape valleys (local minima), probabilistically accepting "bad" moves.[1] As we "cool" the temperature (i.e., reduce T), this process becomes more restrictive, focusing on improvements and converging to a low-energy state (a good minimum of the objective function, ideally the global one). The key here is this slow

[1] We generally aim to set a value of T such that the acceptance probability is high, somewhere between 0.85 and 0.99.

cooling: we want to give the algorithm enough time to find a good minimum before it gets "frozen" and stops searching.

With that in mind, let's look at the pseudocode in Algorithm 1. Here, we start with an initial solution $\mathbf{x}_0$ and set an initial high "temperature" T_0. The variable $\mathbf{x}_{current}$ stores the solution where the algorithm is "currently," while $\mathbf{x}_{best}$ stores the best solution found throughout the search history so far.

Algorithm 1 Simulated Annealing

```
Require: Objective function f(x) to minimize
Require: Initial solution x_0
Require: Initial temperature T_0 > 0
Require: Final temperature T_final ≥ 0 (stopping criterion)
Require: Cooling schedule function COOLDOWN(T) (e.g., T ← αT)
Require: Neighbor generation function GENERATENEIGHBOR(x)
Ensure: Best solution found x_best
Initialization:
x_current ← x_0
x_best ← x_current
T ← T_0
Main Loop:
while T > T_final do
    x_neighbor ← GENERATENEIGHBOR(x_current)
    Δf ← f(x_neighbor) − f(x_current)
    if Δf < 0 then                                   ▷ Neighbor solution is better
        x_current ← x_neighbor
    else               ▷ Neighbor solution is worse or equal, accept with probability
        if RANDOMREAL(0, 1) < exp(−Δf/T) then
            x_current ← x_neighbor
        end if
    end if
                                             ▷ Update best solution found so far
    if f(x_current) < f(x_best) then
        x_best ← x_current
    end if
    T ← COOLDOWN(T)                                  ▷ Update temperature
end while
return x_best
```

In the main loop, we simulate the slow cooling process and continue as long as the temperature T is greater than a predefined final value T_{final}.

Within each iteration of this loop (or, often, after a fixed number of attempts within each temperature level), we generate a new candidate solution $\mathbf{x}_{neighbor}$ by applying a small random perturbation to the current solution $\mathbf{x}_{current}$. Then, we calculate the change in cost Δf, and compare the objective function value f at the neighbor with the value at the current solution.

Now, if the neighbor is strictly better ($\Delta f < 0$), we accept it as the new $\mathbf{x}_{current}$. Otherwise (if it's worse or equal, $\Delta f \geq 0$), we still have a chance to accept it depending on the current temperature T. At the beginning of the search, T is higher, and we can accept solutions that might seem bad but could lead us to a global minimum instead of a local one. Toward the end of the search, with a lower T, we can focus more on local exploration.

Thus, the advantage of this technique lies in its ability to handle complex and non-differentiable objective functions, as well as its capacity to escape local optima. However, we'll need to invest some time in tuning the parameters, as the solution quality and the speed at which we find a solution are strongly related to these values.

Currently, SA is less used compared to EAs, but that doesn't mean it isn't useful. From a theoretical standpoint, the concepts of local search improvement, escaping local optima, and control parameters like T help us better understand other meta-heuristics. Additionally, we can still use it for industrial problems such as routing and scheduling.

4.3 Tabu Search (TS)

Here, the keyword is *tabu*. Consider this: I (meaning, one of the authors of this book) had to balance various demands of personal and professional life to finish this book on time. Consequently, there wasn't much time left for cooking—especially at a time in our lives when we had just moved to a new neighborhood. So, I often had to use a food delivery app to choose what my wife and I would have for dinner.

Naturally, we want a delicious meal that maximizes our satisfaction over time ($f(\mathbf{x})$). In this case, $\mathbf{x}$would be the entire sequence of dishes we'd order over several days.

If I were lazy, I could easily order a hamburger from a famous local burger joint every day. Their burger is good, and the price is fair. The problem with this is a certain complacency: how can one be sure this is the optimal choice without exploring other options? Notice how this example resembles a local optimum: the solution works, but it wouldn't be the best by a long shot.

Therefore, another option I could take would be the following: every day I could open the app and choose other similar suggestions available in the area. What if there's an even better burger place? Or, what if there's a place right next door that serves sensational pizza? On the other hand, we might still feel afraid or too lazy to try something new. To solve this, I could have a rule in my head conveniently called a tabu rule: "I cannot order the same type of food nor from the same restaurant as I did in the last t_{tenure} times." For example, if $t_{tenure} = 5$, then I can't order from the restaurants of the last five orders, nor the type of food I ordered in the last five orders. See how this forces me to be more creative and explore more?

Of course, it's possible to get unlucky and order from bad places, but at least we tried something different and actually know whether that burger joint was indeed the best or not. Additionally, we could even create a rule to bypass that tabu: what if a really good promotion appears for a place that was in the tabu memory? In this case, we could ignore that tabu rule and go to this "forbidden" place anyway. This is called an *aspiration criterion*.

With this analogy of how the tabu search metaheuristic works, let's now analyze its operation in Algorithm 2.

The pseudocode in Algorithm 2 implements the tabu search metaheuristic. Let's detail its operation, focusing especially on the variables and methods that are new compared to a simple local search: first, we define an initial solution $\mathbf{x}_0$ as the current solution $\mathbf{x}_{current}$ and also the best solution $\mathbf{x}_{best}$ found so far. We also store the value of $f(\mathbf{x}_{best})$ as f_{best}.

Initially, the tabu list TL starts empty, and its size depends on t_{tenure}, as mentioned earlier. That said, we have a main loop that executes for a maximum number of iterations set by k_{max}. In each iteration, we generate a set of all neighboring solutions to the current solution, and we calculate the objective function value for each of them. This resembles the rule of looking for nearby restaurants in the example we mentioned. Here, we look for the best neighbor $\mathbf{x}_{next}$ that is not prohibited by TL. In the process of searching for the best neighbor, we identify the differences between $\mathbf{x}_{current}$ and the neighbor to record what justified that move.

Furthermore, we might have a check for the aspiration criterion to allow for exceptional cases, such as a move that results in a solution better than the current $\mathbf{x}_{best}$, even if it is tabu.[2] Thus, we would accept a candidate $\mathbf{x}_{next}$ if it has the lowest value among the candidates, provided it is not tabu, or if it satisfies the aspiration criterion.

Remember we mentioned earlier that tabu search allows us to escape local optima, right? Well, it's perfectly possible that $f(\mathbf{x}_{next}) \geq f(\mathbf{x}_{current})$ or, in other words, that $\mathbf{x}_{next}$ is worse than or equal to $\mathbf{x}_{current}$. Even so, we will update $\mathbf{x}_{current}$ with $\mathbf{x}_{next}$, and this is what allows for this escape.

Important
Note that we will only update $\mathbf{x}_{best}$ and $\mathbf{f}_{best}$ if $\mathbf{x}_{next}$ is also better than it. Otherwise, we will only update $\mathbf{x}_{current}$. Even if $\mathbf{x}_{next}$ is worse than $\mathbf{x}_{current}$, we update our move just to test new locations, but without forgetting the best solution found so far.

Finally, we update the tabu list with the move that resulted in the update of $\mathbf{x}_{current}$ and, when applicable, remove the oldest moves: that is, those that have exceeded t_{tenure}.

Now, the performance of tabu search can be quite sensitive to the choices made in its key components. By far, the most critical parameter is t_{tenure}: a very short value doesn't necessarily prevent returning to recently visited local optima, and a very long value can be too restrictive and lead to search stagnation. That said, an initial suggestion would be to test small fixed values, such as [5, 10, 15, 20, 25].

The choice of $\mathbf{x}_0$ also influences the results. Thus, it might be worthwhile to choose a solution that is simple but makes sense for the problem or, when this isn't possible, use a multi-start strategy as mentioned in Sect. 3.4.

[2] This is the most common rule. After all, this ensures that a historical improvement is never lost due to short-term memory.

Algorithm 2 Tabu Search

```
Require: Objective function f(x) to minimize
Require: Initial solution x_0
Require: Neighborhood generation function GenerateNeighbors(x)
Require: Tabu tenure t_tenure
Require: Maximum number of iterations k_max
Require: (Optional: Aspiration Criteria functions, Move Attribute functions)
Ensure: Best solution found x_best
Initialization:
k ← 0                                  ▷ Iteration counter
x_current ← x_0
x_best ← x_current
f_best ← f(x_best)                     ▷ Objective value of the best solution
TL ← ∅                                 ▷ Tabu List (e.g., stores move attributes and their expiration)
Main Loop:
while k < k_max do
    N(x_current) ← GenerateNeighbors(x_current)
    x_next ← null
    f_next ← ∞
    for all y ∈ N(x_current) do
        move_attributes ← GetMoveAttributes(x_current, y)
        is_tabu ← CheckTabu(move_attributes, TL)
        aspirates ← CheckAspiration(f(y), f_best)      ▷ e.g., f(y) < f_best
        if (¬is_tabu or aspirates) then
            if f(y) < f_next then
                x_next ← y
                f_next ← f(y)
            end if
        end if
    end for
    if x_next = null then
        ▷ No admissible move found (e.g., all neighbors lead to deterioration and are tabu without aspiration). Diversify or stop.
        break
    end if
    x_move_origin ← x_current          ▷ Store current solution before moving
    x_current ← x_next                 ▷ Make the move
    if f(x_current) < f_best then
        x_best ← x_current
        f_best ← f(x_current)
    end if
    ▷ Update Tabu List with the move that was just made
    attributes_of_move_made ← GetMoveAttributes(x_move_origin, x_current)
    UpdateTabuList(TL, attributes_of_move_made, t_tenure)
    k ← k + 1
end while
return x_best
```

Furthermore, note that our algorithm doesn't necessarily specify what we store in TL. Indeed, there are several different implementations for this logic. The most common is to prohibit attributes of a move rather than prohibiting the entire solution. Consider, for example, a traveling salesman problem. In this problem, the TS algorithm might test some combinations of navigating through cities that are very promising, and discarding an entire solution that seems very good just because it's on the tabu list seems too restrictive. On the other hand, we could just prohibit the last change in the navigation plan. In a problem with 1000 cities, it might be that the last solution suggested as a change was to go, at a certain step, from city code 890 to city code 572. Thus, we could prohibit only this last change (890 $\rightarrow$ 572) so we can focus on other parts of the problem. This approach, therefore, focuses on preventing specific recent changes rather than entire solutions.

Now, when is it preferable to use TS? Well, it's less random than SA and, therefore, tends to work better for combinatorial problems or in situations where we need to escape local optima *when compared to SA itself*. However, it's more sensitive depending on the size of the tabu list: notice that it's the list that regulates which moves would be prohibited or accepted. So, for each problem, it's important to experiment with the best parameters. That said, modifying t_{tenure} is not a guarantee that we'll certainly find the global optimum. In other words, we'd have a higher chance, but not a certainty, of converging to the global optimum.

4.4 Variable Neighborhood Search (VNS)

Variable Neighborhood Search (VNS) is based on the premise that a local optimum concerning one definition of "neighborhood" might not be a local optimum if we change how we define what is "neighboring." We know this sentence might sound confusing, so allow us to use another example to explain it.

Your residence might have rooms of different sizes and complexities, right? Meaning, a bathroom might be smaller than a kitchen, which might be smaller than a living room, and so on. And yet, a small bathroom can be much messier and have more boxes and drawers full of objects than a living room.

In this context, let's imagine you're looking for your wallet and your cell phone. You might have an intuition about where they could be, like the desk where your computer is (and which, by the way, is extremely messy). After all, you were sitting there recently for a good while, and you always carry your wallet and cell phone together.

Not forgetting we're talking about optimization, let's say your desk is the current solution. However, if you were very stubborn, you could stay in this spot forever, opening and rummaging under every corner and drawer without finding the wallet.

In that case, we could shift the focus of the search a bit: instead of the desk, we could check the corners of the office chair. Or, perhaps, the cabinet next to this desk. And, let's say you got lucky and found the wallet in a corner of the cabinet, but not the cell phone. Well, it would be intuitive to conduct a thorough search of the cabinet (i.e., a smaller search radius) since it's a place that was promising for our search, right? Only if we don't find anything in this area of your house would we move on to the other rooms, with a larger search radius.

So, let's analyze the pseudocode of the VNS metaheuristic in Algorithm 3. Similar to the other cases, we start with an initial solution $\mathbf{x}_0$ as the current solution $\mathbf{x}_{current}$ and also the best solution $\mathbf{x}_{best}$ found so far. We also store the value of $f(\mathbf{x}_{best})$ as f_{best}, just as we did in the tabu search algorithm.

We also have a variable to store the external stopping criterion t, such as the maximum number of iterations allowed. That said, we have two loops in this algorithm: one for the general search, and another for the neighborhoods.

Notice: at the beginning of each outer loop, we reset the index k to 1. It's this index that selects which structure we'll use for perturbation, where we start with a simpler one and move toward a more complex one. In our analogy of searching for the wallet and cell phone, the first "neighborhood" could be a drawer (smaller search space), while a larger neighborhood could be an entire room (larger search space). Whenever we fail in a search in neighborhood k, we'll move to the next neighborhood $k + 1$.

Now, notice that we have a step called *shaking*. Instead of exploring the entire neighborhood (like searching every corner of a drawer), we pick a random point $\mathbf{x}'$ in this neighborhood to try to find something different from $\mathbf{x}_{current}$. Then, we perform a local search on $\mathbf{x}'$ to find the local optimum $\mathbf{x}''$ *in that region.*

Here, it might be that $f(\mathbf{x}'') < f(\mathbf{x}_{current})$. In this case, we update $\mathbf{x}_{current}$ and reset k to 1. We'll also update $\mathbf{x}_{best}$ if applicable. This is similar to the example we gave of finding the wallet in the cabinet: well, if we found something good there, it's likely we'll find more things if we search properly in the same place, right?

On the other hand, if we have $f(\mathbf{x}'') \geq f(\mathbf{x}_{current})$, it means the local search found nothing in that location. Then, we increase k and try something in a larger scope of the neighborhood (and, returning to our wallet and house example: instead of searching the drawer, we'd search for a random point on the entire right side of the cabinet, then the entire cabinet, and then the whole room).

In the end, we'll have the best solution we found throughout the entire process: $\mathbf{x}_{best}$.

Now, why should you care about this algorithm? Well, if you suspect that your problem has a structure where different scales of change can unlock improvements, VNS might be a good alternative. After all, the algorithm itself operates this way. Thus, problems similar to the traveling salesman problem, or problems involving clustering, vehicle routing, and scheduling could be cases where we could use this, assuming we don't have too much noise in our problem.

4.5 Subtypes of Evolutionary Algorithms (EAs)

We've already talked about EAs in Sects. 3.1.3.3 and 3.2.7, right? Meaning, we already know it performs a search based not on an initial solution $\mathbf{x}_0$, but on a diverse population of solutions that are crossed, tested, and selected from generation to

Algorithm 3 Variable Neighborhood Search (VNS)

Require: Objective function $f(\mathbf{x})$ to minimize
Require: Set of k_{max} neighborhood structures $N_k, k = 1, \ldots, k_{max}$
Require: Initial solution $\mathbf{x}_0$
Require: Local search procedure LOCALSEARCH($\mathbf{x}_{start}$, f)
Require: Stopping criterion (e.g., maximum iterations t_{max})
Ensure: Best solution found $\mathbf{x}_{best}$

Initialization:
$\mathbf{x}_{current} \leftarrow \mathbf{x}_0$
$\mathbf{x}_{best} \leftarrow \mathbf{x}_{current}$
$f_{best} \leftarrow f(\mathbf{x}_{best})$ ▷ Store objective value of best solution
$t \leftarrow 0$ ▷ Iteration counter
Main Loop:
while stopping criterion not met (e.g., $t < t_{max}$) **do**
    $k \leftarrow 1$ ▷ Initialize neighborhood structure index
    while $k \leq k_{max}$ **do**
        // Shaking Step
        Generate a random solution $\mathbf{x}' \in N_k(\mathbf{x}_{current})$ (from k-th neighborhood of $\mathbf{x}_{current}$)
        // Local Search Step
        $\mathbf{x}'' \leftarrow$ LOCALSEARCH($\mathbf{x}'$, f) ▷ Find local optimum from $\mathbf{x}'$
        // Move or Not Step
        $f'' \leftarrow f(\mathbf{x}'')$
        if $f'' < f(\mathbf{x}_{current})$ **then**
            $\mathbf{x}_{current} \leftarrow \mathbf{x}''$ ▷ Update incumbent solution
            if $f'' < f_{best}$ **then** ▷ Update global best if incumbent improved it
                $\mathbf{x}_{best} \leftarrow \mathbf{x}_{current}$
                $f_{best} \leftarrow f''$
            end if
            $k \leftarrow 1$ ▷ Success! Restart search from the first neighborhood
        else
            $k \leftarrow k + 1$ ▷ Failure! Try the next neighborhood structure
        end if
    end while
    $t \leftarrow t + 1$
end while
return $\mathbf{x}_{best}$

generation. Here, we have components of fitness, selection, variation (crossover and mutation), and elitism (i.e., only the best solutions persist to the next generation of individuals).

We also know that it's for this reason that we can perform a global search, having a lower risk of getting stuck in some local optimum when compared to other techniques. However, these techniques can also require more computational resources.

Now, what subtypes of EAs exist? It's expected that there are various variants and academic productions proposing new types of algorithms, but there are some "families" or subtypes of EAs that are more popular. So, we'll focus on these. We'll use the pseudocode we saw in Sect. 3.1.3.3 as the basis for our explanation. To make it easier for you to visualize, we've placed the pseudocode again in Algorithm 4.

Algorithm 4 Generic Evolutionary Algorithm (EA)

```
Require: Population size N (i.e., μ)
Require: Number of offspring to generate N_offspring (i.e., λ)
Require: Fitness function f(x) (conventionally, higher is better; adapt if minimizing)
Require: Parent selection method SelectParents(P, num_needed)
Require: Recombination method Recombine(p_1, p_2) and probability p_c
Require: Mutation method Mutate(ind) and probability p_m
Require: Survivor selection method SelectSurvivors(P_current, O_offspring, N_target)
Require: Termination condition (e.g., max generations G_max)
Ensure: Best individual found x_best
Initialization:
P(0) ← InitializePopulation(N)
EvaluateFitness(P(0), f)
t ← 0
x_best ← FindBestIndividual(P(0))                ▷ Overall best individual found so far
f_best_so_far ← f(x_best)
Main Evolutionary Loop:
while Termination condition not met (e.g., t < G_max) do
                                                 ▷ 1. Variation: Generate N_offspring Offspring
    Offspring(t) ← ∅
    for k ← 1 to ⌈N_offspring/2⌉ do              ▷ Assuming recombination produces two offspring
        (parent_1, parent_2) ← SelectParents(P(t), 2)
        if RandomReal(0, 1) < p_c then
            (child_1, child_2) ← Recombine(parent_1, parent_2)
        else
            child_1 ← parent_1; child_2 ← parent_2    ▷ Clones if no crossover
        end if
        child_1 ← Mutate(child_1, p_m)               ▷ Mutation applied based on p_m
        child_2 ← Mutate(child_2, p_m)
        Add child_1 to Offspring(t)
        if |Offspring(t)| < N_offspring then         ▷ Ensure N_offspring are added
            Add child_2 to Offspring(t)
        end if
    end for
    if |Offspring(t)| > N_offspring then
        Trim Offspring(t) to size N_offspring
    end if
                                                 ▷ 2. Evaluate New Offspring
    EvaluateFitness(Offspring(t), f)
                    ▷ 3. Survivor Selection (Form Next Generation Parent Population)
    P(t + 1) ← SelectSurvivors(P(t), Offspring(t), N)
                    ▷ e.g., (μ + λ) selects N best from P(t) ∪ Offspring(t)
                    ▷ e.g., (μ, λ) selects N best from Offspring(t) (requires N_offspring ≥ N)
                                                 ▷ 4. Update Overall Best Solution
    x_best_in_gen ← FindBestIndividual(P(t + 1))
    if f(x_best_in_gen) > f_best_so_far then         ▷ Assumes f is maximized
        x_best ← x_best_in_gen
        f_best_so_far ← f(x_best)
    end if
    t ← t + 1
end while
return x_best
```

4.5.1 *Genetic Algorithms (GAs)*

This is perhaps the most famous type of EA, and the central idea of this type of algorithm is to mimic the concepts of genetics and natural selection. In fact, during our explanations of EAs, we even alluded to these terms.

We've adapted that EA pseudocode for a Genetic Algorithm (GA) in Algorithm 5. The GA places an emphasis on recombination (crossover) as the primary variation operator. See: mimicking the logic of reproduction, we want to combine the variables of two "parent" solutions to find even better "children" (or offspring).

Let's consider an example of crossover with two continuous variables: x_1 and x_2. A good solution might have the values [2.0, 4.0], and another good solution might have the values [0.5, 8.0]. Crossover could mix the two as a simple average or a weighted average. For example, let's suppose we consider 30% of the first solution and 70% of the second solution. Thus, we would have $0.3 \times [2.0, 4.0] + 0.7 \times [0.5, 8.0] = [0.6, 1.2] + [0.35, 5.6] = [0.95, 6.8]$ as the value associated with the offspring.

Mutation still exists, but it plays a supporting role to crossover. After all, large mutations are rarer in nature compared to reproduction. Thus, it has a lower probability p_m of occurring. Its main idea is to introduce diversity in the generation of new solutions.

Furthermore, notice that in the EA implementation, we only mentioned selecting parents and surviving individuals. Here, we have the concept of a **tournament**. The idea simulates a small championship: to choose a parent, we randomly draw a small group of k individuals from the population (where k, the "tournament size," is a GA parameter, typically 2 or 3[3]). Within this group of k competitors, we compare their fitness values (the quality of the solution they represent). The individual with the best fitness among them "wins" the tournament and is selected as a parent. We repeat this process until we have the necessary number of parents (which is usually the same as the population size N) for the next phase.

4.5.2 *Evolution Strategies (ES)*

The focus of Evolution Strategies (ES) is on optimization problems with continuous variables. This doesn't mean we can't use GAs for this, but ES were classically

[3] $k = 2$ gives more chances to diverse, but not necessarily good, solutions to be selected. A larger k favors the best individuals in the population, which can speed up convergence but also increase the chances of premature convergence.

Algorithm 5 Genetic Algorithm (GA)

```
Require: Population size N
Require: Fitness function f(x) (assumed to be maximized)
Require: Parent selection method SelectParents(P, num)
Require: Crossover method Crossover(p_1, p_2) and probability p_c
Require: Mutation method Mutate(ind) and probability p_m
Require: Number of generations G_max
Require: Number of elite individuals k_elite (e.g., 0, 1, ...)
Require: (Implicit: Chromosome representation and initialization InitializePopulation(N))
Ensure: Best individual found x_best
Initialization:
P(0) ← InitializePopulation(N)
EvaluateFitness(P(0), f)
t ← 0
x_best ← FindBestIndividual(P(0))                       ▷ Overall best individual found so far
f_best_so_far ← f(x_best)
Main Evolutionary Loop:
while t < G_max do
                                                        ▷ 1. Elitism (Carry over best k_elite individuals)
    P(t + 1) ← ∅
    if k_elite > 0 then
        Elites ← Select the k_elite best individuals from P(t)
        P(t + 1) ← P(t + 1) ∪ Elites
    end if
                                                        ▷ 2. Offspring Generation
    Offspring(t) ← ∅
    num_offspring_needed ← N − |P(t + 1)|               ▷ Remaining spots to fill
    for i ← 1 to ⌈num_offspring_needed/2⌉ do            ▷ Generate offspring in pairs
        (parent_1, parent_2) ← SelectParents(P(t), 2)
        if RandomReal(0, 1) < p_c then
            (child_1, child_2) ← Crossover(parent_1, parent_2)
        else
            child_1 ← parent_1; child_2 ← parent_2      ▷ Clones
        end if
        child_1 ← Mutate(child_1, p_m)                  ▷ Mutation applied with probability p_m
        child_2 ← Mutate(child_2, p_m)
        Offspring(t) ← Offspring(t) ∪ {child_1}
        if |Offspring(t)| < num_offspring_needed + num_offspring_needed mod 2 then ▷ Ensure not to add too many if num is odd
            Offspring(t) ← Offspring(t) ∪ {child_2}
        end if
    end for
    if |Offspring(t)| > num_offspring_needed then      ▷ Trim if an extra offspring was made for odd num_needed
        Trim Offspring(t) to size num_offspring_needed (e.g., remove last added)
    end if
                                                        ▷ 3. Evaluate Offspring
    EvaluateFitness(Offspring(t), f)
                                                        ▷ 4. Form Next Generation Population
    P(t + 1) ← P(t + 1) ∪ Offspring(t)                  ▷ Add offspring to fill remaining spots
                                                        ▷ 5. Update Overall Best Solution
    x_best_in_gen ← FindBestIndividual(P(t + 1))
    if f(x_best_in_gen) > f_best_so_far then            ▷ Assumes f is maximized
        x_best ← x_best_in_gen
        f_best_so_far ← f(x_best)
    end if
    t ← t + 1
end while
return x_best
```

designed for combinatorial or discrete problems. However, today we also have GAs for continuous variables.

That said, do you remember that GAs had a greater focus on crossover and less on mutation? Well, here it's the opposite. The mutation step in ES usually adds random noise to the components of the parent solution to generate an offspring. In fact, crossover takes such a backseat that in some implementations, this phase no longer exists. And when we do have it, we often use it to calculate the averages of the parents, for both the variables $\mathbf{x}$and the strategy parameters $\boldsymbol{\sigma}$.

In Algorithm 6, we can also see some additional parameters. As an example, the standard deviation σ and the covariance matrix $\mathbf{C}$ of the Gaussian distribution guide the strength and direction of the mutation, respectively. These are values that also change and are encoded along with the solution itself.[4]

Furthermore, we have another difference in selecting survivors. In one strategy, the μ best individuals are selected from the union of the μ parents and λ offspring, thus ensuring elitism. The name of this strategy is $(\mu + \lambda)$-ES. In the (μ, λ)-ES strategy, which is used by the pseudocode we demonstrate here, the μ best are selected only from the λ offspring (requiring $\lambda \geq \mu$), completely discarding the parents.

4.5.3 Differential Evolution (DE)

The strategy of Differential Evolution (DE) is a bit different from what we've seen with GAs and ES. It's good for global optimization problems with continuous variables, and here we have the concept of differential mutation.

Consider this: in GA, we had a dependency on crossover. In ES, we would add small random noise. On the other hand, as we can see in the pseudocode Algorithm 7 for DE, we have a new variable called $\mathbf{v}_i$ for each individual $\mathbf{x}_i$ of the current population. This vector $\mathbf{v}_i$ stores the vector difference between at least two other individuals chosen randomly from the population.

We also have a differential weight factor called F. It's usually a value between 0.4 and 1.0, and it controls the amplitude of the perturbation. Ultimately, we want a mutation that is not random, but adaptive.

However, this doesn't mean we'll only focus on mutation, as was the case with ES. We have a crossover controlled by a parameter called CR. It stores a rate that is typically a value between 0.5 and 1.0. Crossover occurs between the original vector $\mathbf{x}_i$ and $\mathbf{v}_i$ to create a new vector, called $\mathbf{u}_i$.

The selection step is relatively simple: $\mathbf{u}_i$ will take the place of $\mathbf{x}_i$ in the next generation if it is better or equal in terms of fitness (or, mathematically speaking, $f(\mathbf{u}_i) \leq f(\mathbf{x}_i)$). Otherwise, the target vector $\mathbf{x}_i$ survives.

---

[4] That is, mutation is so important for ES that it even occurs on its parameters.

Algorithm 6 Evolution Strategy (ES)

Require: Number of parents μ
Require: Number of offspring λ ($\lambda \geq \mu$)
Require: Number of parents for recombination ρ ($1 \leq \rho \leq \mu$)
Require: Objective function $f(\mathbf{x})$ to minimize, $\mathbf{x} \in \mathbb{R}^n$
Require: Initial parent population $P(0) = \{(\mathbf{x}_i, \boldsymbol{\sigma}_i)\}_{i=1}^{\mu}$ (object variables and strategy parameters)
Require: (Implicit: Recombination and Mutation operators, self-adaptation learning rates)
Require: Number of generations G_{max}
Ensure: Best solution's object variables $\mathbf{x}_{best}$ and its fitness f_{best}

Initialization:
$t \leftarrow 0$.
EVALUATEFITNESS($P(t)$, f) ▷ Evaluate $f(\mathbf{x}_i)$ for all $(\mathbf{x}_i, \boldsymbol{\sigma}_i) \in P(t)$
$\mathbf{x}_{best} \leftarrow$ FINDBESTSOLUTIONPART($P(t)$) ▷ Best $\mathbf{x}$ from initial population
$f_{best} \leftarrow f(\mathbf{x}_{best})$
Main Evolutionary Loop:
while $t < G_{max}$ **do**
  $Offspring(t) \leftarrow \emptyset$ ▷ Offspring population for current generation
  for $j \leftarrow 1$ **to** λ **do** ▷ Generate λ offspring
    // Parent Selection for Recombination
    $ParentsForJ \leftarrow$ SELECTPARENTSFOROFFSPRING($P(t)$, ρ) ▷ Select ρ parents
    // Recombination
    $\mathbf{x}_{parental} \leftarrow$ RECOMBINEOBJECTVARIABLES($ParentsForJ$)
    $\boldsymbol{\sigma}_{parental} \leftarrow$ RECOMBINESTRATEGYPARAMETERS($ParentsForJ$)
    // Mutation of Strategy Parameters (Self-Adaptation)
    $\boldsymbol{\sigma}'_j \leftarrow$ MUTATESTRATEGYPARAMETERS($\boldsymbol{\sigma}_{parental}$)
      ▷ e.g., $\sigma'_{j,d} = \sigma_{parental,d} \cdot \exp(\tau' N(0,1) + \tau N_d(0,1))$
    // Mutation of Object Variables
    $\mathbf{x}'_j \leftarrow$ MUTATEOBJECTVARIABLES($\mathbf{x}_{parental}$, $\boldsymbol{\sigma}'_j$)
      ▷ e.g., $x'_{j,d} = x_{parental,d} + \sigma'_{j,d} N_d(0,1)$
    Add $(\mathbf{x}'_j, \boldsymbol{\sigma}'_j)$ to $Offspring(t)$.
  end for
  EVALUATEFITNESS($Offspring(t)$, f) ▷ Evaluate $f(\mathbf{x}'_j)$ for all offspring
  // Survivor Selection for (μ, λ)-ES
  $P(t+1) \leftarrow$ select best μ individuals from $Offspring(t)$ based on $f(\mathbf{x})$.
    ▷ Parents $P(t)$ are discarded.
  ▷ Update Global Best (Tracking)
  $\mathbf{x}_{best\_in\_gen} \leftarrow$ FINDBESTSOLUTIONPART($P(t+1)$)
  if $f(\mathbf{x}_{best\_in\_gen}) < f_{best}$ **then**
    $\mathbf{x}_{best} \leftarrow \mathbf{x}_{best\_in\_gen}$
    $f_{best} \leftarrow f(\mathbf{x}_{best})$
  end if
  $t \leftarrow t + 1$.
end while
return $\mathbf{x}_{best}$, f_{best}.

Algorithm 7 Differential Evolution (DE) - DE/rand/1/bin variant

Require: Population size N
Require: Objective function $f(\mathbf{x})$ to minimize, $\mathbf{x} \in \mathbb{R}^n$
Require: Problem dimensionality n
Require: Differential weight $F \in (0, 2]$
Require: Crossover rate $CR \in [0, 1]$
Require: Number of generations G_{max}
Require: (Implicit: Variable limits for initialization and boundary constraints)
Ensure: Best individual found $\mathbf{x}_{best}$ and its fitness f_{best}

```
Initialization:
t ← 0
P(t) ← {x_1(t), ..., x_N(t)} by generating random vectors within bounds.
for all individual x_i(t) ∈ P(t) do
    Evaluate f(x_i(t))
end for
x_best ← FindBestIndividual(P(t))
f_best ← f(x_best)
Main Evolutionary Loop:
while t < G_max do
    P(t+1) ← ∅                              ▷ Initialize population for the next generation
    for i ← 1 to N do                         ▷ For each target vector x_i(t)
        // Mutation (DE/rand/1)
        Choose three distinct random indices r1, r2, r3 from {1, ..., N}, all different from i.
        x_r1 ← P(t)[r1]; x_r2 ← P(t)[r2]; x_r3 ← P(t)[r3].
        Mutant vector v_i ← x_r1 + F · (x_r2 − x_r3).
        v_i ← EnsureBounds(v_i)                  ▷ Apply boundary constraints if necessary
        // Crossover (Binomial)
        Trial vector u_i ← x_i(t)            ▷ Initialize trial with a copy of the target vector
        j_rand ← RandomInt(1, n).
        for j ← 1 to n do                                  ▷ For each dimension
            if RandomReal(0, 1) < CR or j = j_rand then
                u_{i,j} ← v_{i,j}                   ▷ Take component from mutant vector
            end if
        end for
        // Selection
        Evaluate f(u_i).
        if f(u_i) ≤ f(x_i(t)) then
            x_i(t+1) ← u_i.
        else
            x_i(t+1) ← x_i(t).
        end if
        P(t+1) ← P(t+1) ∪ {x_i(t+1)}
    end for
                                          ▷ Update Global Best and Generation
    x_best_in_gen ← FindBestIndividual(P(t+1))
    if f(x_best_in_gen) < f_best then
        x_best ← x_best_in_gen
        f_best ← f(x_best)
    end if
    P(t) ← P(t+1)                ▷ Current population becomes the new population
    t ← t + 1.
end while
return x_best, f_best.
```

When to choose a family? We've discussed GA, ES, and DE. If we choose an Optimization software that includes EAs, it's likely we'll find implementations of these three families. Therefore, the question you might have is: when should we opt for one family of EA algorithms over the others?

Frequently, the choice will be based on the type of decision variables. Let's start with GA: they are strong when the variables of our problem are discrete, binary, or combinatorial. Think, for example, of problems where you need to choose a set of items, define a sequence, or turn options on/off. In practice, we could use GAs for problems like feature selection in machine learning algorithms, or in logistics problems, for example.

However, remember: GAs also work for continuous problems, but they are more flexible compared to ES and DE when not all variables are continuous.

On the other hand, we might prefer ES and DE for optimization problems with continuous variables. Think, for example, of adjusting model parameters, designing controllers, and tuning engine parameters. Now, DE has fewer parameters to adjust than ES. So, it might be more interesting to start testing with DE first. On the other hand, if you believe that the search intensity needs to adapt throughout the process, it might be worth testing ES.

4.6 Swarm Intelligence (SI)

EAs are not the only types of bio-inspired algorithms. In this case, we also have Swarm Intelligence (SI) algorithms.

Here, the main focus is on basing ourselves on collective behaviors found in nature, such as a flock of birds, a school of fish, a pack of wolves, or an ant colony.

Consider this: in EAs, we have the premise of selection and reproduction to produce better individuals that can converge to the best solution, right? Meaning, looking at the pseudocodes and our explanations, you know that we recombine information between individuals until we produce something within a few generations. And, generally, we do this in pairs: a "couple" of good solutions can produce a "child" that isn't necessarily better in one generation, but another "couple" could. On the other hand, this can take time for information about an exceptionally good region of the search space, discovered by one or a few individuals, to effectively spread throughout the entire population.

That is, the propagation of "good genes" depends on selection (that individual needs to be chosen for reproduction) and crossover (the combination needs to be successful and not destroy the good characteristic). If convergence is slow, or if it seems the algorithm is "stuck" without being able to quickly refine a good solution found, perhaps the way information is shared and combined in EAs is not the most efficient for that specific problem.

This is why another strategy exists. In SI, we have many relatively simple individuals (also called *agents*) that have ways to transfer information among themselves. Thus, we would have a coordinated behavior of these individuals, like a flock of birds or a school of fish moving together to converge on a promising point (in the case of animals, toward food or rest. In our case, toward the global optimum.).

Therefore, we'd like to discuss with you the two most popular examples: Particle Swarm Optimization (PSO) and Ant Colony Optimization (ACO). But, before that, we'd like you to analyze the generic swarm intelligence algorithm in Algorithm 8 and compare it with the generic EA algorithm we saw in Sect. 4.5. First, the terminology: we no longer speak of "population" and "individuals," but of a "swarm" and "agents." Also, notice that we mentioned we have an "agent state." In addition to the position $\mathbf{x}$of each agent, we can store more information such as velocity or memory. We'll see more about this shortly. Also, note that we no longer talk about selecting parents to generate offspring, but we have a step to update the general state that will influence all agents.

In fact, we have a new step specifically created to update the state of each agent instead of having a logic for selecting survivors and mutation. Thus, the population of the next iteration is formed by the same agents, but in new positions.

4.6.1 Particle Swarm Optimization (PSO)

Just as with some EAs, explaining bio-inspired algorithms like PSO using analogies is easier than explaining some other techniques we've seen so far, like Simplex, for example.

Consider this: PSO was inspired by the movement of groups of birds or fish searching for food. Imagine the following situation: a flock of birds flies over a hill looking for the spot with the most food. No bird knows where the food is, and a single bird would take a very long time to find the best spot. However, in a flock, they have some advantages. First, each bird knows where it is at that specific moment, and also knows the most promising spot it has flown over so far. After all, if everything goes wrong, it could return to that point.

Furthermore, because it's in a flock, each bird knows the best position its flock (i.e., itself and its colleagues) has found so far.

And, as you know, birds are always moving in the air. Therefore, each bird not only has a position but also its own velocity. This velocity is modified depending on inertia (i.e., depending on the velocity at the previous moment), the attraction to the best global position found by the group, and the attraction to the best personal position it has already found.

With this, the movement of each particle and the group tends to be more natural, like the idea of birds and fish. After all, these animals don't have a perfect mental map or a centralized leader dictating the best direction. Their intelligence arises from what they see and where the other nearby particles are going.

Now, how do we translate this into mathematical terms? Consider the following:

Algorithm 8 Swarm Intelligence

Require: Swarm size N
Require: Objective/Fitness function $f(\mathbf{x})$
Require: Agent state representation and initialization procedure INITIALIZESWARM(N)
Require: Agent state update procedure UPDATEAGENTSTATE(state, SharedInfo, Neighbors)
Require: Procedures for shared information management: INITIALIZESHAREDMEMORY(...), PREPARESHAREDINFORMATIONFORITERATION(...), UPDATESHAREDMEMORYAFTERITERATION(...)
Require: Maximum number of iterations G_{max}
Ensure: Best solution/state found by the swarm $\mathbf{x}_{best}$

Initialization:
$t \leftarrow 0$
$S(t) \leftarrow$ INITIALIZESWARM(N) ▷ Create N agents with initial states $\mathbf{x}_i(t)$
EVALUATESWARM($S(t)$, f) ▷ Evaluate $f(\mathbf{x}_i(t))$ for all agents
$\mathbf{x}_{best} \leftarrow$ FINDBESTPOSITIONINSWARM($S(t)$) ▷ Initial global best position
$f_{best} \leftarrow f(\mathbf{x}_{best})$
INITIALIZESHAREDMEMORY($S(t)$, $\mathbf{x}_{best}$, f_{best}) ▷ e.g., $pbest$ for PSO, $pheromones$ for ACO
Main Loop:
while $t < G_{max}$ **do**
  PREPARESHAREDINFORMATIONFORITERATION($S(t)$, $\mathbf{x}_{best}$, SharedMemory) ▷ e.g., make $\mathbf{g}_{best}$ visible to PSO particles; evaporate pheromones for ACO
  $S_{next\_states} \leftarrow \emptyset$ ▷ To store new states for $S(t+1)$
  for all agent $i = 1$ **to** N **do**
    $state_i^{current} \leftarrow$ GETCURRENTSTATEOFAGENT(i, $S(t)$)
    $neighbor\_info \leftarrow$ GETNEIGHBORINFORMATION(i, $S(t)$, SharedMemory) ▷ If applicable
    $state_i^{new} \leftarrow$ UPDATEAGENTSTATE($state_i^{current}$, SharedMemory, $neighbor\_info$) ▷ (Agent moves/updates based on SI rules, e.g., PSO velocity/position update; ACO probabilistic step)
    $f_i^{new} \leftarrow f(state_i^{new})$ ▷ Evaluate the new state
    UPDATEAGENTPERSONALMEMORY(agent i, $state_i^{new}$, f_i^{new}) ▷ e.g., update $\mathbf{p}_{best,i}$ in PSO
    ▷ Dynamically update global best solution
    $f_{personal\_best\_i} \leftarrow$ GETFITNESSOFAGENTPERSONALBEST(i) ▷ Uses updated personal memory
    if $f_{personal\_best\_i} < f_{best}$ **then**
      $\mathbf{x}_{best} \leftarrow$ GETPOSITIONOFAGENTPERSONALBEST(i)
      $f_{best} \leftarrow f_{personal\_best\_i}$
    end if
    $S_{next\_states} \leftarrow S_{next\_states} \cup \{state_i^{new}\}$
  end for
  $S(t+1) \leftarrow S_{next\_states}$ ▷ Update swarm to new states
  UPDATESHAREDMEMORYAFTERITERATION($S(t+1)$, $\mathbf{x}_{best}$, f_{best}, SharedMemory) ▷ e.g., deposit pheromones in ACO based on solutions in $S(t+1)$
  $t \leftarrow t + 1$
end while
return $\mathbf{x}_{best}$

- Each particle i has a current position $\mathbf{x}_i$.
  We also store its best position $\mathbf{p}_{best,i}$ that it, by itself, has found so far.
  We also store the value of the position $f(\mathbf{p}_{best,i})$.
- We store the best position $\mathbf{g}_{best}$ found by any particle so far. This information is available to all particles.
- Each bird has its own velocity $\mathbf{v}_i$, which is influenced by $\mathbf{v}_i$ from the previous iteration, by $\mathbf{p}_{best,i}$, and by $\mathbf{g}_{best}$.

In a simplified way, the velocity and position of each particle i are updated like this:

$$\begin{aligned}
\mathbf{v}_i(t+1) &= w\mathbf{v}_i(t) && \text{(Inertia)}\\
&\quad + c_1 r_1(\mathbf{p}_{best,i} - \mathbf{x}_i(t)) && \text{(Cognitive)}\\
&\quad + c_2 r_2(\mathbf{g}_{best} - \mathbf{x}_i(t)) && \text{(Social)}\\
\mathbf{x}_i(t+1) &= \mathbf{x}_i(t) + \mathbf{v}_i(t+1) && \text{(Position Update)}
\end{aligned}$$

where t is the iteration, w is the inertia weight, c_1 and c_2 are acceleration coefficients (weights for personal and social influence), and r_1, r_2 are random numbers (typically between 0 and 1) that add stochasticity to the flight.

The process repeats for several iterations. Here, we want to combine individual exploration (i.e., the influence of $\mathbf{p}_{best,i}$) and social information (i.e., the influence of $\mathbf{g}_{best}$) so that we can eventually converge to the region of the global optimum. In Algorithm 9, we adapted the generic swarm intelligence pseudocode for PSO. Thus, notice that the main concepts of this algorithm are indeed w, c_1, c_2, and the concepts of particle velocity and position.

Parameters and Usage

Now, the main parameters we use to adjust PSO are the swarm size (which we call N), the inertia weight w, the cognitive coefficient c_1, and the social coefficient c_2 that we saw in its definition.

Let's start with N. This number depends on the size of the problem, naturally. However, for most practical problems, it's common to start with a number between 20 and 50 particles. The smaller N is, the faster the calculations per iteration will be, but there's also a higher risk of premature convergence (or, in other words, a higher risk of the swarm finding a local optimum). On the other hand, the larger N is, the better the diversification in the search and the better the exploration, but it takes longer to finish running the algorithm.

Next, we have w. We usually use a value in the range $[0.4, 0.9]$. The higher the value of w, the more we favor global exploration, and the particles will more frequently explore new areas. A lower value of w implies better local exploration, and the particles will focus more on searching with greater attention to $\mathbf{p}_{best,i}$ and $\mathbf{g}_{best}$. That said, it's common to have implementations where we start with a high value that decreases over the iterations.

Algorithm 9 Particle Swarm Optimization (PSO)

Require: Number of particles N
Require: Objective function $f(\mathbf{x})$ to minimize, $\mathbf{x} \in \mathbb{R}^D$
Require: Problem dimensionality D
Require: Inertia weight w (can be adaptive)
Require: Cognitive coefficient c_1
Require: Social coefficient c_2
Require: Position bounds $[\mathbf{x}_{min}, \mathbf{x}_{max}]$
Require: Velocity bounds $[\mathbf{v}_{min}, \mathbf{v}_{max}]$ (optional)
Require: Number of iterations G_{max}
Ensure: Global best position $\mathbf{g}_{best}$ and its fitness f_{gbest}
Initialization:
$t \leftarrow 0$
for all particle $i = 1, \ldots, N$ **do**
$\mathbf{x}_i(t) \leftarrow \text{RandomUniform}(\mathbf{x}_{min}, \mathbf{x}_{max})$ ▷ Initialize position
$\mathbf{v}_i(t) \leftarrow \text{RandomUniform}(\mathbf{v}_{min}, \mathbf{v}_{max})$ ▷ Initialize velocity (or to **0**)
$f_i(t) \leftarrow f(\mathbf{x}_i(t))$
$\mathbf{p}_{best,i} \leftarrow \mathbf{x}_i(t)$ ▷ Initialize personal best position
$f_{pbest,i} \leftarrow f_i(t)$ ▷ Initialize personal best fitness
end for
$\mathbf{g}_{best} \leftarrow$ particle's $\mathbf{p}_{best,k}$ with best $f_{pbest,k}$ among all $k = 1, \ldots, N$.
$f_{gbest} \leftarrow f(\mathbf{g}_{best})$.
Main Loop:
while $t < G_{max}$ **do**
for all particle $i = 1, \ldots, N$ **do**
$r_1 \leftarrow \text{RandomReal}(0, 1)$ ▷ Random number for cognitive component
$r_2 \leftarrow \text{RandomReal}(0, 1)$ ▷ Random number for social component
// Update Velocity
$\mathbf{v}_i(t+1) \leftarrow w \cdot \mathbf{v}_i(t) + c_1 \cdot r_1 \cdot (\mathbf{p}_{best,i} - \mathbf{x}_i(t)) + c_2 \cdot r_2 \cdot (\mathbf{g}_{best} - \mathbf{x}_i(t))$
(Optional: Clamp $\mathbf{v}_i(t+1)$ to $[\mathbf{v}_{min}, \mathbf{v}_{max}]$)
// Update Position
$\mathbf{x}_i(t+1) \leftarrow \mathbf{x}_i(t) + \mathbf{v}_i(t+1)$
(Optional: Clamp $\mathbf{x}_i(t+1)$ to $[\mathbf{x}_{min}, \mathbf{x}_{max}]$ or handle boundary conditions)
$f_i(t+1) \leftarrow f(\mathbf{x}_i(t+1))$
// Update Personal Best
if $f_i(t+1) < f_{pbest,i}$ **then**
$\mathbf{p}_{best,i} \leftarrow \mathbf{x}_i(t+1)$
$f_{pbest,i} \leftarrow f_i(t+1)$
end if
// Update Global Best (Corrected Logic)
if $f_{pbest,i} < f_{gbest}$ **then**
$\mathbf{g}_{best} \leftarrow \mathbf{p}_{best,i}$
$f_{gbest} \leftarrow f_{pbest,i}$
end if
end for
$w \leftarrow \text{UpdateInertia}(w, t)$ ▷ Optional: e.g., linear decay
$t \leftarrow t + 1$
end while
return $\mathbf{g}_{best}, f_{gbest}$.

And, finally, we have c_1 and c_2. The most common is to have $c_1 = c_2 = 2.0$, but it's also common to see values between $[1.5, 2.5]$ for both coefficients. If we give more weight to c_1 than to c_2, we'll effectively have a search that prioritizes individual experience. When we give more weight to c_2 than to c_1, we give more weight to the group, which, in practice, increases the risk of convergence to a local optimum.

That said, PSO is an algorithm that is easy to implement and has few parameters when compared to other metaheuristics. In fact, if the problem is continuous, we could test PSO and compare it with ES and DE.

4.6.2 Ant Colony Optimization (ACO)

Consider this: the representations of a position $\mathbf{x}_i$ and velocity $\mathbf{v}_i$ of a particle i are, in practice, vectors of real numbers. The update rules we saw for PSO are also based on continuous spaces, and the very idea of these particles' navigation assumes we wouldn't have a discrete set. On the other hand, we still have real-world problems where we need to work with traveling salesman problems, vehicle routing problems, job scheduling problems, quadratic assignment, and other combinatorial optimization problems, don't we? And, for these cases, we also have a swarm intelligence algorithm: ACO.

ACO is inspired by the behavior of ants searching for the shortest path between the anthill and a food source. I don't know if you've ever stopped at some point in your life to watch how ants behave, but they are capable of walking to food that is very far from the anthill and, still, return home. And, notably, they do this without excellent vision, or without having an aerial view to find the path from a privileged position. Therefore, they manage to succeed in this using pheromones: that is, a chemical substance that ants leave along the path and that is used as a reference by other ants.

Let's continue the explanation: the first ants leave the anthill, and each one follows a random path. After all, they are looking for food but have no idea where it might be. As they walk, they deposit pheromones along their route. The other ants that leave subsequently sense pheromones in different directions around the anthill, but they have a higher probability of choosing the path with more pheromones at that moment.[5]

Now, how do these pheromone "trails," which also evaporate over time, help discover the best path? Think of it this way: an ant that, by chance, finds a shorter path (a "shortcut") to the food will be able to make the round trip more times in a given period. Consequently, it and the ants that follow will reinforce this short path with pheromone deposits more frequently. On longer paths, trips are more time-consuming, passages are less frequent, and evaporation has more time to act, reducing the pheromone concentration. Over time, the short path has increasingly

[5] That is, pheromones don't last forever. Thus, at different times, we will see different concentrations of pheromones in the same place.

more pheromones, probabilistically attracting even more ants, which also deposit more pheromones. On the other hand, long paths become less attractive and tend to be abandoned. The result is that the colony, collectively, converges to the predominant use of the most efficient paths.

That said, let's now think about the implementation of ACO. Just as we did with PSO, we also adapted the generic swarm intelligence algorithm for ACO, as you can analyze in Algorithm 10. Here, we have parameters specific to the ant logic we discussed: the influence of pheromone α, the pheromone evaporation rate ρ, and the initial pheromone value τ_0. We also have the influence of the heuristic β.

Heuristic Information η It functions as a clue or a simple rule that an ant can use considering its current location, based on static information about the problem, even before considering the pheromone trails left by other ants, and it is generally constant information. Here, it's a desirable characteristic that we specify for each problem we'd like to solve with ACO. In the case of a job scheduling problem, this could be the inverse of the processing time of the next task. Or, in a traveling salesman problem, this could be a rule where we establish that it's preferable (but not mandatory) to choose a nearby city.

That said, we start the algorithm by initializing the parameters. Among these parameters, a pheromone matrix $\boldsymbol{\tau}$ is responsible for storing the shared memory of the ant colony and would be equivalent to the chemical trail of pheromones along a terrain around the anthill. Its size depends on the problem we are solving. For example:

- If we need to allocate X tasks to Y machines, then the matrix would be $X \times Y$, where τ_{xy} indicates how desirable it would be to allocate task x to machine y.
- If we need to travel through M cities in a situation similar to a traveling salesman problem, we could have an $M \times M$ matrix, where we store all possible connections between a city i and a city j. Here, a strong τ_{ij} indicates a promising chance of connecting these two cities.

In this matrix, the higher the value of a particular cell τ_{ij}, the greater the chances of an ant adopting that alternative as part of its proposed solution. We usually start with a low and standardized value for all cells, like a matrix of zeros.

Now, returning to the input parameters: the influence of pheromone α controls the weight given to the pheromone trail τ_{ij} in an ant's decision. The higher this value, the greater the chances of an ant repeating a more popular path and not trying something new. The lower it is, the more ants will ignore collective learning. We generally have an α value between $[0, 5]$, with 1 being a very common value.

Next, the parameter β controls the weight given to the heuristic information η_{ij}. It's common to use a value between $[1, 5]$, with 2 being a usual value. The higher this value, the more ants will explore only their local area, and the lower this value, the more they will ignore local heuristic information. As you can imagine, it's important to balance α and β and test these parameters for each problem.

Also, notice that we often use our instincts when we have no information about a place, correct? Similarly, ants could use η to get an initial direction before there are pheromones. Thus, it's common to have a β value greater than α.

We also have the parameter related to the evaporation rate ρ. It controls how long the pheromone will persist, and can be a value between (0, 1], but we normally use something between $[0.01, 0.50]$ (often, 0.1). Here, the higher ρ is, the greater the evaporation, and the ants will forget more quickly. A lower ρ implies greater persistence of pheromones, which means, in turn, a more lasting reinforcement of the best paths found and a higher risk of not trying a better solution (lazy ants, perhaps?).

After this parameter initialization, we'll work in a loop. In this loop, we have the following: each of the N ants constructs a candidate solution S_k, one step at a time. The decision-making for each of these steps is guided by the current $\boldsymbol{\tau}$, by η, and by α and β.

Then, we evaluate $f(S_k)$ and store this solution. Next, we update the best solution S_{best} with S_k if this candidate solution is the best of all solutions found so far.

Finally, we reduce the pheromones based on the evaporation rate ρ and then add more pheromone to the locations where solutions recently used those paths. Here, ants that found more promising routes will deposit more pheromone. Then, we'll repeat the same steps until we reach a maximum number of iterations set by G_{max}. In the end, we'll have the best path found and recorded by S_{best}.

4.7 Practical Considerations and Solver Usage

Before moving to the next chapter, it's worth reconsidering the application of these techniques. Meaning, our expectation was that you could have a good foundation to understand how optimal solutions are found by these different techniques.

In fact, we've included several pseudocode examples. These pseudocodes aren't necessarily a guide you'll need to implement manually. When you're working with optimization problems within a company, you won't have to go back to our explanation of a GA or B&B and implement them from scratch. You may not know this, but these implementations depend heavily on the programming language and the software in which we'd use these solutions. The same GA and the same B&B can have very different speeds, error handling, and other methods to improve the quality of their solutions depending on the software we choose, due to these implementation differences.

Throughout the previous texts, you might have come across the term *solver*, right? Well, a solver is a software that has one or more optimization techniques already implemented and ready for use in different situations. This means they should be fast and work well for different types of data. Thus, in practice, we should invest a good amount of time designing our optimization problem (i.e., transforming a real-world

Algorithm 10 Ant Colony Optimization (ACO)

```
Require: Number of ants N, Set of components C, Problem rules/constraints
Require: Influence factors α, β, Evaporation rate ρ ∈ (0, 1]
Require: Heuristic information η_uv, Objective function f(S)
Require: Iterations G_max, Initial pheromone τ_0, Deposit factor Q
Ensure: Best solution found S_best and its value f_best
Initialize τ_uv ← τ_0 for all valid trails (u, v).
S_best ← null; f_best ← ∞.
for t = 0 to G_max − 1 do
    Solutions(t) ← ∅.
    for ant ant_idx = 1 to N do
        S_k ← InitializeEmptySolution().                              ▷ S_k is S_ant_idx
        Ctx_k ← InitializeContext().                                  ▷ Ctx_k is ConstructionContext_ant_idx
        current_comp ← SelectStartComponent(Ctx_k).
        if current_comp ≠ null then
            AddCompToSolution(S_k, current_comp).
            UpdateContext(Ctx_k, current_comp).
        end if
        while not IsSolutionComplete(S_k, Ctx_k) do
            AllowedNxt ← GetAllowedNextComponents(current_comp, Ctx_k).   ▷ AllowedNxt ≡ CandidateNextComponents
            ProbNum ← ∅; P_sum ← 0.                                   ▷ ProbNum ≡ TransitionNumerators
            for all candidate cand ∈ AllowedNxt do                    ▷ cand ≡ cand_comp
                η_curr,cand ← GetHeuristicValue(current_comp, cand, Ctx_k).
                N_cand ← [τ_current_comp,cand]^α · [η_curr,cand]^β.
                ProbNum[cand] ← N_cand; P_sum ← P_sum + N_cand.
            end for
            if P_sum > 0 then
                next_comp ← SelectProbabilistic(AllowedNxt, ProbNum, P_sum).
            else
                next_comp ← SelectFallback(AllowedNxt, Ctx_k).        ▷ e.g., random
            end if
            if next_comp = null then break;
            end if                                                    ▷ Cannot extend solution
            AddCompToSolution(S_k, next_comp).
            UpdateContext(Ctx_k, next_comp).
            current_comp ← next_comp.
        end while
        Add S_k to Solutions(t).
        f_k ← f(S_k).                                                 ▷ f_k is f_ant_idx
        if f_k < f_best then
            S_best ← S_k; f_best ← f_k.
        end if
    end for
    // Pheromone Update
    τ_uv ← (1 − ρ) · τ_uv for all valid trails (u, v).                ▷ Evaporation
    for each solution S_sol (with value f_sol) in Solutions(t) do     ▷ Deposition
        Δτ_sol ← Q/f_sol.
        for each trail (u, v) in S_sol do
            τ_uv ← τ_uv + Δτ_sol.
        end for
    end for
    (Optional: ApplyPheromoneLimits(τ, τ_min, τ_max)).
end for
return S_best, f_best.
```

problem into optimization equations that represent that problem) and preparing the data. After that, we choose the solver, run this solver for our problem, and, ultimately, interpret the solution obtained by the solver.

4.7.1 Languages and APIs

Now, what are the ways to work with a solver? Look, if you're already working in a company with consolidated processes, it will be difficult to propose something new and reinvent the wheel. Meaning, companies won't necessarily use the same software you worked with at University for various reasons: ease of use, ease of finding skilled labor in your region who can work with solvers, information security, cost, and functionalities. We've seen cases, for example, where we had to use simpler tools because the company had already negotiated a contract with a competitor for another software, and the solver was included "for free." In this case, it was difficult to convince managers that they should spend more money on another solver.

On the other hand, you might also find yourself in a situation where you have the freedom to choose a solver. Therefore, it's good for you to know that there are three main types: AMLs, APIs, and libraries.

4.7.1.1 Algebraic Modeling Languages (AMLs)

The main function of AMLs is the abstract definition of the problem. These tools allow you to write your mathematical model (variables, objective, constraints) in a declarative way, close to mathematics, and independent of the solver that will be used to solve it. Then, the AML translates the model for the solver.

Here you describe your optimization problem in a way very close to the mathematical notation we've used throughout this book. It's very likely that you'll need to invest a few minutes to understand the documentation. After all, each solver has its own implementation logic.[6]

Usually, these software already have the entire interface ready for your use, which in itself is a good thing and, at the same time, a bad thing. It's good because it speeds up your work: you write your problem and can test different techniques without major changes in the implementation. Also, you just need to choose the model parameters. And it's bad because it's not so easy to control your entire algorithm. That is, you can't modify the entire logic of a particular technique: at most, you can implement some kind of calculation or additional heuristic only in the points allowed by the solver itself.

Some of them are standalone software. Others are libraries (also called *packages*) for existing programming languages. Let's look at some of the most popular ones:

---

[6] Take Excel, for example. If here we write something like 10^5, there we would write `=10^5`. In the case of solvers, we'll need to adapt to what each one uses as a base.

- Pyomo (Python Optimization Modeling Objects): It's a library for the Python programming language[7] and it's free. Being in Python, it's easy to integrate this library with other Python libraries such as Pandas (for table manipulation), NumPy (for vector manipulation), and Matplotlib (for creating visualizations).
- PuLP (Python Linear Programming): This is another Python library, and it's also free. It specializes in LP and MILP problems.

4.7.1.2 Application Programming Interfaces (APIs)

APIs, on the other hand, focus on the optimization techniques themselves. In the first weeks of a Computer Science degree, we learn that we communicate with computers through interfaces. Meaning: we type things on a keyboard and use a mouse, and we receive a result on the screen. Or, perhaps, we speak a command to our cell phone, and it returns some information on the screen. Thus, all these interactions with computers need well-defined inputs and outputs. A website or an application is good when we intuitively understand how to navigate through them to perform some operations, right? That said, there are also standardized ways for two systems (or software) to communicate with each other: these are APIs.

API stands for Application Programming Interface. They serve as a "menu of services" or a "contract" that software (in our case, the solver) offers so that other programs (like the rest of your code written in a programming language) can interact with it in a standardized way.

The great advantage of using an API is abstraction: your code doesn't need to know how the solver implements the Simplex or B&B algorithm internally, but only needs to know how to use the API to send the problem and receive the solution. Thus, we can work on our code without worrying about what exists inside the API, and vice versa.

That said, we also have optimization APIs. Some of the most popular are

- CPLEX: CPLEX is one of the most recognized Optimization platforms in the world, and it works especially well for LP, MILP, and QP. The focus is on the performance and robustness of the solution algorithm, and it's a commercial solution from IBM.
- Gurobi: The main competitor to CPLEX for LP, MILP, and QP. It's known for top-notch performance, excellent support, and is widely used for software written in Python.
- OR-Tools: It's a set of optimization tools from Google, and it works for different programming languages. For instance, when we were writing this book, there was support for Python, Java, C#, and C++, all of which are languages used by companies.

---

[7] Currently, Python is one of the first programming languages that students in some Engineering courses and most Computer Science courses learn at University. It's an easier language to start with, and it's also the basis for many of a Data Science team's tasks.

4.7.1.3 Libraries

This might get a bit confused with the solutions we've already seen. There are packages (or libraries) in programming languages that implement algorithms or standardized interfaces for optimization tasks, but are not necessarily an AML.

- Pymoo: It's also developed in Python and focuses on multi-objective optimization problems. However, it also has specific algorithms for single-objective optimization problems like Nelder-Mead, pattern search, GA, DE, and ES.
- DEAP: It's a library designed to enable tests with EAs. In fact, it's possible to implement several of the EA types we've seen in this book, but at a lower level. Meaning: Instead of just saying something like "run an ES algorithm for this problem," you'd have to implement all the ES steps, just as we did in the pseudocode.
- SciPy Optimize: It's part of the SciPy library in Python. It has implementations of algorithms like Nelder-Mead, BFGS, L-BFGS-B, and SLSQP. However, it's made to work with smaller problems.
- MATLAB Optimization Toolbox: It's a proprietary toolbox for MATLAB. In it, we have a set of functions and applications within the MATLAB environment to model and solve LP, MILP, QP, non-linear optimization, among others.

4.7.2 The Importance of Scaling

Let's suppose you're working on an optimization problem and, even though you're using an appropriate solver, the problem isn't converging to a good solution. You might even question if you did something wrong along the way or, perhaps, if you accidentally mistyped something.

Look: Algorithms are studied by Computer Science which, like optimization, is one of the areas of Mathematics. This means that thinking about mathematical representation to facilitate the algorithm's work is as important as creating a representation that is faithful to the real-life problem you're trying to develop.

Thus, it's important for you to understand that optimization algorithms work better when constraints and decision variables operate on similar magnitudes, especially for gradient-based techniques or those that use concepts like distance and velocity, as is the case with PSO and GD.

You may have noticed during the contextualizations of the techniques we discussed that the value of some parameters greatly influences the result, right? Similarly, we also talked about concepts of local and global minima. Let's observe Fig. 4.1 for illustrative purposes, for example. The length of this fissure might be a small value, especially considering a scale of meters or kilometers. On the other hand, the height is quite disproportionate considering the length. Imagine someone kicking a ball in this region: the risk of losing the ball because it accidentally fell into the fissure is high considering its height, isn't it?

Fig. 4.1 An example of a fissure. *Source* Pexels

Similarly, an algorithm could get lost if the scale between different variables isn't standardized. Therefore, it's not ideal when we mix distant numerical magnitudes, such as a variable that can range from [0, 1000000] and another from $[10^{-6}, 10^{-2}]$. This occurs because we could generate ill-conditioned matrices or even have some problems defining parameters. It would be beneficial, in fact, if we could standardize or normalize the values. For example, before sending our continuous optimization problem to a solver, we could put all variables on a scale of $[-1, 1]$ or $[0, 1]$. Similarly, constraint equations could have coefficients added to make them more balanced.

Then, at the end of the algorithm, we could convert the solution $\mathbf{x}_{best}$ to a numerical scale that we humans can better understand.

Important
We mentioned in subsubsection 4.7.1.1 that it's important to know the documentation of the solver you'd be using in your day-to-day work, right? Well, there are solvers that already do this scaling automatically for you. Some of them require you to explicitly activate this setting but, still, it would be one less task for you to perform manually.

4.7.3 Interpreting the Results

Let's assume you managed to convert your real problem into a mathematical optimization formulation with all variables and constraints. And then, you also managed to choose the best technique and ran your solver. And, it finished successfully. Now, what do you do with the results?

As we mentioned earlier, each solver has its own terminology and presentation format. Some of them may have all the items below, and others may have only some of them.

Let's start with the status. When a solver concludes, it may also inform how it finished its execution. You can look for terms like:

- **Optimal** (sometimes also called **Converged**): It means the solver found a solution that satisfies the mathematical conditions of optimality and also determined that no other feasible solution is better. The solver doesn't guarantee it found a global optimum, but it might have found only a local optimum. This is when you should analyze $f(\mathbf{x})$ to judge this (or, perhaps, run another technique).
- **Infeasible**: The algorithm didn't find any solution that satisfies all problem constraints. In this case, it's worth analyzing your optimization problem to see if there wasn't an interpretation error while you were declaring your constraints.

  In the case of metaheuristics and heuristics like Nelder-Mead, this could also indicate that some parameter of your model might need to be changed. For example, the limit of iterations, generations, or population members might be too low to allow your algorithm to find a solution in time.
- **Unbounded**: The algorithm determined it can optimize your problem to infinity (meaning, $f(\mathbf{x})$ tends to infinity). It might be interesting to review the problem constraints. We notice this mainly in cases where an upper or lower bound was missing for one or more variables.
- **Feasible** (sometimes called **Suboptimal** or **Limit**): The algorithm found a feasible solution, but was interrupted by a limit you imposed, such as a maximum number of iterations or a maximum execution time for the optimization. Thus, your algorithm found a solution that is feasible, but didn't have time to prove it's optimal.
- **Error**: Perhaps one of the ugliest messages we could receive (and one of the most common, at first). It could be a typo (such as a misplaced comma or an incorrect sign), a definition error (sometimes you defined a negative value where only positive values were accepted), or something contextual (like something in your problem that prevents a certain calculation from being executed in the solver).

If you have an optimal or feasible solution, interpret the objective function value $f(\mathbf{x}^*)$ that the solver found. Does this value seem plausible given the scale of your problem? Let's go back to the donut shop example: would a profit of only \$2 be something you expect after a long day of work, or should it be a much larger value?

Then, analyze your solution vector $\mathbf{x}^*$. Also, analyze these values and see if they make sense. A donut shop needs to produce a balanced quantity that makes sense in the real world: an unrealistic solution, such as producing −30 chocolate donuts and 50 million coconut donuts in a single day, would be undesirable.

Important
Don't blindly trust the solver's "optimal" output. A model with errors can have a mathematically optimal solution that makes no practical sense. Always validate $\mathbf{x}^*$ in the context of the real problem. Simulate, analyze, and present to domain experts.

Also, check the values of the constraints. This might be counter-intuitive at first: after all, a feasible solution already respects the constraints. However, it might be worth seeing how far we are from the limit values you imposed for each constraint to see how close or distant you are from not respecting them, considering the nuances of real life.

4.8 Comparative Analysis of Optimization Algorithms

Throughout the presentations of the algorithms, we've already discussed the cases in which they could be a good option, as well as their pros and cons. That said, Table 4.1 is a summary of this information.

When analyzing, we'd like you to consider two things. First, performance and complexity are relative and depend on the implementation of each solver and the optimization problem you're implementing. And, second, note that there isn't a single algorithm that is the best for all problems. Therefore, it's good for you to experiment with different methods and parameters frequently.

4.9 Conclusion

As shown in this chapter, metaheuristics offer a robust and flexible toolkit for solving single-objective optimization problems that are challenging for traditional analytical or exact methods due to scale, non-convexity, non-differentiability, or discrete variables. Because many complex industrial problems meet these criteria, a dedicated chapter on these techniques is important.

That said, metaheuristics represent a shift from seeking absolute optimal solutions, which can be time-consuming and resource-intensive, to finding "good enough" solutions within reasonable computational timeframes. This shift is relevant to many real-world engineering and data science challenges where spending extra time and resources for marginal improvements is not justifiable.

Many techniques in this chapter are inspired by nature and physics, necessitating a conceptual explanation before mathematical formalization. Thus, for each technique,

Table 4.1 General characteristics of single-objective optimization algorithms

Method/Family	Common problems	Main pros	Main cons	Practical performance	Complexity (Concept/Impl.)
Gradient-Based (Unconstrained/Subproblems)					
GD	Quadratic Min.	Simple, Base for others	Slow, Sensitive to α, Local optimum (non-conv)	Slow/Moderate	Low
SGD	ML training (deep learning, large scale)	Fast per iteration (Big Data), escapes shallow local min.	Noisy, tuning α_k/batch, doesn't converge exactly	Very Good (Big Data)	Medium
Newton	Scientific/Eng. optim. (small/medium n)	Fast local convergence (quadratic)	Cost/Inversion of Hessian, requires stable H, sensitive to $\mathbf{x}_0$	Excellent (local)	High
Quasi-Newton	General unconstrained NLP, ML training (medium)	Good balance speed/cost, superlinear conv. (L-BFGS scales well)	More complex than GD	Very Good	Medium
Derivative-Free (Unconstrained/Subproblems)					
Nelder-Mead	Param. tuning (n small), Simple black-box optim.	Simple, No derivatives needed, Robust to initial noise	Slow (esp. n>10), No theoretical guarantee, Can stagnate/fail	Variable/Slow	Low
Pattern search	Black-box optim., Eng. (n small/medium), Robust	No derivatives needed, theoretical guarantees (GPS), systematic	Slow (esp. high n), Tuning of pattern/step	Slow/Moderate	Medium
Bundle methods	Large-scale non-smooth convex optim. (Lagrangian relax.)	Solid theory (convex), Better than subgradient, Large scale	Complex, mainly convex	Good (non-smooth conv.)	High
For Specific Constrained Problems/Transformations					
Penalty	(Technique w/ others)	Simple concept, general	Tuning ρ, Ill-conditioning, residual violation	(depends on solver)	Low (concept)
Barrier (IPM)	LP/QP/conv./smooth NLP large scale	Efficient, robust, high precision, global optimum (convex)	Requires feasible $\mathbf{x}_0$, Internal complexity	Excellent	High
SQP	General smooth NLP (small/medium n)	Effective, good local theory, handles non-linearities well	Complex, QP subproblem expensive, local optimum (non-conv)	Very Good	High

(continued)

Table 4.1 (continued)

Method/Family	Common problems	Main pros	Main cons	Practical performance	Complexity (Concept/Impl.)
Simplex	LP	Global optimum LP, Efficient (practice), basis for MILP	Only LP, worst-case exp. (rare)	Excellent (practice)	Medium/High
QP Solvers (IPM/ActiveSet)	QP (esp. Convex)	Global optimum (convex), very efficient for QPs	Specific to QP	Excellent (QP)	High
B&B / B&C	IP / MIP (esp. MILP)	Global optimum IP/MIP (guaranteed), standard framework	Can be slow (exp.), requires relaxation solver	Variable (Good w/ Cuts)	High
Metaheuristics (Usually Unconstrained + CHTs for Constraints)					
SA	General global optim. (Comb./Cont.), Scheduling, Design	Simple concept, Easy to escape local (initially)	Slow fine convergence, Tuning T, No optimum guarantee	Variable/Good	Medium
TS	Combinatorial optim. (Scheduling, VRP, QAP), Continuous	Effective (combinatorial), Memory avoids cycles	Tuning t_{tenure}/attributes, can stagnate	Very Good (combinatorial)	Medium
VNS	Combinatorial optim. (VRP, clustering, scheduling)	Systematic, simple concept, uses local search well	Requires good N_k, LS cost can be high	Very Good (combinatorial)	Medium
GA (EA)	General optim. (cont./disc./comb.), robust, black-box	Versatile (represent.), global exploration, parallelizable	Tuning params., slow fine conv., no optimum guarantee	Good/Very Good	Medium/High
ES (EA)	Continuous optim. (non-linear, non-conv.)	Self-adaptation, effective for complex continuous (CMA-ES)	Complex (esp. CMA-ES), less trivial for discrete	Very Good/Excellent (continuous)	High
DE (EA)	Continuous global optim.	Simple, few params., robust, effective for continuous	Can stagnate, less focused on discrete	Very Good (continuous)	Medium
PSO (SI)	Continuous global optim., Param. tuning, eng.	Simple, fast initially, few params.	Premature convergence, tuning w, c_1, c_2	Good/Very Good (continuous)	Low/Medium
ACO (SI)	Combinatorial/Graph optim. (TSP, VRP, routing)	Effective for path problems, positive feedback	Specific (graphs), tuning $\alpha, \beta, \rho, \ldots$	Excellent (graphs)	Medium/High

we explain the underlying concepts and provide pseudocode with practical explanations of its operation. The success of these algorithms often depends on parameter tuning, which can itself be an optimization task.

Furthermore, no single technique is universally superior: each has specific use cases. Therefore, engineers and data scientists must choose the technique most suitable for their specific needs. Nevertheless, many core metaheuristic concepts can be adapted for scenarios with multiple conflicting goals, as discussed in the next chapter.

Chapter 5
Multi-objective Optimization

Abstract This chapter focuses on Multi-Objective Optimization (MOO), where problems require the simultaneous optimization of two or more, often conflicting, objectives. To this end, we first introduce the concept of Pareto dominance and the Pareto front, representing the set of optimal trade-off solutions among the available objectives. MOO techniques are then broadly categorized into two types. First, a priori methods are presented, where decision-maker preferences are incorporated before optimization. Examples include the Weighted Sum method, the ϵ-Constraint method, Goal Programming, and Utility Functions. Second, a posteriori methods are discussed, which generate a set of Pareto-optimal solutions for subsequent selection by the decision-maker. Examples of these techniques include Multi-Objective Evolutionary Algorithms (MOEAs) such as NSGA-II, NSGA-III, SPEA2, MOEA/D, and MOPSO. The chapter also explains the role of Multi-Criteria Decision-Making (MCDM) techniques in understanding, selecting, and ranking solutions from the Pareto front. Similar to the previous chapter, each strategy's core principles, mechanisms, and key parameters are explained, balancing technical depth with conceptual overview.

You probably feel more comfortable with optimization if you have progressed from the start of this book up to this chapter. We've already covered the general concept in Chap. 1, how it appears in nature and our daily lives, and then how we formalize it mathematically in Chap. 2. Later, in Chap. 3, we opened the toolbox and learned to use algorithms to find the best solution when we have a single well-defined objective. Furthermore, we explored metaheuristics for this same type of problem in Chap. 4. We could even say you're ready to work with the vast majority of real-world optimization problems, couldn't we?

However, as life loves to present us with more complex challenges, we soon realize that focusing on a single objective is often a simplification. Consider various aspects of life where success is multi-faceted. It's highly unlikely that success is measured by a single metric. Perhaps you chose a cell phone that is inexpensive *and* fast. A home that is spacious *and* in a great location. A car that is fast *and* safe *and* cheap. Or maybe you want to live in a city that has good mobility *and* clean air *and* affordable housing. A pattern emerges here: we are constantly trying to reconcile desires that compete with each other.

W. R. Monteiro and G. R. Meza, *A Practical Guide to Optimization in Engineering and Data Science*, https://doi.org/10.1007/978-3-032-04633-8_5

The truly interesting problems, and those whose solutions will generate the most impact, will be the more complex ones. And the optimization problems that fit this description are, for the most part, Multi-Objective Optimization (MOO) problems with two or more conflicting objectives. Sometimes, we don't just want to find a single best possible solution, but rather to understand the map of the best possible answers and help choose the most appropriate one for each context.

Important
We assume that you understand the basic optimization concepts we've covered in previous chapters, such as the meaning of mathematical terms like $\mathbf{x}$, f, and constraints. We also assume you understand the difference between single-objective optimization and MOO.

5.1 Pareto Optimality

Remember when we said that optimization is the art of doing the best you can with what you have available in Chap. 1? Well, in Multi-Objective Optimization, the concept is a bit more subtle: how do we compare two solutions when one is better on objective A but worse on objective B? That's where the concept of dominance, which we saw in Sect. 2.3.3, comes in. So you don't have to go back there, we'd like to represent Fig. 5.1 just so you can recall what we mentioned in that section.

Just to recap: if we have two options (solutions $\mathbf{x}_1$ and $\mathbf{x}_2$) and $\mathbf{x}_1$ is as good as $\mathbf{x}_2$ in all aspects (objectives f_i) and strictly better in at least one aspect, then there's no reason to even consider $\mathbf{x}_2$. We say that $\mathbf{x}_1$ *dominates* $\mathbf{x}_2$. To help explain this, we'd like to invoke another analogy: in this case, a simple GPS problem.

5.1.1 Decision-Makers and the Pareto Front

You likely used some app to find the best route to a place, right? Some people use Google Maps, others use Waze, others use Apple Maps, and some use Gaode and Baidu Maps, and so on. Regardless of your app, let's imagine you need to leave your home for the airport and are using one of these apps that offers you different routes. You have two objectives (both to be minimized):

- f_1: Estimated travel time (in minutes).
- f_2: Estimated travel cost (in \$, considering tolls and perhaps an estimate for fuel/wear and tear).

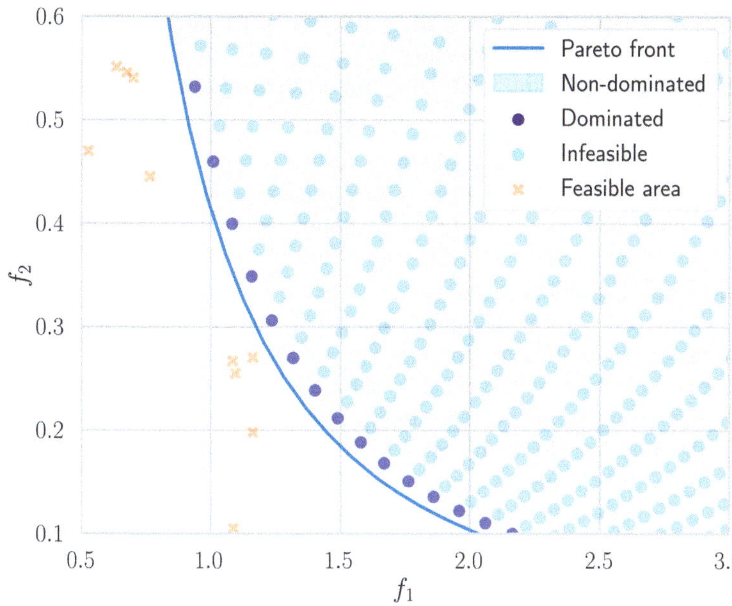

Fig. 5.1 Example of Pareto optimality for a two-objective minimization problem. It shows both infeasible, dominated, and non-dominated solutions. The feasible area and the actual Pareto front are also displayed. *Source* The author, 2025

And the app, behind the scenes, has calculated four possible routes for you:

- Route α: Time 30 min, cost \$15. Fast, but with an expensive toll.
  Mathematical representation: $\mathbf{f}_\alpha = [30, 15]$
- Route β: Time 50 min, cost \$5. Slow, no tolls, and therefore very cheap. But the path is longer.
  Mathematical representation: $\mathbf{f}_\beta = [50, 5]$.
- Route γ: Time 45 min, cost \$10. Average time and cost considering α and β.
  Mathematical representation: $\mathbf{f}_\gamma = [45, 10]$.
- Route δ: Time 40 min, cost \$8. Good balance between time and cost.
  Mathematical representation: $\mathbf{f}_\delta = [40, 8]$.

Remember we want to minimize time and cost. Now, let's compare for dominance:

- α versus β: α is better in time ($30 < 50$), but worse in cost ($15 > 5$). $\rightarrow$ *Non-dominated* with respect to each other.
- α versus γ: α is better in time ($30 < 45$), but worse in cost ($15 > 10$). $\rightarrow$ *Non-dominated* with respect to each other.
- α versus δ: α is better in time ($30 < 40$), but worse in cost ($15 > 8$). $\rightarrow$ *Non-dominated* with respect to each other.
- β versus γ: β is worse in time ($50 > 45$), but better in cost ($5 < 10$). $\rightarrow$ *Non-dominated* with respect to each other.

- β versus δ: β is worse in time ($50 > 40$), but better in cost ($5 < 8$). $\rightarrow$ *Non-dominated* with respect to each other.
- γ versus δ: δ is better in time ($40 < 45$) **and** better in cost ($8 < 10$). $\rightarrow$ *δ dominates γ*.

Thus, observe that γ is dominated by δ, since δ is superior in all objectives (less time and lower cost). Therefore, there is no rational reason for the app to show route γ if δ is available. Routes α, β, and δ are not dominated by any of the other presented options. Each of them represents a different trade-off: α prioritizes time, β prioritizes cost, and δ offers a middle ground.

Here, α, β, and δ form the approximation of the Pareto front for this problem.[1] Pareto-optimal solutions represent the equilibrium points where any attempt to improve one objective will inevitably cause a loss in another objective. They are the best possible compromises that an optimization algorithm has allowed us to find.

Now, which one would be the best? Notice that the choice becomes subjective. I, for example, tend to prefer faster routes. You, on the other hand, might prefer cheaper routes. However, it could be that I'm trying to save money and left home well in advance. Therefore, for a specific case, I might prefer a more economical route.

Do you see how this depends on each case and becomes a decision for someone interested in that problem? This is why you will sometimes see the term *decision-makers*: it's the person, or team of people, who would choose the solution from the Pareto front approximation that makes the most sense for a given occasion.

Important
We will use this example as a basis for explaining some of the techniques we will see later. For simplicity, we will refer to this case as "the GPS problem."

Each of the optimization techniques (especially the a priori methods) will not necessarily find the same routes α, β, γ, and δ that we saw earlier, but for simplicity and learning purposes, we will consider how some techniques would compare these proposed routes.

5.1.2 Frontier Challenges

Finding or approximating the Pareto front in real problems can be more difficult than it seems. Also, the difficulty does not increase linearly. Therefore, let's think about some challenges we might face depending on the type of problem.

Let's start with the shape. We saw in Sect. 2.2.3 that we can have convex and non-convex problems. Let's look at Fig. 5.2 as an example.

[1] In real problems, we hardly ever know what the true Pareto-optimal solutions would be. Hence, it's more common to refer to these approximations found by our algorithms simply as "Pareto fronts," without the word *approximation*.

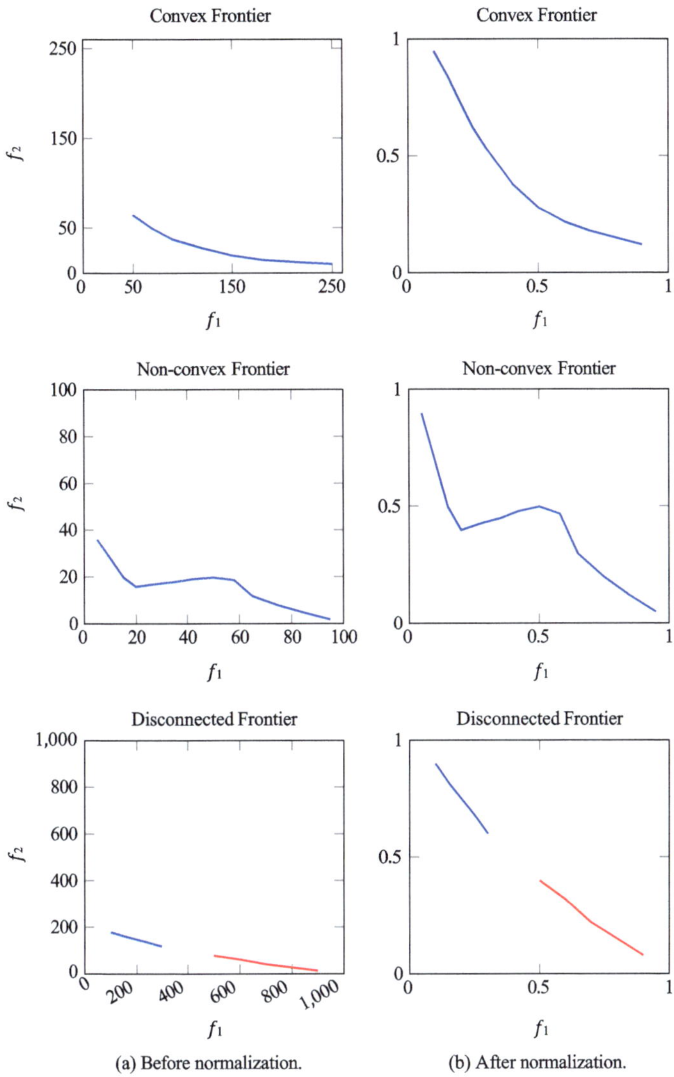

Fig. 5.2 The shape of the frontier (left) dictates the difficulty and choice of method. Non-convexity requires more robust methods. The scale of the objectives (right) can distort perception and must be normalized for many methods. *Source* The author, 2025

Convex problems (i.e., those that resemble a smooth curve projecting "outward") tend to be easier to solve and allow for the adoption of simpler optimization techniques, such as the weighted sum method we'll see shortly.

Then, we have non-convex problems. These have concave parts that indent into the feasible region. These problems are a bit more complex and are commonly found in real-world problems permeated by various non-linearities in the objective functions or constraints.

Tips

Non-linear objective functions or constraints often generate non-convex frontiers. If your problem has quadratic terms, exponentials, or complex logic, assume you are dealing with non-convexity. In such cases, metaheuristics based on evolutionary algorithms (EAs) are generally more suitable than simpler techniques.

Now, we have another problem related to normalization. We briefly mentioned this earlier, but it's worth expanding on it a bit more. You see, some MOO methods rely on reference points to calculate the distance between one solution and another, as is the case with some EAs. And, in practice, these techniques use Euclidean distance[2] or other similar techniques to find the proximity of these points.

The problem with using these distances is that when variables (or objectives) have very different numerical scales, calculations that combine these variables (like Euclidean distance) are dominated by the variable with the largest magnitude, ignoring the relative importance of the others. Imagine, for example, comparing travel time in seconds (ranging from 1200 to 7200) with cost in money (ranging from \$5 to \$50). If we simply calculate Euclidean distances directly, the objective with the largest numerical magnitude (time, in this case) will completely dominate the result, even if both are equally important to you. The cost would become almost irrelevant in practice.

Therefore, we almost always need to bring all objectives to a common scale, such as $[0, 1]$ or $[-1, 1]$, before applying weights or calculating distances. That said, there are cases where we use normalizations like min-max in MOO problems, as illustrated below:

$$f_i^{\text{norm}}(\mathbf{x}) = \frac{f_i(\mathbf{x}) - f_i^{\text{ideal}}}{f_i^{\text{nadir}} - f_i^{\text{ideal}}} \tag{5.1}$$

where f_i^{ideal} is the best value found for f_i and f_i^{nadir} is the worst value on the frontier for f_i. Thus, we ensure that all f_i^{norm} are on the same numerical scale $[0, 1]$.

[2] This is the "straight-line" distance between two points, as you would measure with a ruler on a map or in space.

Attention
There are some models where normalization is mandatory, even if all your variables are on the same scale and order of magnitude. This is the case for some EA applications and some other classical methods, such as the weighted sum.

Tips
Models that depend on calculating distances between solutions will give poor answers if we don't normalize the data first. And it's common for more than one person to work on an optimization problem within a company over the years. Therefore, there's a risk that someone might end up replacing a model that doesn't depend on these distances with one that does in the future. Thus, consider normalizing your variables whenever possible before optimizing.

Furthermore, we have cases where our headache is dimensionality. The "curse of dimensionality" is a known issue in machine learning (ML) problems, but it also applies to optimization cases.

One of the reasons we study MOO algorithms with only two or three objectives is visualization: you can understand a two-dimensional graph very well, or sometimes a well-formatted three-dimensional graph in a printed book or PDF file, right? Now, with four or more dimensions, things start to become harder to think about intuitively.[3] Besides being harder to think about, it's also more difficult for one solution to dominate the others, and the Pareto front is not so easy to keep visualizing.

There's also a scenario we heard about in one of those work meetings: "*You said these seven objectives are conflicting even among themselves. Are you sure? Couldn't they just be different angles of two larger objectives?*" Here we return to the Pareto front: if we simplify the problem, it's easier to optimize, analyze the results, and obtain a good solution. So, sometimes it might be worthwhile to reduce the number of objectives if your problem allows for it, right?

There's also a specific case for us: simulation. Let's think about two scenarios to make this clearer:

- Training ML problems: Let's suppose we're trying to optimize the hyperparameters of an ML model. Consequently, each solution would be a candidate ML model, right? Now, depending on the problem, we'd have to train each of these candidate models using a pre-existing database so we can evaluate its performance, correct? But what if we have such a large volume of data that each ML model would take several minutes or hours to train? If we're impatient, we wouldn't be able to

[3] Sometimes *you* can visualize spaces with four or more dimensions without problems, but this doesn't mean that all decision-makers in your company can do so with the same ease.

wait that long for a response from the optimization algorithm until it tests a large number of solutions in the feasible space.
- Product simulations: Similarly, we could imagine we're designing an object that needs to be simulated, like a wing or an airfoil that needs to undergo aerodynamic tests to evaluate one or more optimization objectives. Likewise, we'd have to spend several minutes or hours to see the result of each candidate solution. Therefore, we might not be able to test a few thousand solutions in a timely manner.

In these cases, it wouldn't be very interesting to adopt methods that depend on many evaluations, as is the case with some EAs. As an option, we could use techniques that use Bayesian approaches or other methods aimed at reducing the number of comparisons and tests.[4]

That said, it's for these reasons that we will hardly ever find the entire true Pareto front. We want solutions that are a good approximation of the frontier, naturally. That's why Multi-Objective Optimization has a set of goals for its execution instead of just attempting to find the best solution for all objectives.

The first goal is **convergence**. While the Pareto-optimal curve is often unknown in real-world scenarios, MOO algorithms are expected to find a set of solutions that are members of this curve or as close as possible to it.

The second is **diversity**: the solutions found should represent all of the Pareto-optimal curve as much as possible instead of forming a cluster of close solutions while missing the other parts of the curve.

The third is **relevance**: the decision-maker can have a particular interest in a specific region of the Pareto-optimal curve instead of all of it. An example of this interest is in the form of a bias toward a subset of objectives. Since decision-makers are expected to visualize all solutions and choose the one that better fits their needs, it is also relevant to present solutions as close as possible to the user preferences.

5.2 When Do We Fit in "Multi-objective"?

Let's use the Pareto front and its challenges as our starting point. Decision-makers might be people who want to access and compare all the optimal solutions presented by a Pareto front, but they might also be people in a hurry or who don't have much knowledge or desire to compare all these solutions, right? For instance, presenting a hundred candidate solutions from a Pareto front during a brief business meeting is often impractical.

Or, perhaps, you'd just like to reuse the concepts and solvers we saw in Chap. 3 in some way. Therefore, one of the most common ways to divide MOO algorithms is by the moment when decision-makers... well, make their decisions.

The first situation is when decision-makers state what they want before starting the optimization process with an algorithm. You'll find mentions of this type of

---

[4] This is one of the reasons why data scientists might use Bayesian Optimization techniques to optimize hyperparameters of an ML model.

strategy under the *a priori methods* name. The idea here is to have only a single Pareto-optimal solution produced instead of a Pareto front.

The other alternative is the *a posteriori methods*. Here, the MOO algorithm will produce a Pareto front with several Pareto-optimal solutions. After that, decision-makers will use their experience to choose the most suitable solution for that occasion.

Still, there's a third alternative: *interactive methods*. Here, the optimization algorithm runs for a shorter period, presenting some solutions it found to the decision-makers. With this partial information, decision-makers can opt to provide feedback so the algorithm can focus more or less on certain regions of the Pareto front. Then, the algorithm runs for a while longer and produces more solutions. The decision-makers then conduct a new analysis and produce more feedback until a good solution is found.

5.3 A Priori Methods

Thus, let's first concentrate on a priori methods. Here, we want to use the decision-makers' preferences before running an optimization algorithm. This means, in practice, that we will convert a MOO problem into one or more single-objective optimization problems. By doing so, we can use the techniques we saw in Chaps. 3 and 4 to find a solution.

We see some advantages in these methods. Firstly, we can reuse much of our knowledge about how single-objective optimization techniques work. Secondly, we have a simpler problem to execute and analyze: since we convert the problem to have only a single objective, we also get a single answer. The issue with this is that decision-makers will need to think in advance about their preferences and priorities among the different objectives before they can visualize a Pareto front. Let's take the GPS example we saw earlier: decision-makers would need to say, for example, that time would have a weight of 75% in the decision-making process and cost 25%. However, why couldn't time have a weight of 74% and cost 26%? Or 76% and 24%? These differences could change the priority of the results, and this would be a second disadvantage: decision-makers wouldn't have access to solutions that could potentially be better compared to the solution produced by the single-objective optimization algorithm due to the very nature of this technique.

That said, let's consider some popular a priori methods.

5.3.1 Weighted Sum Method

This is the simplest a priori method. Here, for each objective i, we assign a weight w_i attributed by the decision-makers. Thus, an MOO problem with k objectives could be written as:

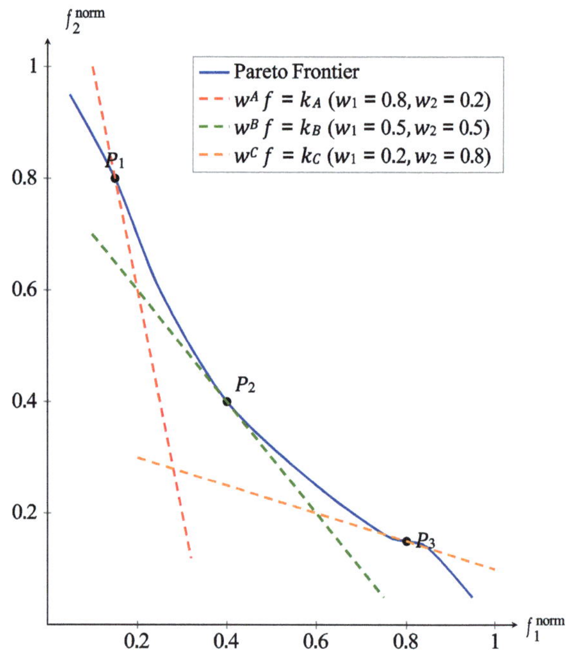

Fig. 5.3 The Weighted Sum Method finds the point on the frontier that first touches a line with a slope defined by the weights **w**. In this figure, each weight vector $\mathbf{w} = [w_1, w_2]$ defines a line $w_1 f_1 + w_2 f_2 = k$ with slope $-w_1/w_2$. The optimal solution for each **w** is found where this line first touches the Pareto frontier. *Source* The author, 2025

$$\min f(\mathbf{x}) = \sum_{i=1}^{k} w_i f_i(\mathbf{x}) \tag{5.2}$$

where:

- $w_i \geq 0$ is the weight assigned to the i-th objective $f_i(\mathbf{x})$.
- $\sum_{i=1}^{k} w_i = 1$, assuming we understand the weights as percentages of importance for each of the objectives. Therefore, the weights would sum to 100%.

Attention
This is a clear example where we need to normalize the objectives $f_i(\mathbf{x})$ to the same scale before applying the weights. Otherwise, the objectives with a larger numerical magnitude will dominate this sum regardless of the chosen weights.

The fact that this technique is simple does not imply that it has no disadvantages. When working with non-convex problems, it's common to have some parts of the feasible space that are particularly interesting right in these non-convex parts. That is, it's common to witness situations in these locations where small deteriorations in one objective cause large improvements in other objectives (Fig. 5.3).

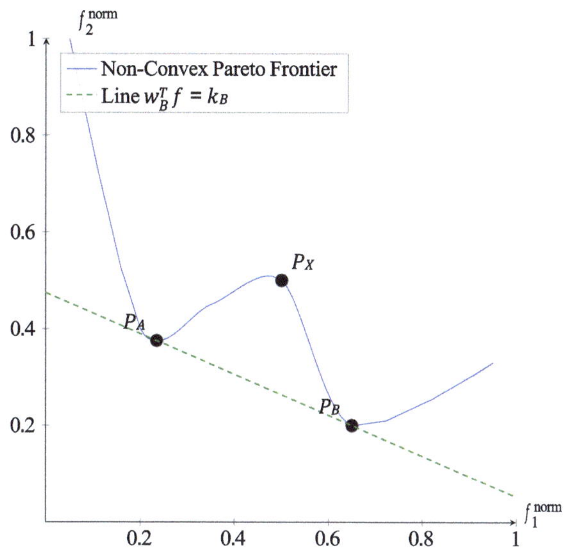

Fig. 5.4 The key limitation of the Weighted Sum: inability to find solutions in non-convex parts of the Pareto Frontier, as weighted sum lines can only be tangent. See how we are unable to find any points in the concave region. These points (e.g., P_X) are on the frontier but may not be found by the sample weighted sum dotted line. *Source* The author, 2025

The problem is that this method cannot access these areas, and a large portion of the optimization problems we will encounter in our careers are non-convex. Therefore, if we apply this method to non-convex problems, we risk completely ignoring very good solutions. For this reason, this method is useful for convex problems and for understanding the general concept of a priori (Fig. 5.4).

Now, how would this method look in the GPS example we saw earlier? Remember we saw that we had the non-dominated routes $\alpha = [30, 15]$, $\beta = [50, 5]$, and $\delta = [40, 8]$, where f_1 is time and f_2 is cost.

First, we would need to normalize f_1 and f_2 to the same numerical scale, like $[0, 1]$. Then, the choice of weights would determine the result. If we gave a very high weight to minimizing time ($w_1 = 0.9$, $w_2 = 0.1$, for example), the algorithm would return a route like α. If we prioritized cost ($w_1 = 0.1$, $w_2 = 0.9$), a route similar to β would be chosen. On the other hand, more balanced weights ($w_1 = 0.5$, $w_2 = 0.5$) could lead to a route like δ or some other point on the frontier if it were continuous and convex between these points.

However, a challenge arises: how can a decision-maker precisely define numerical weights reflecting their subjective preferences without first understanding the available trade-offs (e.g., the specific time and cost of options like α, β, δ)? It's very difficult to translate a subjective preference into numerical weights without knowing the pros and cons of each choice. How, for example, do you quantify "time is a little more important today"? This is why it's common to have to test different weights to explore the frontier, and even then, we can fail with non-convex problems.

5.3.2 ϵ-Constraint Method

If in the weighted sum method we combined the objectives into one, here we choose one objective to remain as the objective to be optimized, while the others are converted into constraints.

Here, we choose *one* objective f_m that we really want to optimize. For all other f_j objectives, we simply inform the maximum limit ϵ_j that we are willing to tolerate. Thus, we have a single-objective constrained optimization problem:

$$\begin{aligned} \min_{\mathbf{x}} \quad & f_m(\mathbf{x}) \\ \text{subject to} \quad & f_j(\mathbf{x}) \leq \epsilon_j, \quad \forall j \neq m \\ & g_i(\mathbf{x}) = 0, \quad i = 1, \ldots, m \quad \text{(Original equality constraints)} \\ & h_l(\mathbf{x}) \leq 0, \quad l = 1, \ldots, p \quad \text{(Original inequality constraints)} \end{aligned} \tag{5.3}$$

Important
It is common to encounter people who only know about single-objective optimization who confuse MOO with this technique. We have heard several times phrases like, "*Yes, I know MOO! You choose one objective to optimize and put the rest as constraints.*" In fact, when we later look at the a posteriori methods, you will see that this view is too simplistic for all existing scenarios.

Let's suppose we have two objectives to minimize: $f_1(\mathbf{x})$ and $f_2(\mathbf{x})$. In the ϵ-constraint method, we could choose one of these objectives (say, $f_1(\mathbf{x})$) and rewrite the problem as:

$$\begin{aligned} \min_{\mathbf{x}} \quad & f_1(\mathbf{x}) \\ \text{subject to} \quad & f_2(\mathbf{x}) \leq \epsilon_2 \\ & \text{(Original problem constraints)} \end{aligned} \tag{5.4}$$

In this case, ϵ_2 is an upper limit (a maximum value) that decision-makers define for the second objective. The algorithm then finds the solution that minimizes f_1 without letting f_2 exceed the value ϵ_2.

Using this method has some advantages. First is non-convexity: we can now find solutions in non-convex parts of the Pareto front, since the optimization algorithm focuses on only one of the objectives. Also, we notice greater ease for decision-makers to map and prioritize each objective. Notice that they are forced to choose only one objective as the main one, and still need to define what would be the maximum acceptable value for the other constraints (Fig. 5.5).

Let's return once more to our GPS example from Sect. 5.1.1, with some simulations summarized in Table 5.1. Let's assume that you, as the decision-maker, choose time f_1 as the primary objective. In this case, we would try to minimize this objective and nothing else. Now, cost f_2 becomes a constraint.

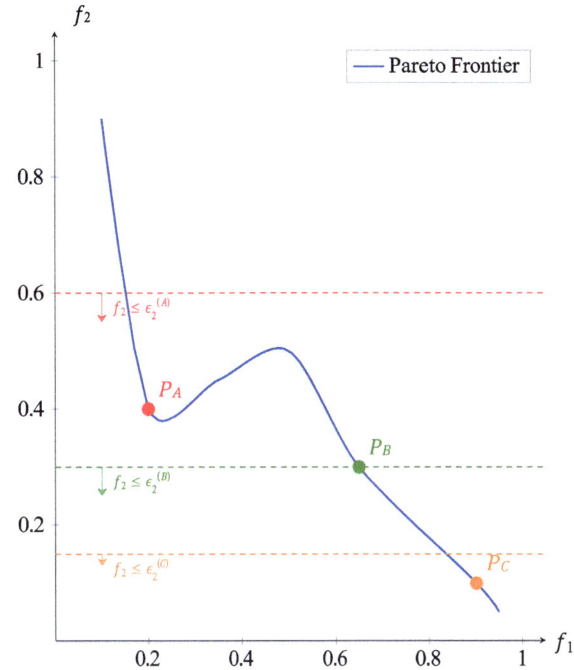

Fig. 5.5 The ϵ-Constraint Method: Optimizes one objective (f_1) subject to limits on the second objective (ϵ_2). By varying the ϵ_2, the frontier is mapped, even in non-convex parts. The solution points P_A, P_B, and P_C are examples of valid solutions considering the different constraints $\epsilon_2^{(A)}$, $\epsilon_2^{(B)}$, and $\epsilon_2^{(C)}$, respectively. *Source* The author, 2025

Table 5.1 Example of applying the ϵ-constraint method to the GPS problem (Minimize Time f_1)

Cost limit ϵ_2	Constraint ($f_2 \leq \epsilon_2$)	Feasible routes (Respect $f_2 \leq \epsilon_2$)	Best f_1 (Time) (Among feasible)	Resulting route [Time, Cost]
\$90	$f_2 \leq 90$	α, β, δ	min (30, 50, 40) = 30	α [30, 15]
\$10	$f_2 \leq 10$	β, δ	min (50, 40) = 40	δ [40, 8]
\$7	$f_2 \leq 7$	β	min (50) = 50	β [50, 5]
\$2	$f_2 \leq 2$	None	–	(Infeasible)

Let's suppose you don't want to spend more than \$90. This is a relatively high value. In this case, we'd have min f_1 subject to $f_2 \leq 90$. Here, routes α, β, and δ respect this constraint. And, among them, the route with the smallest f_1 is route α. Therefore, it would be the chosen route even though it has the highest cost among all.

Now, what if we set a more restrictive value for ϵ_2? You could say, for example, that you don't want to spend more than \$10. Then, we would define $\epsilon_2 = 10$, and we would have the problem min f_1 subject to $f_2 \leq 10$. The routes that satisfy $f_2 \leq 10$ are β (cost 5) and δ (cost 8). Between them, the one that minimizes f_1 is δ (time 40). Therefore, the result would be Route δ.

And, if we were even more restrictive with $\epsilon_2 = 7$, only route β would satisfy the constraint, and it would be the solution found (minimizing f_1 within the remaining

Table 5.2 Example of optimal solutions for each objective function in a problem with $k = 3$. The first row shows solution $\mathbf{x}^{(1)*}$, the second row shows solution $\mathbf{x}^{(2)*}$, and the third row shows solution $\mathbf{x}^{(3)*}$. The best solutions we originally found are in red

	f_1	f_2	f_3
min f_1	$f_1(\mathbf{x}^{(1)*})$	$f_2(\mathbf{x}^{(1)*})$	$f_3(\mathbf{x}^{(1)*})$
min f_2	$f_1(\mathbf{x}^{(2)*})$	$f_2(\mathbf{x}^{(2)*})$	$f_3(\mathbf{x}^{(2)*})$
min f_3	$f_1(\mathbf{x}^{(3)*})$	$f_2(\mathbf{x}^{(3)*})$	$f_3(\mathbf{x}^{(3)*})$

feasible set). Notice that the difficulty here lies in choosing ϵ values that are both realistic (not making the problem infeasible, like asking for $\epsilon_2 \leq 2$) and that explore the desired trade-off region.

5.3.2.1 Payoff Table, Ideal and Nadir

You must have noticed that choosing the correct value of ϵ completely changes the result, right? Therefore, we can use a payoff table to help us in this task to find a realistic value for each ϵ_j.

The first step we take in this table is to go back to our initial MOO problem: that is, before choosing one objective to remain as the main objective and modifying the others as constraints. In this original problem of k objectives, we solve the optimization problem min $f_i(\mathbf{x})$ considering each objective $i = 1, \ldots, k$ and all original constraints. With this, we will find the optimal solution $\mathbf{x}^{(i)*}$ for each objective.

After that, for each of these optimal solutions $\mathbf{x}^{(i)*}$, we will calculate its values for the other objectives $f_j(\mathbf{x}^{(i)*})$ $(j = 1, \ldots, k;\ j \neq i)$. This gives us an idea of how each of these generated optimal solutions considers the other objectives.

In fact, to facilitate this notion, we construct a table with k rows and k columns. This is the payoff table, an example of which can be seen in Table 5.2.

Notice that we have the best possible values for each objective in the form of $[f_1(\mathbf{x}^{(1)*}), f_2(\mathbf{x}^{(2)*}), f_3(\mathbf{x}^{(3)*})]$. We call this the *ideal point*. Let's go back to the GPS case: route α is the fastest at 30 min, and route β is the cheapest with a cost of \$5. In this case, the ideal point would be an imaginary route that takes 30 min and costs \$5.

The *nadir point*, on the other hand, is the vector of worst objective values found across the entire true Pareto-optimal set. One way to obtain its *approximation* is to take the worst value from each column of the payoff table. In the GPS case, the slowest route we had found was β with a duration of 50 min, and the most expensive is α with a cost of \$15. Therefore, the nadir point would be the imaginary route that takes 50 min and costs \$15.

Using the ideal and nadir points, we define the relevant range for each ϵ_j (that is, for each objective j that we are not optimizing) as a value between $[f_j^{ideal}, f_j^{nadir}]$. Thus, notice that it's not a random value we choose to define ϵ_j, but rather something within this range. In fact, we could even divide this range into smaller intervals to test performance. Let's return to the GPS example: if we want to optimize time f_1, cost f_2 would be the constraint. Here, the value of ϵ_2 would be a value defined in the range [5, 15]. We could divide this into two intervals ([5, 10] and [11, 15]), three intervals ([5, 8], [9, 12], and [13, 15]), or more.

5.3.3 *Goal Programming (GP)*

The focus of this strategy is on defining goals g_i for each objective f_i. Here, we want to minimize g_i, not f_i. That is, we want to minimize deviations from predefined goals set by decision-makers, not in relation to the objectives themselves.

Now, how do we work with g_i? First, we add constraints for an objective $f_i(\mathbf{x})$ as follows:

$$f_i(\mathbf{x}) - d_i^+ + d_i^- = g_i, \quad \text{with } d_i^+, d_i^- \geq 0 \tag{5.5}$$

Here, we have two new deviation variables: d_i^+, which defines how much our objective $f_i(\mathbf{x})$ exceeds the goal g_i (overshoot), and d_i^-, which defines how much the objective $f_i(\mathbf{x})$ falls short (undershoot) of the goal g_i. With this, we want to get closer to these goals. In fact, it's common to use this technique if decision-makers need to think in terms of specific goals for each objective. Something like "I'd like my GPS route to give an option with a cost of up to \$X," "production needs to be at least Y units," or "I'd like the waiting time to be around Z minutes." Notice that in all these phrases, we don't focus on finding the best value for each objective, but rather on minimizing unwanted deviations from the goals. Thus, we could think of this technique when we face problems of resource allocation to achieve departmental goals, or when we need to plan production with volume and budget targets (Fig. 5.6).

That said, we have two main routes for this type of technique: weighted GP and lexicographic GP.

5.3.3.1 Weighted Goal Programming

We use this technique when we need to assign different penalties to different types of deviations. Here, we try to minimize a weighted sum of all unwanted deviations at the same time. For this, decision-makers need to assign weights w_i^+ and w_i^- that represent the penalty for each goal g_i. The objective function takes the following form:

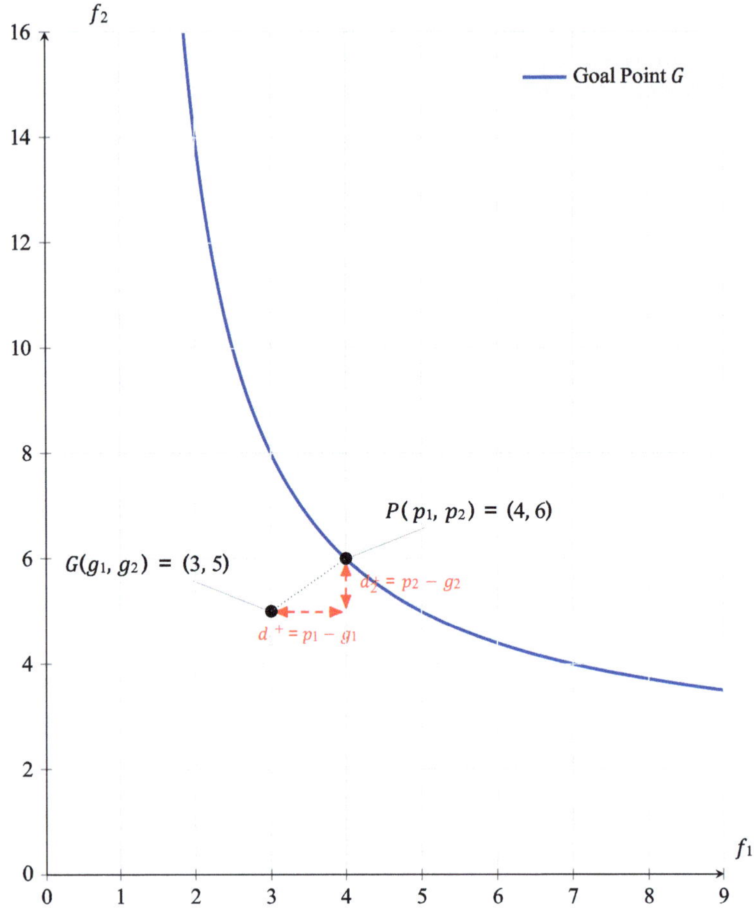

Fig. 5.6 Goal Programming: here, we are looking for the solution P on the frontier that minimizes deviations from the goals (G) defined by the decision-makers. *Source* The author, 2025

$$
\begin{aligned}
&\min_{\mathbf{x},\mathbf{d}^+,\mathbf{d}^-} \quad G(\mathbf{x}, \mathbf{d}^+, \mathbf{d}^-) = \sum_{i=1}^{k} (w_i^+ d_i^+ + w_i^- d_i^-) \\
&\text{subject to} \\
&\qquad f_i(\mathbf{x}) - d_i^+ + d_i^- = g_i, \quad \forall i = 1, \ldots, k \\
&\qquad g_j(\mathbf{x}) = 0, \quad \forall j = 1, \ldots, m \\
&\qquad h_l(\mathbf{x}) \leq 0, \quad \forall l = 1, \ldots, p \\
&\qquad d_i^+, d_i^- \geq 0, \quad \forall i = 1, \ldots, k \\
&\qquad \text{(Any other constraints on } \mathbf{x}\text{, such as } \mathbf{x} \geq 0)
\end{aligned}
\tag{5.6}
$$

In this format, we need to optimize the values of the original variables **x**, but also the values of all deviation variables d_i^+ and d_i^-. These variables are also defined by the constraint $f_i(\mathbf{x}) - d_i^+ + d_i^- = g_i$. This forces that, for any solution **x** found, the values of the deviations d_i^+ and d_i^- correspond exactly to the difference between the value of the objective $f_i(\mathbf{x})$ and the goal g_i. This is what ensures that, by minimizing the objective function G, we are effectively trying to make $f_i(\mathbf{x})$ approach g_i. We also maintain the original constraints of the MOO problem, such as $g_j(\mathbf{x}) = 0$ and $h_l(\mathbf{x}) \leq 0$.

Finally, notice that the original objective function $f_i(\mathbf{x})$ for an objective i becomes a constraint: after all, it needs to respect d_i^+ and d_i^-. Therefore, the single-objective optimization algorithm will find the combination of $\mathbf{x}$, d_i^+, and d_i^- that minimizes the weighted sum of all unwanted deviations G, respecting both the original constraints and the new goal constraints that define the deviations. The value of $f_i(\mathbf{x}^*)$ in the optimal solution will be the final result for the original objective i.

Let's think once more about the GPS problem: let's consider that we, the decision-makers, have defined the goals $g_1 = 35$ (time in minutes) and $g_2 = \$7$ (cost, in \$). For both, exceeding the goal is not good (d_1^+ and d_2^+ should be minimized), but finding something below these values is interesting ($w_1^- = 0$, $w_2^- = 0$).

Here, we could think of two scenarios as an example:

- **Equal penalties** ($w_1^+ = 1$, $w_2^+ = 1$). Thus, we want to minimize $G(\mathbf{x}) = d_1^+ + d_2^+$. Reusing those same Pareto-optimal routes we saw throughout this chapter, we would have:

  G_α (for [30, 15]): $d_1^+ = 0, d_2^+ = 8 \rightarrow G = 0 + 8 = 8$. This happens because 30 is less than 35 (good) and $15 - 7 = 8$ (i.e., deviation d_2^+ of 8).

  G_β (for [50, 5]): $d_1^+ = 15, d_2^+ = 0 \rightarrow G = 15 + 0 = 15$. This happens because we had a deviation in d_1^+ ($50 - 35 = 15$), but no deviation for d_2^+ (5 is less than our goal of \$7).

  G_δ (for [40, 8]): $d_1^+ = 5, d_2^+ = 1 \rightarrow G = 5 + 1 = 6$. This happens because we had two deviations: $d_1^+ = 40 - 35 = 5$ and $d_2^+ = 8 - 7 = 1$. This would be the best route as it has the lowest $G(\mathbf{x})$ value.

- **Different penalties** ($w_1^+ = 5$, $w_2^+ = 1$. That is, time is much more important than cost). Here, we have $G(\mathbf{x}) = 5d_1^+ + d_2^+$. Following the same goals of $g_1 = 35$ and $g_2 = \$7$, we would have:

  G_α: $5(0) + 8 = 8$. This is the best route for having the lowest $G(\mathbf{x})$.

  G_β: $5(15) + 0 = 75$.

  G_δ: $5(5) + 1 = 26$.

Note that, for these two examples, we always seek a single equilibrium point that balances all unwanted deviations according to the penalties defined by the decision-makers.

5.3.3.2 Lexicographic Goal Programming (Preemptive Goal Programming)

If we need to prioritize some objectives (and goals) over others, then we could use a lexicographic GP technique. In this technique, we group goals into priority levels. Here, the algorithm minimizes deviations from the highest priority level first. Then, it minimizes deviations from the second level, ensuring that the optimization of the first level is not worsened, and so on.

Therefore, decision-makers no longer establish weights, but rather absolute priority levels for achieving one or more goals. Thus, an algorithm will try to satisfy the highest priority goals as best as possible first, disregarding lower priority goals at that moment. Here, we end up with a sequence of single-objective optimization problems. Let's see:

1. Decision-makers establish L goals, where L is a non-negative scalar. It can be equal to the number of objectives (i.e., one goal equals one objective), or a smaller value (where we could group more than one objective for a goal).
   Here, for each goal, we will have a priority level P. Therefore, we will have a sequence of priorities $P_1, P_2, \dots, P_L$, where we will solve P_1 first, then P_2, and so on.
2. Since we will have L goals, we will also have L single-objective optimization problems. Just as we introduced an objective function G in the previous section, here we will also reintroduce G, but with a small change: $G_r = \sum_{i \in P_r} (w_i^+ d_i^+ + w_i^- d_i^-)$ will measure the deviation for the goals at priority level r. Here, the weights w can be 1 if we want to maintain the same weight for all priorities.
   Important: Notice that we defined $\sum_{i \in P_r}$, and not $\sum_{i \in P_L}$. This means we are considering more priority levels as we increase the value of r.
3. Then, for the first priority ($r = 1$), we will look for the smallest possible deviation D_1^* for this priority. Or, mathematically speaking, we will look for $D_1^* = \min G_1(\mathbf{d}^+, \mathbf{d}^-)$ while respecting the original constraints and, also, trying to ensure that the deviations are not negative.
4. After finding D_1^*, we will advance to the next priority $r = 2$. To find D_2^*, we will need to respect the original constraints, the non-negativity constraints of the goals, and also ensure that the optimal deviation levels of higher priorities are not worsened.
5. Similarly, we will advance through the other priorities until the process is complete.

That said, let's think about the formulation of the optimization problem. We would have for each r:

$$\min_{\mathbf{x},\mathbf{d}^+,\mathbf{d}^-} \quad G_r(\mathbf{d}^+, \mathbf{d}^-) = \sum_{i \in P_r} (w_i^+ d_i^+ + w_i^- d_i^-) \quad \text{(Objective for Priority } r\text{)}$$

subject to

$$\begin{aligned}
&f_i(\mathbf{x}) - d_i^+ + d_i^- = g_i, \quad \forall i = 1, \ldots, k \quad \text{(Goal Constraints)} \\
&g_j(\mathbf{x}) = 0, \quad \forall j = 1, \ldots, m \quad \text{(Original Equality Constraints)} \\
&h_l(\mathbf{x}) \leq 0, \quad \forall l = 1, \ldots, p \quad \text{(Original Inequality Constraints)} \\
&G_s(\mathbf{d}^+, \mathbf{d}^-) = D_s^*, \quad \forall s = 1, \ldots, r-1 \quad \textbf{(Achievement Constraints)} \\
&d_i^+, d_i^- \geq 0, \quad \forall i = 1, \ldots, k
\end{aligned} \tag{5.7}$$

Notice that we introduced achievement constraints here. They are very important for this method and ensure that we don't ruin what we've done so far. I mean, it would be useless to find a good solution for the highest priority only to spoil everything later when modifying something less relevant, right? Therefore, we force the current solution to at least maintain the best deviation level we found for previous priorities.

In the end, the value of $f_i(\mathbf{x}^*)$ in the optimal solution obtained by solving priority level L will be the final result for the original objective i.

For comparison and example purposes, let's return to our hypothetical GPS case: here, we could establish two priorities since we have two objectives. We could even consider two different situations. Just as we did in Sect. 5.3.3.1, let's consider we have the goals $g_1 = 35$ (time in minutes), $g_2 = 7$ (cost, in \$). Let's see:

1. Scenario 1: The first priority is cost, and the second priority is time.

   We minimize d_2^+ (excess cost). Looking at the routes, the minimum d_2^+ is 0, achieved by route β. So, $D_1^* = 0$.

   Now that we've solved the first priority, let's minimize d_1^+ (excess time). To do this, we need to obey $d_2^+ \leq 0$ (i.e., $f_2 \leq 7$). The only route that satisfies $f_2 \leq 7$ is β. Its d_1^+ is 15.

   Thus, the optimal solution is β. We perfectly achieved the cost goal (priority 1), and did the best possible for the time goal (exceeding by 15 min) without violating priority 1.
2. Scenario 2: Priority 1 is time, and priority 2 is cost.

   We minimize d_1^+ (excess time). The minimum d_1^+ is 0, achieved by α. So $D_1^* = 0$.

   Now that we've solved the first priority, let's minimize d_2^+ (excess cost) subject to $d_1^+ \leq 0$ (i.e., $f_1 \leq 35$). The only route that satisfies $f_1 \leq 35$ is α, and its d_2^+ is 8.

   Here, the optimal solution is α. We perfectly achieved the time goal (priority 1), and did the best possible for the cost goal (exceeding by \$8) without violating priority 1.

5.3.3.3 Practice

Here, the question to ask decision-makers is: are goal deviations something that needs to be balanced simultaneously, or according to a hierarchy? If we need to solve simultaneously, then it would be a case for considering weighted GP. In fact, using weighted GP can be a good option in situations where we can allow a slightly larger deviation in a less important goal to further reduce the deviation in a more important goal. Or, perhaps, there are cases where we want to find an equilibrium point that minimizes all deviations simultaneously. We see this, for example, in situations where we need to plan a factory's production, or allocate the budget among different departments.

Otherwise, it might be a case for lexicographic GP. We use it when some goals absolutely must be satisfied before we think about other goals. These are cases where we have less flexibility to trade one goal for another, such as in the manufacturing of pharmaceutical and food products, or in situations where regulatory and safety prerequisites are mandatory.

5.3.4 Lexicographic Method

The logic of this method is quite similar to lexicographic GP, which we saw in Sect. 5.3.3.2, but there are still subtle differences. That's why we decided to dedicate extra space to talk about it.

In lexicographic GP, we seek to find a solution that comes as close as possible to achieving the established goals, respecting the hierarchy of importance of these goals. The result depends heavily on the values of the goals g_i, defined by the decision-makers.

In the lexicographic method, however, we want to find the solution that achieves the best absolute possible value for the highest priority objective, then the best absolute value for the second (without sacrificing the first), etc. It does not depend on predefined goals, but only on the ranking of the objectives f_i.

Thus, we adapt equation Eq. 5.7 for each priority r to take the following form in the lexicographic method:

$$
\begin{aligned}
&\min_{\mathbf{x}} \quad f_r(\mathbf{x}) \quad \text{(Objective for Priority r)}\\
&\text{subject to}\\
&\qquad g_j(\mathbf{x}) = 0, \quad \forall j = 1, \ldots, m \quad \text{(Original Equality Constraints)}\\
&\qquad h_l(\mathbf{x}) \leq 0, \quad \forall l = 1, \ldots, p \quad \text{(Original Inequality Constraints)}\\
&\qquad f_s(\mathbf{x}) \leq f_s^*, \quad \forall s = 1, \ldots, r-1 \quad \textbf{(Lexicographic Priority)}
\end{aligned}
\tag{5.8}
$$

Here, we are directly minimizing the objective function $f_r(\mathbf{x})$, which is the function corresponding to the current priority level r. We also no longer have deviation variables, and optimization occurs directly on $\mathbf{x}$.

Now, the same maxim about lexicographic GP also applies to this "pure" lexicographic method: decision-makers need to establish a strict and absolute order of priority among the objectives.

5.3.5 *Utility Functions*

In the literature and in university classes, it's common to find the definition that utility functions are the most complete a priori method. After all, we define all the decision-makers' preferences in a single-objective function, which we'll call U.

In this case, U takes as input the values of all objectives for $\mathbf{x}$ and returns only a single scalar value that represents how useful[5] that solution is for the decision-makers. Thus, the original multi-objective problem transforms into a single-objective optimization problem with the following characteristic:

$$
\begin{aligned}
&\max_{\mathbf{x}} \quad U(f_1(\mathbf{x}), f_2(\mathbf{x}), \ldots, f_k(\mathbf{x})) \\
&\text{subject to} \\
&\qquad g_j(\mathbf{x}) = 0, \quad \forall j = 1, \ldots, m \quad \text{(Original Equality Constraints)} \\
&\qquad h_l(\mathbf{x}) \leq 0, \quad \forall l = 1, \ldots, p \quad \text{(Original Inequality Constraints)}
\end{aligned}
\tag{5.9}
$$

Now, how do we define U? This is the most important question and is where we usually stumble when trying to implement this in practice. We believe you'll understand this by understanding the complexity of the steps below. Let's go:

1. First, decision-makers need to determine a utility function $u_i(f_i)$ for each objective i. Here, the idea is to understand how utility changes as we change each objective.

   Let's think about the GPS case: does a reduction in travel time from 70 to 60 min have the same relevance as a reduction from 30 to 20 min? Sometimes, when we leave for the airport with some time to spare, a reduction from 30 to 20 min doesn't make much difference—after all, we'd still spend a good amount of time waiting in the lounge. Now, if we arrive just in time, the difference between 70 and 60 min could be the difference between missing the flight or not. See how these ten minutes between these two examples have different importance? Here, the utility function could be non-linear.

   Or, perhaps, a reduction in travel cost from \$10 to \$8 is as relevant as a reduction from \$8 to \$6: the \$2 difference will have the same weight for you. Here, the utility function can be linear.

[5] Hence the name *utility functions*. We also call them *value functions*, because the scalar value represents the relevance of that solution.

2. Then, we evaluate the trade-offs. We'll need to understand from the decision-makers how they would trade one objective for another. This allows us to determine the correct mathematical form to combine the individual utilities $u_i(f_i)$ into the global function U.

   Common forms are additive ($U = \sum w_i u_i(f_i)$) or multiplicative, but they are only valid under certain independence assumptions.

   To exemplify the independence assumptions, let's think about the GPS case: does the preference between different levels of travel cost change according to travel time?
3. Finally, we determine the weights that reflect the relative importance of the variation ranges of each individual utility $u_i(f_i)$.

Independence Assumptions Let's say that after reading this book and learning many things about Optimization, you have two job offers. Company λ offered you a high salary (f_1) (good), but you need to spend a lot of time commuting (f_2) to get there (bad). Company ζ offered you a low salary (bad), but it's practically next door to your house (good). Please keep these two situations in mind.

The idea of independence assumptions is to understand if the decision-makers' preference changes as we achieve the other objectives. With that, let's look at the main types of independence assumptions.

In *preferential independence*, consider the following question: would you still choose Company λ if the growth potential (f_3) in both companies was high? And would you still choose it if the growth potential in both companies was low? If your choice doesn't change no matter the value of f_3, then we would say that $[f_1, f_2]$ are preferentially independent of f_3.

In *utility independence*, the question is a bit different, as it's a stronger condition related to your attitude toward risk. Imagine we fix the commute (f_2) at a "medium" level and the growth potential (f_3) at a "low" level for two other companies: ι and κ. Now, you have two options regarding salary (f_1):

- Conservative choice: receive a guaranteed "medium-high" salary from company ι.
- Riskier choice: company κ hasn't revealed its salary range. You have a 50% chance of a "very high" salary (better than what company ι offers) and a 50% chance of a "medium-low" salary, lower than company ι.

Which option would you choose? Your choice here reflects your risk preference regarding salary, considering the current level of commute and growth.

Now, imagine we change the context: the commute (f_2) remains "medium," but the growth potential (f_3) becomes "high." We present you with exactly the same choice about salary (f_1): the certainty of "medium-high" versus the gamble of "very high" and "medium-low." Did your choice between certainty and the salary lottery change just because the growth prospect improved?

If your preference between the lottery and certainty for salary (f_1) does not depend on the level at which we fixed the commute and growth, then we say that the salary objective (f_1) is utilitarily independent of the growth objective (f_3). To be able to use the simpler forms of the utility function U (like the additive $U \approx \sum w_i u_i(f_i)$), we generally need this utility independence to be mutual among the objectives.

Remember we mentioned that these functions would be among the most complete considering all a priori methods, right? This is good, as we could transform a complex MOO problem into a single-objective problem that could be solved with the techniques we saw in previous chapters and, still, with the benefit of capturing all preferences and trade-offs. We could relatively easily achieve a solution that is preferred by decision-makers.

However, it's very difficult to use in practice due to the complexity of defining U—perhaps due to the inherent subjectivity of humans. It might be easy for you to answer "*How many more minutes of commuting would you trade for a \$100 increase in your monthly salary?*" or "*How much would you like to cut from your salary to have to commute 20 min less to work?*," but it's not a simple question for decision-makers to answer and consistently translate into a mathematical equation. In fact, constructing a good function U can take so much time that, in practice, we opt for other techniques (Fig. 5.7).

To exemplify this little stress, let's return to the GPS example. We could think of utility functions like:

- $U = f_1 - 0.5 \times (f_2)^2$. Here, dissatisfaction with a route increases linearly with time, and non-linearly with cost. We also have a scaling coefficient to adjust aversion to cost (0.5).
- $U = f_1 - 3f_2$. Here, time and cost are linear, but we assume that increasing the cost by \$1 is 3 times worse than increasing travel time by 1 min.
- $U = (f_1)^2 - 5f_2$. Here, dissatisfaction with time grows quadratically and dissatisfaction with cost grows linearly, but with a weight of 5.
- $U = f_1 - f_2 - 0.1 \times f_1 \times f_2$. In this case, we have linear dissatisfaction with time and cost and an additional penalty that is greater when both are high. That is: a long and expensive trip is particularly bad.

Now, let's consider the first utility function U: how would the three solutions $\alpha = [30, 15]$, $\beta = [50, 5]$, and $\delta = [40, 8]$ be evaluated by this function? Let's see:

- $U_\alpha = 30 - 0.5 \times (15^2) = 30 - 112.5 = -82.5$
- $U_\beta = 50 - 0.5 \times (5^2) = 50 - 12.5 = 37.5$
- $U_\delta = 40 - 0.5 \times (8^2) = 40 - 32 = 8$

Thus, the solution that maximizes U would be β. However, the challenge of consistently listing decision-makers' preferences remains.

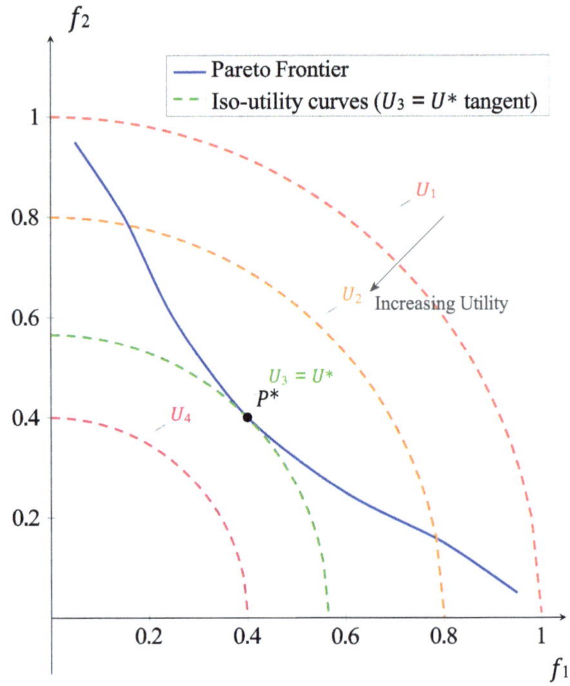

Fig. 5.7 Optimization with a Utility Function: Finds the point on the Pareto Frontier that maximizes the decision-maker's global satisfaction function U, corresponding to the tangency point with the highest iso-utility curve. *Source* The author, 2025

5.4 A Posteriori Methods

If in a priori methods we needed the decision-makers' opinions before optimization, here we leave this work for the end. After all, there are situations where we want to visualize some different options before making a final decision, and having access to only one solution can hide some other good opportunities along the way. For example, what if you're buying a cell phone and only realize after completing your purchase that there was another cell phone 30% cheaper on sale, but with a slightly inferior battery to the option you bought? Maybe you wouldn't even have bought the cell phone you chose if you had known this in advance, right?

Therefore, the idea of a posteriori methods is to generate an approximation of the Pareto front for the decision-makers. This way, they can choose more calmly which solution(s) best suit their needs. This, in fact, resembles those situations we saw in Sect. 5.1.1 where a map application shows two or more optimal route options so that we, the decision-makers, can choose what makes the most sense for us by comparing the pros and cons of each option. Since the goal is to generate a diverse set of solutions, many of these methods handle the non-convex parts of the frontier well, overcoming the limitation we saw in the Weighted Sum method, for example.

However, the fact that we have more than one solution to choose from can become a problem when we have tens, hundreds, or even thousands of solutions on the approximated Pareto front. Furthermore, generating this frontier may require more computational resources than in a single-objective problem.

5.4.1 Particularities

The a posteriori approach is divided into a few steps, as shown in Fig. 5.8. The first is the Multi-Objective Problem (MOP) definition. In contrast to the objective functions in SOO or the output from a priori approaches, it returns an objective vector instead of a single value.[6] When modeling the problem, we need to clearly define the objective functions $f_i(\mathbf{x})$ that we want to optimize for each objective i, as well as the constraints that $\mathbf{x}$ must satisfy.

In a posteriori techniques, objective vectors are optimal if none of their values can be improved without causing harm to the others. In the GPS example, the solution

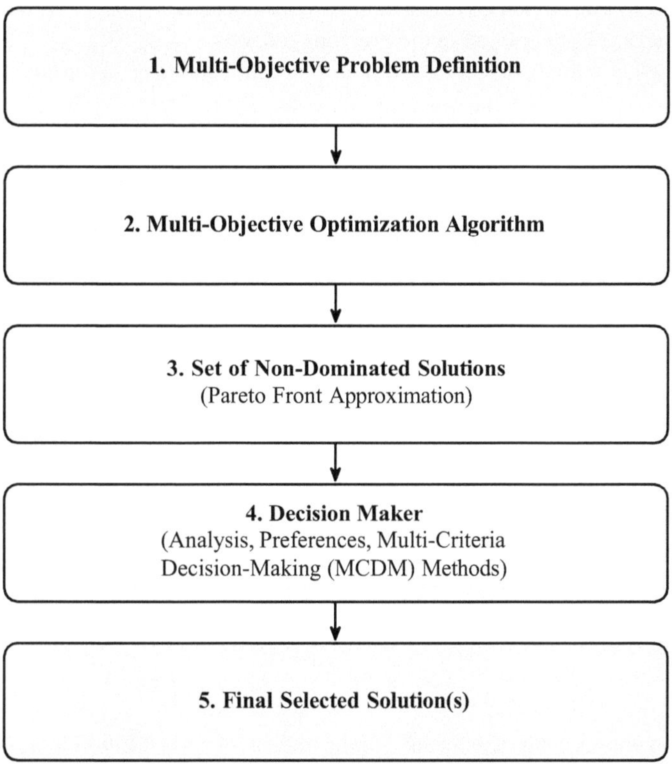

Fig. 5.8 Typical pipeline of a posteriori method. *Source* The author, 2025

[6] This is why we refer to the outputs from Multi-Objective Optimization problems as $\mathbf{f}(\mathbf{x})$ in Sect. 2.2.5.

$\mathbf{f}_\delta = [40, 8]$ would be considered optimal if we couldn't decrease the time from 40 min without compromising the cost of $8.

The next step after defining the MOP is to optimize it via MOO algorithms. When there are multiple conflicting solutions in play, no unique solution that is the best considering all objectives will often exist. With that in mind, an MOO algorithm will output several feasible solutions. Depending on the technique adopted, some of these feasible solutions are repeated copies, or some of them are dominated by other solutions.

Therefore, after obtaining this initial set of solutions from the MOO algorithm, we often apply a domination filter. This process is based on the concept of Pareto dominance, which we have seen in Sect. 2.3.3: a solution a is said to dominate another solution b if a is better than or equal to b in all objectives, and strictly better in at least one objective.

The domination filter systematically compares every solution in the generated set against all others. Any solution found to be dominated by another is removed until only the non-dominated solutions remain. As a result, there will be multiple equally good solutions that could be presented to the decision-maker—hence the Pareto front approximation we mentioned in the previous sections.

Here, what matters to us is both the set of candidate solutions $\mathcal{P}_{approx} = \{\mathbf{x}^{(1)}, \mathbf{x}^{(2)}, \ldots, \mathbf{x}^{(N)}\}$ and the corresponding set of the Pareto front approximation $\mathcal{PF}_{approx} = \{\mathbf{f}(\mathbf{x}^{(1)}), \mathbf{f}(\mathbf{x}^{(2)}), \ldots, \mathbf{f}(\mathbf{x}^{(N)})\}$. Ideally, we want $\mathcal{PF}_{approx}$ to be as close as possible to the true Pareto front $\mathcal{PF}_{true}$. Also, we want $\mathcal{PF}_{approx}$ to fulfill the goals of convergence, diversity, and relevance as seen in Sect. 5.1.2.

These solutions can then be filtered, post-processed, and presented to the decision-maker through the multicriteria decision-making step. This step enables a holistic visualization of all feasible solutions that help with data management, evaluation, and preliminary exploration. Furthermore, this step can help provide humans with valuable feedback on whether the solutions are in the right direction and if the solutions have good quality. To this end, decision-makers can also tune the MOP and the MOO algorithm and re-run them according to their needs and choose what is (or are) ultimately the most preferred solution(s) from $\mathcal{P}_{approx}$ for their implementation under their own work context.

That said, we will move on to Multi-Objective Optimization techniques, considering that we have already covered the definition of an MOP in Sect. 2.2.5. The techniques we will see are evolutionary algorithms. This is the most popular family of algorithms for solving Multi-Objective Optimization problems via a posteriori methods.

Tips

In case you missed it, we introduced the main concepts of Evolutionary Algorithms with Single-Objective Optimization problems in Sects. 3.1.3.3 and 3.2.7. Those concepts will also be the basis for the explanations of the methods we will see in this section.

As we've seen before, the foundation of these algorithms is biological: we use the evolution of species through natural selection as inspiration to produce good individuals that adapt to a given ecosystem. In our case, we generate good solutions that adapt to our optimization objectives and constraints.

Now, if a single-objective EA seeks the individual with the best scalar fitness (i.e., a single numerical value $f(\mathbf{x})$), how do we adapt this to find a set of solutions on an approximated Pareto front with several conflicting objectives? Well, let's go back to the concept of biology: after a few generations of a species in an ecosystem, we will find a population that is best adapted to live there, correct? And it would also be correct to admit that this same population has its own differences among its members: we have individuals who are taller than others, or stronger, or with fewer health problems. That is, this same population would not be composed of identical clones, but rather of individuals who, although optimized for that environment, still present differences among themselves.

Similarly, an EA for Multi-Objective Optimization will have a population that will converge to the Pareto front and spread across it, maintaining diversity among the conflicting objectives.

Therefore, let's return to the generic evolutionary algorithm we saw in Sect. 4.5. For readability, we've placed the pseudocode again below in Algorithm 1.

What should we modify here to transform this EA into a Multi-Objective EA (or, as they are popularly known, MOEAs)? We've put the modifications of this algorithm in Algorithm 2. Firstly, we no longer have a single fitness function f as input, but rather a vector of objective functions $\mathbf{f}(\mathbf{x})$. We also no longer have a single output solution $\mathbf{x}_{best}$, but rather the non-dominated solutions $\mathcal{P}_{approx}$ and their corresponding values of the Pareto front approximation $\mathcal{PF}_{approx}$ considering the objectives established by $\mathbf{f}$. We will use $\mathcal{PF}_{approx}$ for the Multi-Criteria Decision-Making (MCDM) step, which we will discuss in Sect. 5.5.

Furthermore, we are evaluating the fitness of more than one objective—hence, the result of this function is now a vector $\mathbf{f}$ for each $\mathbf{x}$. In fact, this change from evaluating a single value to multiple values (one for each objective) is repeated throughout the code. The routine for selecting "parent" solutions also becomes more complex.

If for single-objective problems we would select parents based on the best value of a single fitness function, in MOO the selection needs to balance two inherently conflicting objectives for the population itself. We want the parents to be non-dominated solutions so they can be as close as possible to the true Pareto front. But, on the other hand, we don't want the parents to be very similar solutions, lest they produce offspring that are also similar. Otherwise, we would tend to converge to a single point: something we really don't want to happen in MOO problems.

Similarly, we have to change the way we select the individuals that will advance to the next generation. In this case, we stop evaluating for a single objective and start evaluating for several conflicting objectives. Different MOEA implementations will have different strategies to address this point.

Algorithm 1 Generic Evolutionary Algorithm (EA)

```
Require: Population size $N$ (i.e., $\mu$)
Require: Number of offspring to generate $N_{offspring}$ (i.e., $\lambda$)
Require: Fitness function $f(\mathbf{x})$ (conventionally, higher is better; adapt if minimizing)
Require: Parent selection method SELECTPARENTS($P$, num_needed)
Require: Recombination method RECOMBINE($\mathbf{p}_1, \mathbf{p}_2$) and probability $p_c$
Require: Mutation method MUTATE(ind) and probability $p_m$
Require: Survivor selection method SELECTSURVIVORS($P_{current}, O_{offspring}, N_{target}$)
Require: Termination condition (e.g., max generations $G_{max}$)
Ensure: Best individual found $\mathbf{x}_{best}$
Initialization:
$P(0) \leftarrow$ INITIALIZEPOPULATION($N$)
EVALUATEFITNESS($P(0), f$)
$t \leftarrow 0$
$\mathbf{x}_{best} \leftarrow$ FINDBESTINDIVIDUAL($P(0)$)                ▷ Overall best individual found so far
$f_{best\_so\_far} \leftarrow f(\mathbf{x}_{best})$
Main Evolutionary Loop:
while Termination condition not met (e.g., $t < G_{max}$) do
                                                        ▷ 1. Variation: Generate $N_{offspring}$
    $Offspring(t) \leftarrow \emptyset$
    for $k \leftarrow 1$ to $\lceil N_{offspring}/2 \rceil$ do       ▷ Assuming recombination produces two offspring
        (parent$_1$, parent$_2$) $\leftarrow$ SELECTPARENTS($P(t)$, 2)
        if RandomReal(0, 1) $< p_c$ then
            (child$_1$, child$_2$) $\leftarrow$ RECOMBINE(parent$_1$, parent$_2$)
        else
            child$_1 \leftarrow$ parent$_1$; child$_2 \leftarrow$ parent$_2$      ▷ Clones if no crossover
        end if
        child$_1 \leftarrow$ MUTATE(child$_1$, $p_m$)                    ▷ Mutation applied based on $p_m$
        child$_2 \leftarrow$ MUTATE(child$_2$, $p_m$)
        Add child$_1$ to $Offspring(t)$
        if |Offspring($t$)| $< N_{offspring}$ then                  ▷ Ensure $N_{offspring}$ are added
            Add child$_2$ to $Offspring(t)$
        end if
    end for
    if $|Offspring(t)| > N_{offspring}$ then
        Trim $Offspring(t)$ to size $N_{offspring}$
    end if
                                                        ▷ 2. Evaluate New Offspring
    EVALUATEFITNESS($Offspring(t), f$)
                                    ▷ 3. Survivor Selection (Form Next Gen. Parent Pop.)
    $P(t+1) \leftarrow$ SELECTSURVIVORS($P(t), Offspring(t), N$)
                    ▷ e.g., $(\mu + \lambda)$ selects $N$ best from $P(t) \cup Offspring(t)$
        ▷ e.g., $(\mu, \lambda)$ selects $N$ best from $Offspring(t)$ (requires $N_{offspring} \geq N$)
                                                        ▷ 4. Update Overall Best Solution
    $\mathbf{x}_{best\_in\_gen} \leftarrow$ FINDBESTINDIVIDUAL($P(t+1)$)
    if $f(\mathbf{x}_{best\_in\_gen}) > f_{best\_so\_far}$ then             ▷ Assumes $f$ is maximized
        $\mathbf{x}_{best} \leftarrow \mathbf{x}_{best\_in\_gen}$
        $f_{best\_so\_far} \leftarrow f(\mathbf{x}_{best})$
    end if
    $t \leftarrow t + 1$
end while
return $\mathbf{x}_{best}$
```

In the end, we return the best non-dominated solutions extracted directly from the final population.

Algorithm 2 Generic Multi-Objective Evolutionary Algorithm (MOEA) Framework

```
Require: Population size N, Offspring count N_offspring
Require: Objective functions f(x) = [f_1(x), ..., f_k(x)]
Require: Methods: SelectParentsMOO, Recombine, Mutate
Require: Methods: SelectSurvivorsMOO, UpdateArchive (opt. with N_archive)
Require: Termination condition (e.g., t < G_max)
Ensure: Approximation of Pareto Set P_approx and Pareto Front PF_approx
t ← 0; P(0) ← InitializePopulation(N); EvaluateObjectives(P(0), f)
A(0) ← ∅
if using archive then
    A(0) ← UpdateArchive(A(0), P(0), N_archive)
end if
while Termination condition not met (e.g., t < G_max) do
    MatingPool ← SelectParentsMOO(P(t), A(t), N_offspring)
    Offspring(t) ← ∅
    for i ← 1 to ⌈N_offspring/2⌉ do
        Select parent_1, parent_2 from MatingPool
        if RandomReal(0, 1) < p_c then
            (child_1, child_2) ← Recombine(parent_1, parent_2)
        else
            (child_1, child_2) ← (parent_1, parent_2)                ▷ Clones
        end if
        child_1 ← Mutate(child_1, p_m); child_2 ← Mutate(child_2, p_m)
        Add child_1 to Offspring(t);
        if |Offspring(t)| < N_offspring then
            Add child_2 to Offspring(t)
        end if
    end for
    if |Offspring(t)| > N_offspring then
        Trim Offspring(t) to size N_offspring
    end if
    EvaluateObjectives(Offspring(t), f)
    P(t + 1) ← SelectSurvivorsMOO(P(t), Offspring(t), N, A(t))
    if using archive then
        A(t + 1) ← UpdateArchive(A(t), P(t) ∪ Offspring(t), N_archive)
        A(t) ← A(t + 1)
    end if
    P(t) ← P(t + 1); t ← t + 1
end while
if using archive and A(t) ≠ ∅ then
    P_approx ← ExtractNonDominated(A(t))
else
    P_approx ← ExtractNonDominated(P(t))
end if
PF_approx ← {f(x) | x ∈ P_approx}
return P_approx, PF_approx
```

That said, let's see how the most well-known MOEAs work, starting, perhaps, with the most popular of all: NSGA-II.

5.4.2 NSGA-II (Non-dominated Sorting Genetic Algorithm II)

Anyone who has studied MOEAs in practically any university on the planet has seen, used, or heard of NSGA-II. It's popular both for generating diverse solutions along the Pareto front and for being scalable. One of the biggest fears some professors have is teaching something that is only useful in the classroom. That is: it would be useless for you to learn a technique in this book that you couldn't apply in your job because performance would worsen too much with a real problem, right? Well, when we say a problem is *scalable*, we mean it can be used for both small problems and for problems with hundreds or thousands of variables and more complex constraints.

5.4.2.1 Elitism

A third reason behind NSGA-II's popularity is elitism. Look: EAs can have selection, crossover, and mutation operators as we've seen before, right? The idea of these operators is to help us explore different combinations of the feasible search space to find different solutions. Let's think about our lives: sometimes, we need to experiment with something to see if it's good or bad. This applies to foods, places, and experiences, for example. Therefore, these operators help us in this experimentation process.

However, there's also a risk in implementing these algorithms. In the pseudocodes we've seen so far, it wasn't very common to save everything that happened in previous generations in a history or something similar. This means it was possible to find an excellent solution in a generation t and, due to bad luck, this solution might not be chosen as a parent or might be ruined by some wrong crossover or mutation. Remember this is possible, since these operators have a random component. Therefore, this solution would disappear for generation $t + 1$, which would be a waste.

Now, how would we avoid this? Well, the idea would be to preserve the best solutions (in our case, non-dominated solutions) and pass them on to the next generation. This is called **elitism**. However, if we preserve the best solutions rigidly, we'll reduce the population's diversity. We would quickly stagnate the search, or have some trouble controlling the population size depending on how many non-dominated solutions we found.

At the end of each generation t, we don't immediately discard the parent population $P(t)$. In the case of this algorithm, we create a temporary combined population $R(t)$ that unites the parents $P(t)$ and the offspring $Offspring(t)$ generated by the selection, recombination, and mutation steps. If the size of population $P(t)$ was N individuals and if we created N offspring, then the size of $R(t)$ is $2N$. Do you remember that we need to filter N individuals for the next generation? Well, we assign a rank to all solutions in a population *before* executing the selection, recombination, and mutation operators. In this ranking, non-dominated solutions have the value F_1, solutions that are dominated only by rank F_1 receive the value F_2, and so on. Figure 5.9 illustrates this in practice.

From this ranking, we select all solutions with rank F_1 for the next generation. If we have enough space (i.e., if rank F_1 has fewer than N solutions), we select all solutions from the next rank, and so on. Eventually, we'll reach a point where there isn't enough space to include all solutions from a certain rank F_{last}. In this case, we also use crowding distance to maximize diversity: individuals from F_{last} with the largest crowding distance (i.e., those that are not extremely similar to other individuals) are selected to fill the remaining spots.

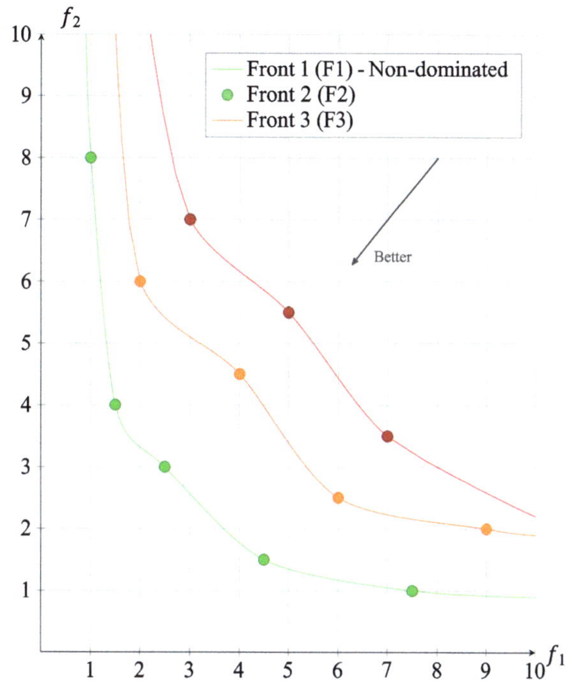

Fig. 5.9 Non-Dominated Sort: Partitioning the population into hierarchical fronts (F1, F2, F3...). Being in a lower numbered front is always better. *Source* The author, 2025

Figure 5.10 illustrates this in practice. Look: on a Pareto front, all solutions on that front are equally "good" from the perspective of Pareto dominance, since none dominates another. However, some solutions may be very close to each other in the objective space, while others may be more isolated. If we need to choose between solutions from the same front, doing so randomly could be risky: we might end up with many solutions clustered in a small portion of the frontier and few or none in other parts.

Here, we use crowding distance to estimate how close (or far) a solution is in relation to its neighbors on that specific front. With this calculation done, we will give preference to solutions that are in regions farther from other solutions, as they contribute more to diversity and a better representation of the entire frontier.

After choosing the N best solutions, we would ignore the remaining ranks. After all, we've already selected all the best solutions (both parents and offspring) for the next generation $P(t+1)$.

Fun fact

If we're talking about NSGA-II and NSGA-III, then we could assume there's an NSGA-I (or just NSGA), right? The answer is yes, this algorithm also exists and, by the way, was introduced by the same researcher: Kalyanmoy Deb. NSGA-I was one of the first MOEAs, but it had its criticisms. For example,

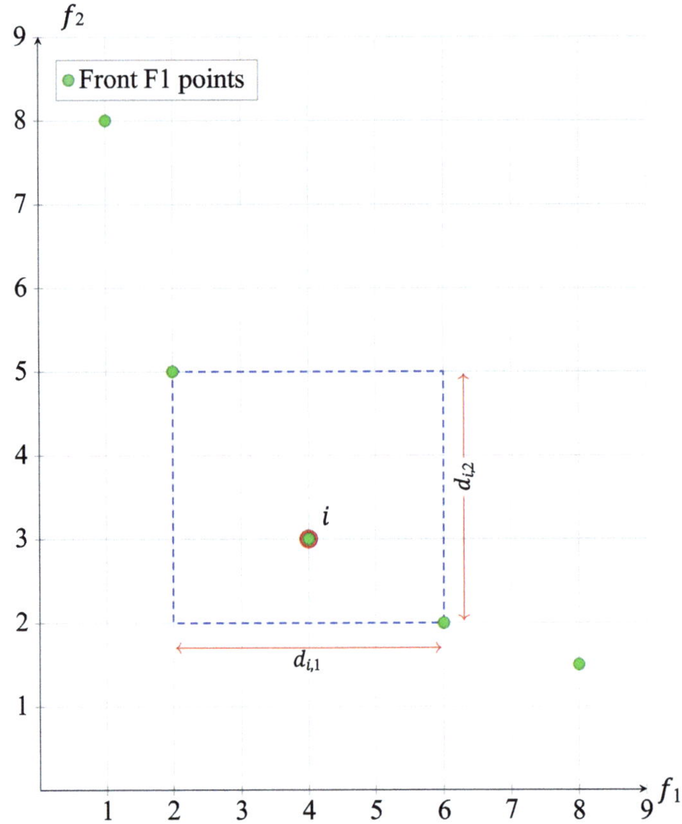

Fig. 5.10 Calculating Crowding Distance (CD) in NSGA-II: Measures how "spaced out" an individual is from its neighbors on the same front. Higher CD = Less crowded = Better for diversity. Here, CD is defined as $CD(i) = d_{i,1} + d_{i,2}$, after normalization. *Source* The author, 2025

it had high computational complexity. That is, the number of iterations grew exponentially ($O(kN^3)$) as the complexity of the problem increased, making it unfeasible for the kind of complex problems we need to solve in companies today. Furthermore, it didn't have the concept of elitism that we have in NSGA-II. Finally, to ensure population diversity, we had to configure another parameter in the algorithm, which becomes an additional responsibility to manage and a potential risk of not generating good solutions. In fact, NSGA-II was created to solve these three problems.

That said, let's look at the pseudocode in Algorithms 3, 5, and 6. Here, we use Algorithm 2 as a basis. Notice that we no longer define the number of offspring $N_{\text{offspring}}$, since the number of offspring is usually the same as the population size N. Also, notice that we have a new operation for population $P(0)$: `FastNonDominatedSort`.

5.4.2.2 Explaining the "Fast"

In this case, we call it "fast" because the challenge of dividing solutions into non-domination ranks like $F_1, F_2, \ldots$ was implemented in a more scalable way compared to a cruder approach.

You see, the simplest way to find F_1 would be to compare each solution p with all other solutions in the population. If no one dominates p, then p belongs to F_1, correct? Now, imagine we need to compare two individuals, p and q, to see who dominates whom. Here we would have:

- If we had two objectives: we would need to compare p_1 with q_1, and p_2 with q_2. So, we would need two comparisons.
- If we had three objectives: we would need to compare p_1 with q_1, p_2 with q_2, and p_3 with q_3. So, we would need three comparisons.
- If we had k objectives: we would need to compare p_1 with q_1, p_2 with q_2, p_3 with $q_3, \ldots, p_k$ with q_k. So, we would need k comparisons.

And, if we have N individuals, we would need to perform approximately $N \times (N - 1)$ comparisons. Therefore, to find F_1 we would need $O(kN^2)$ dominance comparisons.[7] And, in the worst case where each front has only one solution, we would need to repeat this process N times, leading to a total complexity of $O(kN^3)$.

In NSGA-II, however, we have a different strategy: first, for each solution p in $R(t)$, we calculate how many solutions dominate p, and store this count in a domination counter called n_p. We also calculate how many solutions p dominates, and store this in a list S_p. This process has the same $O(kN^2)$ complexity we saw earlier. However, the process of distributing solutions among ranks from n_p and S_p is much faster. Thus, we end up with a complexity of $O(kN^2)$. It might seem simple, but for a two-objective problem, it's the difference between 10,000 and 1,000,000 comparisons in a case where $N = 100$. So, it's quite a lot.

---

[7] The Big-O Notation, like $O(kN^2)$ or $O(N^3)$, is a way to describe how quickly the execution time (or memory usage) of an algorithm grows as the size of the input (like the size of the population being sorted) increases. Here, we want to understand how the algorithm behaves for very large inputs, ignoring constant factors and slower-growing terms. Common values include (ranked from best to worst):

- $O(1)$: Constant time. The time to execute the algorithm doesn't change regardless of how much data we're processing.
- $O(\log N)$: Logarithmic time. The time grows very slowly with size N.
- $O(N)$: Linear time. The time grows proportionally to N.
- $O(N^2)$: Quadratic time. The time grows much faster than N.
- $O(N^3)$: Cubic time. Among all these examples, this would be the worst case. For large populations (N), N^3 becomes computationally very expensive and slow.

Algorithm 3 NSGA-II (Non-dominated Sorting Genetic Algorithm II)

```
Require: Population size N
Require: k objective functions f(x) = [f_1(x), ..., f_k(x)] (to be minimized)
Require: Number of generations G_max
Require: (Implicit: Decision variable ranges, Crossover, and Mutation operators/probabilities)
Ensure: P_approx (Approximation of the Pareto Set), PF_approx (Approximation of the Pareto Front)
  Main Algorithm Steps:
t ← 0
P(t) ← INITIALIZEPOPULATION(N)                                   ▷ See Algorithm 4
EVALUATEOBJECTIVES(P(t), f)                                      ▷ See Algorithm 4
AllFronts ← FASTNONDOMINATEDSORT(P(t))                           ▷ See Algorithm 4
for all front F_i in AllFronts where F_i is not empty do
    CALCULATECROWDINGDISTANCE(F_i)                               ▷ See Algorithm 5
end for
while t < G_max do
    Offspring(t) ← ∅
    for idx = 1 to N do
        parent_1 ← TOURNAMENTSELECTION(P(t))                     ▷ See Algorithm 5
        parent_2 ← TOURNAMENTSELECTION(P(t))                     ▷ See Algorithm 5
        child ← APPLYVARIATION(parent_1, parent_2)               ▷ See Algorithm 4
        Add child to Offspring(t)
    end for
    EVALUATEOBJECTIVES(Offspring(t), f)                          ▷ See Algorithm 4
    R(t) ← P(t) ∪ Offspring(t)
    AllFronts_R ← FASTNONDOMINATEDSORT(R(t))                     ▷ See Algorithm 4
    P(t + 1) ← ∅
    i ← 1
    while AllFronts_R[i] is not empty and |P(t + 1)| + |AllFronts_R[i]| ≤ N do
        CALCULATECROWDINGDISTANCE(AllFronts_R[i])                ▷ See Algorithm 5
        P(t + 1) ← P(t + 1) ∪ AllFronts_R[i]
        i ← i + 1
    end while
    if AllFronts_R[i] is not empty and |P(t + 1)| < N then
        CALCULATECROWDINGDISTANCE(AllFronts_R[i])                ▷ See Algorithm 5
        SORTBYCROWDINGDISTANCE(AllFronts_R[i])                   ▷ Sort descending
        Needed ← N − |P(t + 1)|
        P(t + 1) ← P(t + 1) ∪ AllFronts_R[i][1 ... Needed]
    end if
    P(t) ← P(t + 1)
    t ← t + 1
end while
FinalFronts ← FASTNONDOMINATEDSORT(P(G_max))                     ▷ See Algorithm 4
P_approx ← FinalFronts[1]
PF_approx ← {f(x) | x ∈ P_approx}
return P_approx, PF_approx
```

5.4.2.3 Crowding Distance and Loop

For each individual within each front, we also calculate the crowding distance. These initial non-domination ranks and crowding distances are important for the first parent selection.

The algorithm then enters a loop that repeats for a maximum number of generations G_{max}. In each generation, we create more solutions and choose the most promising ones. To begin, we select N parents from the current population using the `TournamentSelection` function. In this tournament process, we generally choose individuals with the best non-domination rank and, in case of a tie in rank, we would choose the individual with the largest crowding distance to ensure better diversity.

Algorithm 4 NSGA-II Sub-procedures, Part I

```
procedure InitializePopulation(N)                ▷ Implicit: Uses decision variable ranges
    P ← ∅
    for idx = 1 to N do
        Generate individual x_idx randomly
        Add x_idx to P
    end for
    return P
end procedure
procedure EvaluateObjectives(Population, f)
    for each individual x in Population do
        Evaluate f(x) and store with x
    end for
end procedure
procedure FastNonDominatedSort(Population)
    Fronts ← [∅, ...]                             ▷ List of fronts, F_1, F_2, ...
    for each individual p in Population do
        p.S ← ∅                                   ▷ Set of solutions dominated by p
        p.n ← 0                                   ▷ Domination counter for p
        for each individual q in Population, q ≠ p do
            if p dominates q then
                Add q to p.S
            else if q dominates p then
                p.n ← p.n + 1
            end if
        end for
        if p.n = 0 then
            p.rank ← 1
            Add p to Fronts[1]
        end if
    end for
    i ← 1                                         ▷ Front index
    while Fronts[i] is not empty do
        Q_next_front ← ∅
        for each individual p in Fronts[i] do
            for each individual q in p.S do
                q.n ← q.n − 1
                if q.n = 0 then
                    q.rank ← i + 1
                    Add q to Q_next_front
                end if
            end for
        end for
        i ← i + 1
        if Q_next_front is not empty then
            Fronts[i] ← Q_next_front
        end if
    end while
    return Fronts (up to the last non-empty front with assigned ranks)
end procedure
```

Algorithm 5 NSGA-II Sub-procedures, Part 2

```
procedure CalculateCrowdingDistance(Front)                ▷ Assumes k objectives known
    l ← |Front|
    if l = 0 then return
    end if
    for each individual p in Front do
        p.distance ← 0
    end for
    for m = 1 to k do                                         ▷ For each objective
        Sort Front using objective m values in ascending order.
        Front[0].distance ← ∞
        Front[l − 1].distance ← ∞
        if l > 2 then                  ▷ Only calculate for internal points if more than 2 points
            f_m^min ← value of objective m for Front[0]
            f_m^max ← value of objective m for Front[l − 1]
            if f_m^max − f_m^min > 0 then
                for j = 1 to l − 2 do              ▷ For internal points Front[1] to Front[l − 2]
                    obj_val_next ← value of objective m for Front[j + 1]
                    obj_val_prev ← value of objective m for Front[j − 1]
                    Front[j].distance ← Front[j].distance + (obj_val_next − obj_val_prev)/(f_m^max − f_m^min)
                end for
            end if
        end if
    end for
end procedure
procedure TournamentSelection(Population)                  ▷ Uses rank and crowding distance
    Select two individuals p1, p2 randomly from Population
    if p1.rank < p2.rank then return p1
    else if p2.rank < p1.rank then return p2
    else if p1.distance > p2.distance then return p1               ▷ Higher distance is better
    else if p2.distance > p1.distance then return p2
    elsereturn randomly one of p1, p2
    end if
end procedure
procedure ApplyVariation(parent1, parent2)                            ▷ Implicitly uses pc, pm
    child ← ApplyCrossover(parent1, parent2)
    child ← ApplyMutation(child)
    return child
end procedure
```

Next, we apply crossover and mutation operators to the parents to generate a new population of N offspring. For each of the offspring, we calculate the objective vectors $\mathbf{f}(\mathbf{x})$.

Then, the parent population and the offspring population are merged into a single combined population $R(t)$ of size $2N$. We sort this entire population using that same `FastNonDominatedSort`, resulting in fronts $F_1, F_2, \ldots$. As we mentioned earlier, front F_1 contains the best non-dominated solutions found so far, considering both parents and offspring. Elitism, remember?

Then, we select the next population $P(t+1)$ containing N individuals. We add all fronts starting with F_1, then F_2, and so on, until we exceed N individuals. If the last front F_i only partially fits, we use crowding distance again with `CalculateCrowdingDistance` and sort in descending order of distance with `SortByCrowdingDistance` to select only the most diverse individuals from F_i to fill the remaining spots in $P(t+1)$.

After G_{max} generations, we will terminate the loop. The final population will have the N best solutions found, trying to be as diverse as possible for this algorithm. Thus, the approximation of the Pareto set $\mathcal{P}_{approx}$ is normally the first non-dominated front F_1, and the approximation of the Pareto front $\mathcal{PF}_{approx}$ is the set of objective vectors $\mathbf{f}(\mathbf{x})$ corresponding to the solutions in $\mathcal{P}_{approx}$.

5.4.2.4 Variants of NSGA-II

NSGA-II was created at a time when the euro barely existed as a currency, *The Lord of the Rings: The Two Towers* was playing in cinemas, people generally accessed the internet via a computer instead of a cell phone, and people listened to music via discs or the radio. Despite its age, NSGA-II remains a very good algorithm, especially for problems with 2 or 3 objectives.

Therefore, it doesn't necessarily mean that NSGA-II is unbeatable in all aspects. As the number of objectives increases, their effectiveness decreases. When we have many objectives, it becomes much harder for one solution to dominate another, and almost all solutions end up being non-dominated with respect to each other. Think, for example, about defining who are the best basketball players or the best soccer players of all time. Well, it would be relatively easy to find a Pareto front of the best players using criteria like average points or goals scored per game, or average passes per game. Now, when we start adding more criteria like "communication skills," "penalties received," "performance difference to the second-best player on the same team," or even "social projects undertaken," it becomes harder to find any player who is superior or equivalent to all others in all criteria, don't you agree? Just like everything in life, each will have its pros and cons compared to others, which means we will have many non-dominated solutions.

It also becomes complicated to maintain a good spread of solutions along a frontier with 4, 5, 10, or even more dimensions (i.e., objectives). After all, the "volume" grows exponentially in high-dimensional spaces. With a fixed population size N, the space becomes very "empty," and the average distance between points increases. Thus, the idea of crowding distance here loses some of its meaning.

Therefore, over time, variants of NSGA-II were developed. Let's see:

R-NSGA-II

Here, the prefix "R-" means "Reference-Point Based." There are cases where decision-makers have some specific preferences for certain regions of the Pareto front, and it would make sense to focus our search on these preferences.

In this case, R-NSGA-II is a variation of NSGA-II in which we incorporate these reference points, provided previously by decision-makers, to guide the algorithm's selection process, as illustrated in Fig. 5.11.

All the NSGA-II pseudocode we saw earlier is maintained. That is, all that explanation we gave about elitism and the idea of the "fast" non-dominated sort still applies. However, the way we handle diversity and choose solutions now changes. You see, instead of relying solely on crowding distance, R-NSGA-II uses a "preference operator." This operator works as follows:

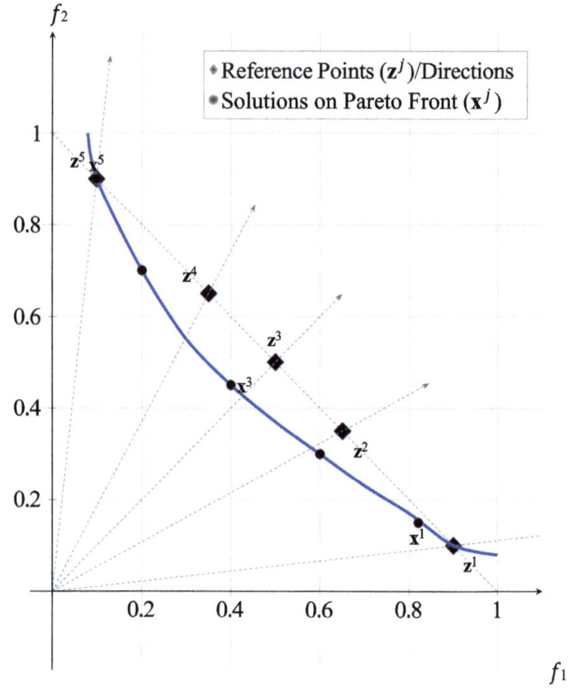

Fig. 5.11 Reference points used by the R-NSGA-II algorithm. *Source* The author, 2025

1. First, for each solution, we calculate the distance[8] between it and the nearest reference point. Here, solutions closer to any of the reference points would have a better rank.
2. Then, solutions that are closer to the reference points have a preference distance value of 1, those that are a bit further away have a value of 2, and so on. In our case, we will use this distance for the tournament process. There, the solution with the smallest preference distance is chosen.
3. We also need to maintain a certain diversity near the reference points. For this, we use the idea of *ϵ-clearing*. Here, the idea is as follows: first, we choose a random solution from a set of non-dominated solutions. All solutions within an ϵ neighborhood (i.e., whose normalized sum of differences in objectives is equal to or less than an ϵ value) would be penalized so that we can reduce clustering among solutions.

The way we've put it, it seems like R-NSGA-II is an evolution of NSGA-II, and it seems like NSGA-II is somewhat obsolete, right? Actually, that's not quite the case. Although R-NSGA-II is a sophisticated adaptation, it doesn't make NSGA-II obsolete. They are different tools for different purposes.

---

[8] Usually Euclidean distance. We can also assign weights to objectives so that one has more weight than another in the optimization process.

If we are starting an optimization project and want to do an initial exploration of possible solutions, having a view of the possible trade-offs, NSGA-II makes sense. In cases like this, we don't know exactly what the distribution of the Pareto front would be like, and it would be nice to have a diversified approximation.

Now, we can also have scenarios where decision-makers already have a clearer idea about which objective values are more interesting to them—either because management already has a target, or because there's a market benchmark, or because there's another solution already implemented, or because we've already done an initial analysis with NSGA-II and know which region of feasible solutions we should focus on. In these cases, we would like to concentrate computational resources to find solutions that are closer to a region of interest.

NSGA-III

Now, both NSGA-II and R-NSGA-II work well when we have few objectives, like two or three. However, things get worse when we deal with many-objective problems. When we have many objectives, the probability of one solution dominating another decreases significantly, and thus, we have a situation where a large part of the population becomes non-dominated with respect to each other. And, with that, we have two more headaches. First, since NSGA-II and R-NSGA-II prioritize non-dominated fronts, selection loses effectiveness because almost all solutions belong to the first front F_1. This makes it difficult to advance toward the optimal Pareto front with each generation. Second, since a large part of the N individuals population will belong to F_1, we will have a large number of solutions for decision-makers to choose from. And, with a large number of solutions, it becomes difficult for us to choose, compare, and prioritize what really matters—especially when these solutions are of low quality.

And, still, we have the problem of crowding distance that we mentioned earlier. After all, the more dimensions we have, the less relevant this metric becomes and, consequently, the less relevant the solutions generated by the algorithm would be.

To solve this, we have NSGA-III. It also has the same basic functioning as NSGA-II. We continue with non-dominated sorting to classify the population into fronts, like $F_1, F_2, \ldots$. We also continue with the elitist approach of combining parents and offspring. However, we replace the crowding distance with an approach based on reference points[9] to maintain diversity, especially when selecting individuals from the last front to be included in the next generation.

Thus, this approach works as follows:

1. First, we generate a set of reference points that are well spread across the hyperplane composed of k dimensions, where each dimension is an objective. The idea of spreading them is to ensure that NSGA-III doesn't get lost in this space and, also, that we have diversity in the solutions. The number of points we generate here normally guides the size of our population.

[9] Which you now know what it is after we went through R-NSGA-II, right?

If decision-makers have a notion about which regions are more interesting, they can define these reference points themselves. If they do so, then NSGA-III is used to concentrate the search in the vicinity of these user-supplied points.

Now, if decision-makers don't know where to go, we can use a method that automatically generates these points. For this, we use approaches like Das-Dennis, which covers the entire objective space uniformly with distributed points.

2. We normalize the objectives of the current population before using the reference points. The idea is to adapt this in each generation, considering the ideal point and extreme points so that NSGA-III works well even if we have objectives with different scales.
3. Then, we associate each solution in the normalized population within a specific generation with each of the reference points. We do this by creating a reference line for each of the reference points, connecting them to the origin of the hyperplane. Then, we calculate the perpendicular distance of each solution to each of these reference lines. We then associate the reference point whose line is closest to each solution.
4. Do you remember that, in NSGA-II, we usually divided solutions into fronts, right? We usually choose all individuals from the first front F_1, then from the second front F_2, and so on. Eventually, we would reach a certain front where we couldn't include all its individuals due to space limitations, as we would have to choose a total of N individuals to be moved to the next generation. Well, here, this logic continues. The difference is that we use a niching method to choose the individuals from this last Pareto front to occupy the next generation. In this case, we prioritize solutions that are associated with reference points that have few or no other members already present in the part of the solution that has already been selected. It might be a simplistic explanation, but consider that we would always prefer to add "representatives" close to a reference point rather than choosing some individual who has other similar solutions (i.e., belonging to the same reference point) already selected. After all, our intention of selecting individuals from a diversity perspective still holds.

Just from the explanation of the differences, you must have realized that it's more complex than NSGA-II, right? Well, that's exactly it. NSGA-III introduces a series of additional steps compared to NSGA-II's crowding distance calculation, mainly in the algorithm phase, where we choose the N individuals for the next generation.

This complexity isn't something that was done just to complicate things for your university exam. In fact, this is why it works better than NSGA-II in problems with many objectives. After all, it's by using reference points that we can maintain diversity in a more controlled and directed way in high-dimensional spaces.

U-NSGA-III

So, NSGA-II works well for few objectives, and NSGA-III was designed to solve problems with many objectives. Now, what if for some reason you wanted to use a single technique without having to worry about whether we're working with a multi-objective problem or a many-objective problem?

Well, the idea of U-NSGA-III goes along these lines. Here, the prefix "U-" means *unified*. It's based on the general structure of NSGA-III, but has some modifications that allow the algorithm to be simplified if we have fewer objectives. In fact, it can even work for single-objective problems.

First, the population size N depended on the number of reference points in the original NSGA-III execution. Here, the number of reference points can be any value less than or equal to N. This ends up being interesting because we can have more individuals than reference points, something especially important in problems with only one or two objectives (after all, sometimes we could have two reference points for a population of 500 or more individuals per generation, don't you think?).

With this, we have a problem: it's quite possible that we'll have more individuals in the population than reference points, and that the niching logic of NSGA-III will need to be modified to address this. Therefore, we introduce here a tournament selection operator before recombination to increase selective pressure: if two parents competing in the tournament belong to different niches (are associated with different reference points), one of them is chosen randomly. Now, if both parents belong to the same niche, we first select by non-domination rank (better rank wins). If the ranks are equal, the parent closest to the reference line associated with the niche is chosen.

R-NSGA-III

We saw R-NSGA-II in the previous pages. It's a technique that uses user-supplied reference points to focus the search on specific regions, but it was originally designed for few objectives. And we also saw NSGA-III, which is great for many objectives. Logically, we might want to combine these two into one, right? This is where R-NSGA-III comes in.

But what about U-NSGA-III?

U-NSGA-III was created years before R-NSGA-III. So, why would we even think of another technique? Let's see: for U-NSGA-III, the main goal is to have a single code for any dimensionality of objectives, usually seeking the optimal solution (when we have only one objective) or a representation of the entire Pareto front (when we have more than one objective). Before U-NSGA-III, you might use a GA for single-objective problems, NSGA-II for multi-objective problems, and NSGA-III for many-objective problems, requiring different implementations for each of these strategies.

However, U-NSGA-III doesn't have the specialized mechanism to generate a focused set of reference points around the decision-makers' desires. U-NSGA-III would use the reference points as a reference set (possibly with few points), while R-NSGA-III would use the user's points to create a larger, clustered set of reference points, allowing NSGA-III's niching to work more effectively to densely explore the vicinity of the specified preferences. Here, the keyword is density.

R-NSGA-III is an adaptation of NSGA-III designed to find Pareto-optimal solutions close to one or more user-supplied reference points. And, similarly to R-NSGA-II, this technique can be used when a previous run of an algorithm (like NSGA-III) showed results across the entire frontier, and the user now wants to explore in more detail (i.e., find more points) or investigate certain regions on the many-objective frontier.

5.4.2.5 Choosing Someone from the NSGA-II Family

Perhaps upon reading about the NSGA-II-based techniques, you felt a bit confused. After all, the general concept and terminology are similar, even though we've discussed the advantages and disadvantages of each. Therefore, allow us to invest some additional time to talk about *when* you should use each of these techniques. Of course, this would also apply to other Multi-Objective Optimization techniques, but here we'd like to clarify when you might use each of these NSGA-II-based techniques.

That said, the questions we could ask would be: "*how many objectives are we optimizing?*" and "*are we interested in any specific region of the Pareto front?*" If we have two or three objectives and want a good approximation of the entire Pareto front, the answer would probably be NSGA-II. Since it uses a crowding distance logic, we achieve diversity along the Pareto front and find it with acceptable quality and speed.

Now, it may be that decision-makers want a more targeted search. They might already have an idea of which regions of the Pareto front are more interesting and, therefore, want to concentrate the search in these areas. If this is the case and we have two or three objectives, then it might be worth choosing R-NSGA-II—after all, this technique introduces reference points.

On the other hand, we could have a scenario where we have four or more objectives. It's for these cases that we would use NSGA-III or R-NSGA-III. Similarly to NSGA-II and R-NSGA-II, we would choose between one and the other based on the definition (or not) of regions of interest on the Pareto front. If we want a good approximation of the Pareto front in general, then we would opt for NSGA-III. Otherwise, for R-NSGA-III. In fact, as we mentioned earlier, we can also adopt R-NSGA-II when, after an initial exploration (with NSGA-III or another technique), we want to investigate certain areas of the many-objective frontier more deeply.

That said, U-NSGA-III can be useful for both few and many objectives, and theoretically would be the best choice in all scenarios. However, in practice, this wouldn't necessarily be true.

The first case would be the learning curve. You see, one of the major problems we might have when starting in the industry is assuming that implementing a more advanced solution is necessarily the best choice. Even though U-NSGA-III was proposed over a decade ago, not everyone with knowledge of Optimization has heard of it. So, sometimes it wouldn't be worthwhile for a company to replace an algorithm like NSGA-II, which is old but well-known, simple, and reliable, with a solution that people don't trust and don't know how it works. For example, for problems with 2 or 3 objectives where there's no clear need for complex reference-point guidance, it might be faster to use NSGA-II in terms of development, testing, and integration.

Or, perhaps, there's a political issue. It might be simpler for you to explain that an algorithm is explicitly "reference-point oriented" than a "unified" algorithm with internal mechanisms that are more complex to justify. Or, maybe, the Optimization library you'd use in the company doesn't even have U-NSGA-III available.

Finally, there's an issue of reliability and maintenance: let's suppose you decided to test U-NSGA-III to replace NSGA-II in a company, and it actually produced a slightly better Pareto front approximation. Now, will people in the future know how to handle all the additional parameters specific to U-NSGA-III? Wouldn't there be a greater risk of someone accidentally tweaking a parameter and losing this performance improvement? So, sometimes a simpler algorithm can also have more predictable behavior.

5.4.3 SPEA2 (Strength Pareto Evolutionary Algorithm 2)

Perhaps SPEA2 and NSGA-II are the most well-known Multi-Objective Optimization algorithms. Therefore, it's likely that a good portion of universities and optimization software work with these two techniques, at least.

SPEA2 is an elitist evolutionary algorithm that uses the concept of an "elite archive." In each generation, we produce new solutions (offspring) from existing solutions (parents) using operators like crossover and mutation, right? And, as you should know by this point in the book, these operators have a random component. Randomness is good because it helps explore new regions of the search space, but it also introduces the risk of ignoring or removing good solutions from the previous generation.

To avoid losing these individuals, we have introduced this elite archive. It's a reserve population where we store a set of the "best" solutions (i.e., non-dominated solutions) found so far throughout all generations.

Furthermore, SPEA2 has a slightly different philosophy for measuring the "value" (fitness) of each solution. If in NSGA-II we used the concept of dividing solutions into fronts or ranks $F_1, F_2, \ldots$, here we use continuous values to evaluate the strength and raw fitness of each individual. Well, before we move on to the pseudocode, it's important to explain these concepts to you. Therefore, let's assume that the candidate solutions for your optimization problem are competitors in a tournament with multiple objectives.

In SPEA2, the "strength" of a solution i, denoted as $S(i)$, directly measures its domination power. It's simply the count of how many other solutions (both in the main population $P(t)$ and in the elite archive $\bar{P}(t)$) are dominated by solution i. In that tournament analogy we mentioned, $S(i)$ stores the number of direct "victories" of solution i over the others.

On the other hand, the raw fitness $R(i)$ complements the strength $S(i)$. If $S(i)$ measured how many solutions i dominated, here we do the opposite: we measure how many solutions *dominate* solution i, using $S(i)$ as a basis. So, let's consider a few hypotheses:

- An individual i is not dominated by anyone. Then, it will have $R(i) = 0$. This is the best possible raw fitness value, indicating it's a current "champion."
- An individual i is dominated by individuals j_1 and j_2. These two individuals have strengths $S(j_1)$ and $S(j_2)$, respectively. Therefore, individual i will have a raw fitness $R(i) = S(j_1) + S(j_2)$.

  Being dominated by a "weak" individual (one that dominates few other solutions and thus has a low $S(j)$) is less bad than being dominated by a "strong" individual (one that dominates many other solutions and thus has a high $S(j)$).

  Therefore, a high $R(i)$ means that solution i is being dominated by much superior solutions.

Important
Therefore, in SPEA2, we want to minimize the raw fitness $R(i)$. Notice that, by being calculated from $S(i)$, it allows us to differentiate solutions between those that are non-dominated ($R(i) = 0$), those that are "almost" non-dominated ($R(i)$ is low, indicating they are dominated by few solutions or by "weak" solutions), and those that are clearly dominated ($R(i)$ is high, indicating they are dominated by many "strong" solutions).

Having this type of control that goes beyond simply differentiating between dominated and non-dominated solutions helps us to have a more precise selection of solutions.

That said, let's consider the SPEA2 algorithm as illustrated in Algorithm 6. Here, notice that we have the calculation of strength $S(i)$, raw fitness $R(i)$, and density $D(i)$ for each individual in the combined population R_t.

To promote diversity, we calculate a density estimate $D(i)$ for each $i \in R_t$. First, we calculate the Euclidean distance in the objective space between i and all other individuals $j \in R_t$. Then, we sort these distances, where σ_i^k is the distance to the kth nearest neighbor and k is a parameter, typically $\sqrt{|R_t|}$. Finally, we determine the density $D(i) = 1/(\sigma_i^k + 2)$. A high density (very close neighbors, small σ_i^k) results in a larger $D(i)$, which is bad—after all, we want diverse solutions, right?

Therefore, the final fitness $F(i)$ is the sum of $R(i) + D(i)$, where smaller values are better. This is the main idea behind SPEA2's fitness strategy, and we use $F(i)$, including for selecting parents to produce new solutions for the current generation.

Also notice that we construct the archive for the next generation $\bar{P}_{t+1}$. First, all non-dominated individuals (those with $F(i) < 1$) from R_t are copied. If the archive is too small, the best dominated individuals—that is, those with the smallest $F(i)$—are added until the size $\bar{N}$ is reached. Or, if the archive is too large, we reduce the archive by removing individuals from the densest regions to favor diversity.

Algorithm 6 Strength Pareto Evolutionary Algorithm 2 (SPEA2)

Require: Main population size N
Require: Archive size $\bar{N}$
Require: Maximum number of generations G_{max}
Require: Vector of k objective functions $\mathbf{f}(\mathbf{x})$
Require: Recombination operator (and prob. p_c), Mutation operator (and prob. p_m)
Ensure: Set of non-dominated solutions A (approximation of Pareto Front)
Initialization:
$t \leftarrow 0$
$P_t \leftarrow$ INITIALIZEPOPULATION(N)
$\bar{P}_t \leftarrow \emptyset$ ▷ Initial empty archive
EVALUATEOBJECTIVES($P_t, \mathbf{f}$)
Main Loop:
while $t < G_{max}$ **do**
  $R_t \leftarrow P_t \cup \bar{P}_t$ ▷ Combined population
  ▷ Ensure all individuals in R_t have up-to-date objective values (if needed)
  // Step 1: Fitness Assignment for individuals in R_t
  for all individual $i \in R_t$ **do**
    $S(i) \leftarrow |\{j \mid j \in R_t \wedge i \succ j\}|$ ▷ Strength: number of solutions i dominates
  end for
  for all individual $i \in R_t$ **do**
    $R(i) \leftarrow \sum_{j \in R_t, j \succ i} S(j)$ ▷ Raw fitness: sum of strengths of i's dominators
    Calculate σ_i^{knn} (distance to knn^{th} nearest neighbor in R_t, e.g., $knn \approx \sqrt{|R_t|}$)
    $D(i) \leftarrow 1/(\sigma_i^{knn} + 2)$ ▷ Density (adding 2 to avoid division by zero)
    $F(i) \leftarrow R(i) + D(i)$ ▷ Final Fitness (to be minimized)
  end for
  // Step 2: Environmental Selection (Build Archive $\bar{P}_{t+1}$)
  $\bar{P}_{t+1} \leftarrow \{i \in R_t \mid F(i) < 1\}$ ▷ Copy all non-dominated individuals
  if $|\bar{P}_{t+1}| < \bar{N}$ **then** ▷ Archive too small
    Sort dominated individuals in R_t (where $F(i) \geq 1$) by $F(i)$ in ascending order.
    Add the first $(\bar{N} - |\bar{P}_{t+1}|)$ individuals from sorted list to $\bar{P}_{t+1}$.
  else if $|\bar{P}_{t+1}| > \bar{N}$ **then** ▷ Archive too large, truncate
    TRUNCATEARCHIVE($\bar{P}_{t+1}, \bar{N}$)
    ▷ (Iteratively remove individuals with smallest knn^{th} neighbor distance σ_i^{knn})
  end if
  // Step 3: Termination Check
  if $t \geq G_{max} - 1$ **then** ▷ Check if this is the last iteration to set final A
    $A \leftarrow \bar{P}_{t+1}$
    break
  end if
  // Step 4: Mating Pool Selection
  $MatingPool \leftarrow \emptyset$.
  for $j = 1$ **to** N **do** ▷ Create mating pool of size N
    Select individual from $\bar{P}_{t+1}$ using binary tournament based on $F(i)$ values.
    Add selected individual to $MatingPool$.
  end for
  // Step 5: Offspring Generation
  $P_{t+1} \leftarrow$ APPLYVARIATION($MatingPool, N, p_c, p_m$) ▷ Generate N offspring
  EVALUATEOBJECTIVES($P_{t+1}, \mathbf{f}$)
  // Prepare for Next Iteration
  $P_t \leftarrow P_{t+1}$
  $\bar{P}_t \leftarrow \bar{P}_{t+1}$
  $t \leftarrow t + 1$.
end while
if $t < G_{max}$ **then** ▷ Loop may have been broken by other criteria if added
  $A \leftarrow \bar{P}_t$ ▷ Assign current archive if loop broke not by G_max from within
end if
return A.

These specific characteristics of SPEA2 make it useful for certain types of problems in the industry. Consider, for example, situations where we'd like to access solutions that are part of the extremes (or "edges") of a Pareto front. It's common to have situations like this when we need, for example, to know the financial strategies with the lowest possible risk or the highest possible return. Or, perhaps, the engineering component that is as light as possible or as stiff as possible. Here, the fact that we have an archive of solutions allows us to access these solutions without the risk of them being eventually removed during the selection, mutation, and crossover stages.

Or, still, let's consider scenarios where we need to evaluate various solutions, considering objectives like cost, time, and quality. Here, a solution that is "dominated" only by exceptionally strong projects (that outperform many others) might be viewed differently by decision-makers than a solution dominated by mediocre projects. And, as you can imagine, the raw fitness logic enables this type of analysis, comparison, and choice.

5.4.4 DEMO (Differential Evolution for Multi-objective Optimization)

Differential Evolution (DE), as we saw earlier in Sect. 4.5.3, is an evolutionary algorithm for single-objective optimization. Therefore, the DEMO algorithm is an adaptation that combines the classic DE operators with selection and diversity maintenance mechanisms (like those used by NSGA-II) for solving Multi-Objective Optimization problems.

The central idea of DEMO is to maintain DE's efficiency in generating new candidate solutions, while explicitly managing convergence toward the Pareto Front and the diversity of the solutions found.

In fact, DEMO has several variants. In Algorithm 7, we've included the implementation of the most well-known one, DEMO/parent. For each individual (target vector) P_i in the current population, a new candidate individual C is generated using DE's differential mutation and crossover mechanism, which we've already seen in Sect. 4.5.3.

After generating and evaluating candidate u, we use Pareto dominance to determine whether parent P_i or candidate u will survive. Candidate u will replace P_i only if it dominates P_i. If u is dominated, then u will be discarded. And, if one solution does not dominate the other, then we add u to the population without eliminating P_i. Thus, both can enrich the set of solutions jointly.

Because of this last case, it's possible to have more solutions than the original population size N_{pop}. When this happens, we remove the worst solutions using the same non-domination rank concept we saw in NSGA-II.

Algorithm 7 Differential Evolution for Multiobjective Optimization (DEMO) - DEMO/parent Variant

Require: Population size N_{pop}, k objective functions $\mathbf{f}(\mathbf{x})$ (to minimize), DE scaling factor $F \in (0, 2]$, DE crossover rate $CR \in [0, 1]$, max generations G_{max}, decision space dimensionality n.
Ensure: Approximation of the Pareto Set $\mathcal{P}_{final}$.

Initialization:
$g \leftarrow 0$.
$P_g \leftarrow$ INITIALIZEPOPULATION(N_{pop}, n). ▷ Each $\mathbf{x} \in \mathbb{R}^n$
EVALUATEOBJECTIVES($P_g, \mathbf{f}$).
Main Loop:
while $g < G_{max}$ **do**
  $P_{temp} \leftarrow$ Copy(P_g). ▷ Working population from P_g
  $P_{added\_candidates} \leftarrow \emptyset$. ▷ For new non-dominated trial vectors
  for $i \leftarrow 1$ **to** N_{pop} **do** ▷ For each target individual $\mathbf{x}_{target}$ in P_g
    $\mathbf{x}_{target} \leftarrow P_g[i]$.
    // DE Mutation (DE/rand/1) and Crossover
    Select three distinct indices $r1, r2, r3$ from $\{1, \ldots, N_{pop}\} \setminus \{i\}$.
    $\mathbf{x}_{r1} \leftarrow P_g[r1]$; $\mathbf{x}_{r2} \leftarrow P_g[r2]$; $\mathbf{x}_{r3} \leftarrow P_g[r3]$.
    Mutant vector $\mathbf{v} \leftarrow \mathbf{x}_{r1} + F \cdot (\mathbf{x}_{r2} - \mathbf{x}_{r3})$.
    $\mathbf{v} \leftarrow$ ENSUREBOUNDS($\mathbf{v}$). ▷ Handle boundary constraints
    Trial vector $\mathbf{u} \leftarrow$ BINOMIALCROSSOVER($\mathbf{x}_{target}, \mathbf{v}, CR, n$).
    EVALUATEOBJECTIVES($\{\mathbf{u}\}, \mathbf{f}$). ▷ Evaluate single trial vector
    // Selection: Target $\mathbf{x}_{target}$ *vs. Candidate* $\mathbf{u}$
    if $\mathbf{u} \succ \mathbf{x}_{target}$ **then** ▷ $\mathbf{u}$ dominates $\mathbf{x}_{target}$
      $P_{temp}[i] \leftarrow \mathbf{u}$. ▷ Replace target in working population
    else if not ($\mathbf{x}_{target} \succ \mathbf{u}$) **then** ▷ $\mathbf{x}_{target}$ does not dominate $\mathbf{u}$
      $P_{added\_candidates} \leftarrow P_{added\_candidates} \cup \{\mathbf{u}\}$.
    end if ▷ Else ($\mathbf{x}_{target} \succ \mathbf{u}$), $\mathbf{u}$ is discarded and $\mathbf{x}_{target}$ from $P_{temp}[i]$ is kept.
  end for
  $P_{extended} \leftarrow P_{temp} \cup P_{added\_candidates}$. ▷ Combine potential survivors and new candidates
  if $|P_{extended}| > N_{pop}$ **then**
    $P_{g+1} \leftarrow$ ENVIRONMENTALSELECTION($P_{extended}, N_{pop}, \mathbf{f}$). ▷ E.g., NSGA-II's non-dominated sort and crowding distance truncation
  else
    $P_{g+1} \leftarrow P_{extended}$.
  end if
  $P_g \leftarrow P_{g+1}$.
  $g \leftarrow g + 1$.
end while
$\mathcal{P}_{final} \leftarrow$ EXTRACTNONDOMINATED(P_g). ▷ E.g., first front from $P_{G_{max}}$
return $\mathcal{P}_{final}$.

5.4.5 MOEA/D (Multi-Objective Evolutionary Algorithm Based on Decomposition)

MOEA/D tackles the problem in a different way. What if, instead of trying to find the entire front as a set, we broke it down (decomposed it) into several single-objective subproblems and solved them simultaneously and cooperatively?

First, the original problem $\min[f_1(\mathbf{x}), ..., f_k(\mathbf{x})]$ is transformed into N single-objective optimization problems $g(\mathbf{x}|\mathbf{w}^i, \mathbf{z}^*)$, where N is the population size and $\mathbf{w}^i = [w_1^i, \ldots, w_k^i]$ is a weight vector associated with the ith subproblem.

As seen in Algorithm 8, the external population EP is initialized empty, and we calculate the neighborhoods $B(i)$ for each subproblem i, based on the proximity of the weight vectors λ. Each $B(i)$ contains the indices of the T closest weight vectors.

We also initialize the reference point z (ideal) with the best values for each objective found in the initial population.

Algorithm 8 Multi-Objective Evolutionary Algorithm based on Decomposition (MOEA/D)

```
Require: MOP: minimize f(x) = (f_1(x), ..., f_k(x))
Require: Number of subproblems N (population size)
Require: Set of N uniformly spread weight vectors Λ = {λ^1, ..., λ^N}
Require: Neighborhood size T
Require: Aggregation function g_agg(x|λ, z) (e.g., Tchebycheff, PBI)
Require: Genetic operators (e.g., ApplyGeneticOperators(parent1, parent2))
Require: (Optional) Repair/Improvement operator
Require: Stopping criterion (e.g., max generations G_max)
Ensure: Set of non-dominated solutions EP (External Population)
EP ← ∅
For each i = 1, ..., N, compute neighborhood B(i) (indices of T closest weight vectors to λ^i).
{x^1, ..., x^N} ← InitializePopulation(N).              ▷ Initial population of solutions
For i = 1, ..., N: FV^i ← f(x^i).                        ▷ FV^i stores objective vector of x^i
Initialize reference point z = (z_1, ..., z_k)^T where z_j = min_{i=1...N}{f_j(x^i)}.
for all i = 1 to N do                                    ▷ Initial EP update
    UpdateEP(EP, x^i, f(x^i))
end for
g ← 0.                                                   ▷ Generation counter
while stopping criterion not met (e.g., g < G_max) do
    for i = 1, ..., N do                                 ▷ For each subproblem i
        // 2.1 Reproduction
        Select parent indices p1, p2 randomly from B(i).
        y_new ← ApplyGeneticOperators(x^p1, x^p2).
        y' ← ApplyProblemSpecificRepairOrImprovement(y_new).
        f(y') ← EvaluateObjectives({y'}, f).             ▷ Evaluate objectives of the new candidate
        for j = 1, ..., k do
            if f_j(y') < z_j then
                z_j ← f_j(y').
            end if
        end for
        for all index idx ∈ B(i) do
            if g_agg(y'|λ^idx, z) ≤ g_agg(x^idx|λ^idx, z) then
                x^idx ← y'.
                FV^idx ← f(y').                          ▷ Update stored objective vector for x^idx
            end if
        end for
        UpdateEP(EP, y', f(y'))

        procedure UpdateEP(EP_current, x_cand, f_cand)
      Remove solutions from EP_current dominated by f_cand.

            if f_cand is not dominated by any solution in EP_current then
        Add x_cand (and its f_cand) to EP_current.
        (Optional: If |EP_current| exceeds limit, truncate EP_current)

            end if

        end procedure
    end for
    g ← g + 1.
end while
return EP.
```

In the main loop, we iterate over each of the N subproblems. For the current subproblem i, two parents (x_k and x_l) are randomly selected from the neighborhood

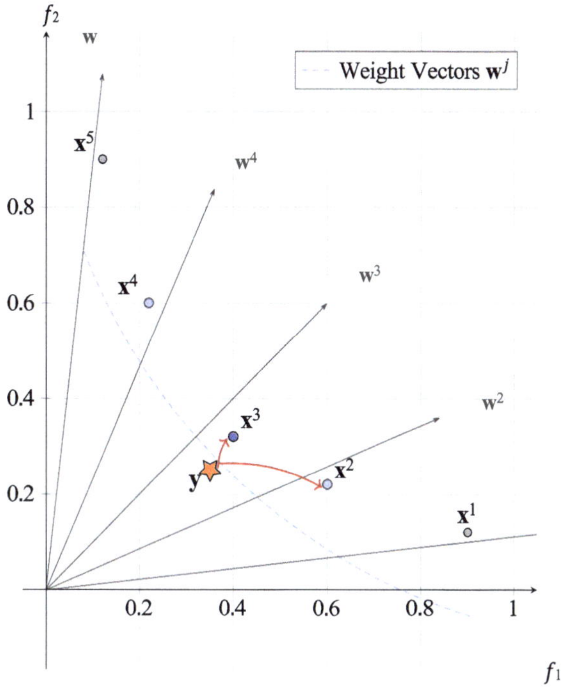

Fig. 5.12 MOEA/D: Decomposition into scalar subproblems $g(\mathbf{x}|\mathbf{w}^i)$ and cooperative evolution through neighborhoods $B(i)$. A solution $\mathbf{y}$ replaces $\mathbf{x}^j$ if $g(\mathbf{y}|\mathbf{w}^j) < g(\mathbf{x}^j|\mathbf{w}^j)$ for $j \in B(i)$. *Source* The author, 2025

$B(i)$. A new offspring y' is generated by applying genetic operators, such as those from Differential Evolution.

The offspring y may undergo a repair process for being infeasible, or a local improvement heuristic, resulting in its adjusted version y'. After evaluating its objectives, we update the reference point z if y' is better for any of the objectives (Fig. 5.12).

Subsequently, we have the main cooperation step. You see, the key to MOEA/D is that the subproblems (and their solutions $\mathbf{x}^i$) do not evolve in isolation. If the offspring $\mathbf{y}$ is better than the current solution $\mathbf{x}^j$ for subproblem j (i.e., $g(\mathbf{y}|\mathbf{w}^j, \mathbf{z}^*) < g(\mathbf{x}^j|\mathbf{w}^j, \mathbf{z}^*)$), then $\mathbf{x}^j$ is replaced by $\mathbf{y}$.

Finally, we update the external population EP so that it only contains non-dominated solutions. Notice that this resembles a situation where we have N friends trying to find the best spot in N different directions at a crowded concert by a major artist. Each friend $\mathbf{x}^i$ focuses on their direction $\mathbf{w}^i$. However, they communicate with the friends closest to them $j \in B(i)$, something like: "Hey, I found a spot $\mathbf{y}$ here that's better for your direction $\mathbf{w}^j$ than the spot $\mathbf{x}^j$ where you are!" This exchange of information (good solutions spread through the neighborhood) helps the entire population converge to the best spot at the concert more efficiently, doesn't it?

5.4.6 MOPSO (Multi-objective Particle Swarm Optimization)

Do you remember the Particle Swarm Optimization (PSO) technique we saw in Sect. 4.6.1? Well, we also have a version of it specifically for Multi-Objective Optimization problems.

PSO particles, inspired by the coordinated flight of birds or the swimming of fish schools, explore the search space and learn from their own experiences and the group's success, considering only one objective. In MOPSO, we extend the fundamental concepts of PSO (i.e., particles with positions, velocities, and memory) to optimize more than one problem at the same time. We still maintain the idea where each particle navigates through the search space, but now its decisions and learning are guided by the principles of Pareto dominance.

Back in Sect. 4.6.1, we told you that in traditional PSO, each particle stores its best position $\mathbf{p}_{\text{best},i}$ that it, by itself, has found so far. We also said that each particle is influenced by the best position $\mathbf{g}_{\text{best}}$ found by any particle up to that point.

In MOPSO, we keep the same concept. However, now $\mathbf{p}_{\text{best},i}$ no longer stores a scalar value (i.e., the value related to a single objective), but rather the best non-dominated solution it has found in its trajectory, which is a vector.[10]

If the particle finds a new position that dominates its $\mathbf{p}_{\text{best},i}$, it updates its memory. If $\mathbf{p}_{\text{best},i}$ dominates the new position, it keeps its old memory unchanged. And, if the two are mutually non-dominated, one of them is randomly chosen to be the new $\mathbf{p}_{\text{best},i}$.

We also change the idea of $\mathbf{g}_{\text{best}}$. After all, on a Pareto front, we wouldn't have just one best candidate solution, right? Instead, we have an archive called the "global repository" (REP) containing all non-dominated solutions found by any particle in the swarm so far. In each iteration cycle, we update the REP, adding new non-dominated solutions and removing solutions that are in this repository but have been outperformed by some other recently found solution.

That said, we have another important change here: if in single-objective optimization we wanted particles to converge to a single point (i.e., the global optimum), here we want particles to spread out along the Pareto front. Therefore, for each particle, we will choose another solution from the REP to influence it (Fig. 5.13).

This choice happens as follows: first, we divide the search space into sectors, or hypercubes.[11] Then, we evaluate how populated each sector (or hypercube) is: sectors with many non-dominated solutions already found are considered dense, while sectors with few are sparse.

After this definition, we choose one of these hypercubes by a weighted roulette system, where sparser sectors have a higher chance of being selected. And, after choosing a hypercube, we choose a random solution from the REP that is in that hypercube. The intention here is to favor regions with fewer solutions so that we

---

[10] After all, we need to represent the values of two or more objectives at the same time.

[11] If it were a two-dimensional space (i.e., with two objectives), we could divide this space into small rectangles. If it were three-dimensional (three objectives), we would divide the space into small cubes. And, if it had n dimensions, we would divide it into n-cubes (or hypercubes).

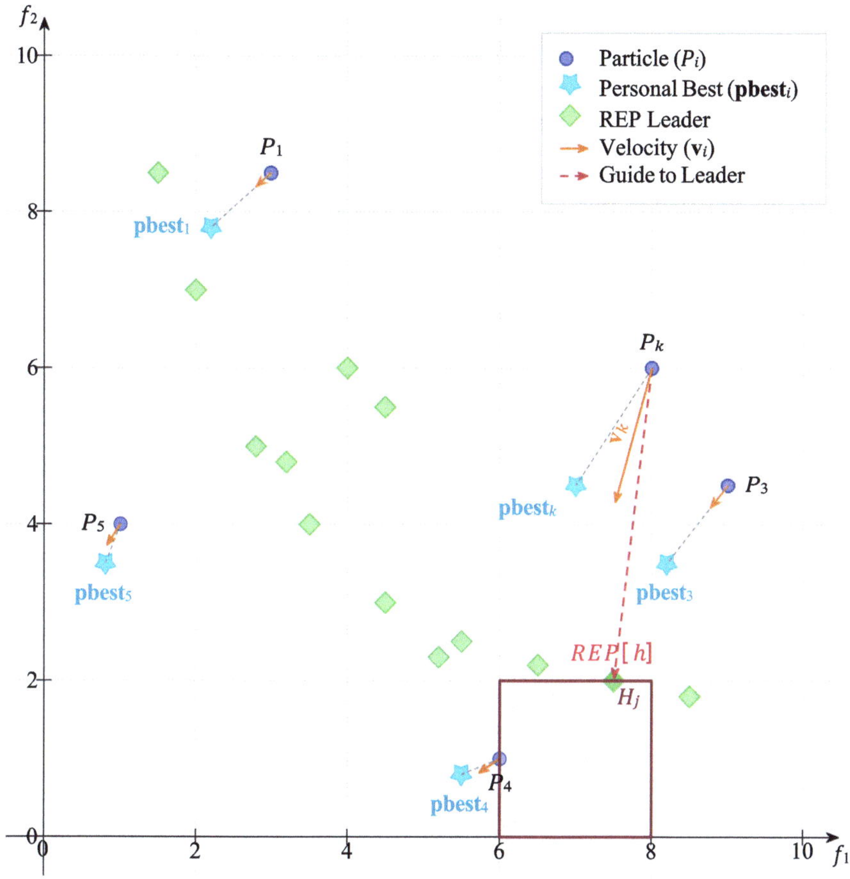

Fig. 5.13 Illustration of the MOPSO mechanism. Particles (circles) navigate the objective space, each maintaining its personal best position (**pbest**$_i$, connected by a dashed line). A repository (REP, squares/diamonds) stores the global non-dominated leaders. To guide its movement, a particle selects a leader $REP[h]$ from the repository, with preference for leaders from less dense regions of the frontier (indicated by the hypercube grid and the arrow from P_k to $REP[h]$). The particle's velocity is then adjusted by its inertia, its **pbest**$_i$, and the selected leader $REP[h]$, driving the swarm toward the Pareto Front. *Source* The author, 2025

ultimately have more diversity. Furthermore, it may happen that we no longer have space in the REP to store more non-dominated solutions. In these situations, we will prioritize solutions that are in the most populated hypercubes.

5.5 Multi-criteria Decision-Making

Now, imagine the following scenario: you've developed an MOO problem and chosen some technique to produce a Pareto front with several non-dominated solutions. However, we have a real problem: in our analogies and examples, or in the tutorials we see out there, we might generate a relatively small Pareto front, containing about three, five, or ten solutions. In such cases, it's relatively simple to compare these solutions and choose the best one: after all, the number of solutions found is sufficiently low to recognize the differences between each of them to make a final choice.

That said, in real life, decision-makers are unlikely to be very happy having to choose one solution from tens, hundreds, or thousands of candidate solutions—all of them being fully valid and non-dominated. Remember: even though the choice of the best solution is somewhat subjective, it would be interesting if we could somehow assist in the decision-making process. Fortunately, we can do this with MCDM. The idea of MCDM techniques is to help us process and choose one or more solution(s) from those available on a Pareto front.

We don't know if you're already working in the industry or not, but sometimes a professional's day-to-day life can be stressful. Just as a university student needs to manage various deadlines and responsibilities, in the industry we also have projects, deadlines, responsibilities, an endless sea of meetings, and people wanting to clarify doubts and align some points. In this context, some professionals may feel overwhelmed by the workload and stress. Therefore, it's in our interest to reduce our cognitive load when possible, and to focus only on those decisions that are strategic and truly matter. In our optimization case, this means it would be interesting to facilitate the work of decision-makers and avoid confusion due to excess—after all, the chances of choosing a not-so-good option due to lack of attention (and being held responsible for it, leading to even more stress) are higher when we have thousands of solutions than if there were a few relevant solutions that had already been filtered in advance.

That said, we were already doing a certain MCDM process earlier. You see: at the beginning of the chapter, we mentioned that decision-makers' preferences are incorporated into different parts of the process depending on the technique. In a priori techniques, preferences are included before the start of the optimization process. Therefore, the result of these techniques is a single solution that takes such preferences into account.

Now, what we'll see here are techniques intended for a posteriori methods. Regardless of the chosen technique, in the end, we'll have an approximation of the Pareto front. With these solutions in hand, we can use an MCDM technique from this section to analyze and select the solutions that are most aligned for a given situation. Or, perhaps, we could even combine more than one technique for the same problem.

5.5.1 Visual Techniques

Let's start with the most intuitive approach. Often, the easiest (and most intuitive) way to choose a solution is through visual analysis. There's a reason why managers prefer dashboards created with tools like Looker, PowerBI, or Tableau. There's also a reason why we, along with authors of other books and guides, prefer to include images with textual explanations. Furthermore, there's a reason why some people like to watch a demonstration of how others solve a problem before attempting it themselves. Indeed, sometimes just looking at something allows us to intuitively understand what's happening.

Therefore, one possibility in MCDM is to opt for using graphics to aid in decision-making. However, it's not just a matter of generating some images and showing them to decision-makers: there's a significant difference between producing a visualization and producing a visualization *that is functional*. The effectiveness of a chart will depend on how we present it. In fact, there are some common mistakes that people who aren't graphic design specialists (like us, the authors, and potentially you, if you're also not from a design background) might make when trying to produce these visualizations. These mistakes, in practice, can lead to suboptimal solution choices or even misinterpretations.

Important
There's an interesting parallel here between the worlds of machine learning and Optimization. Just as much of machine learning research is dedicated to techniques rather than data, in Optimization, we dedicate a significant portion of research to techniques and not as much to visualizations. However, it's no use having the best technique in the world if people can't make good decisions based on the algorithms' output. That's why we'd like you to understand that choosing a good MCDM technique is as important as choosing a good Optimization algorithm.

A frequent error is information overload, where we try to cram too much data (like thousands of solutions, for example) into a single chart. This could result in a cluttered graph that's difficult to interpret. Or, perhaps, we might run the risk of inferring that the people viewing the graph have the same context as we do, who produced the Optimization algorithm and the MCDM graphics. Consider this: what other people might call `cost`, you might call f_1 in your graph. Wouldn't it make sense to use a more user-friendly name for everyone (like `cost`) along with an intuitive graph title and legends?

Or, perhaps, we could pay additional attention to colors and scales. Sometimes we might opt for a color scheme that isn't easily distinguishable for someone who might have color blindness, or a text font size that's hard to see on the wall of a physical meeting room. Or, maybe, we could use a numerical scale with distorted proportions that might lead to incorrect conclusions. Therefore, remember: the idea behind these visual MCDM techniques is to help the decision-makers, who are the people with the most knowledge in the context of the problem you're trying to solve. Thus, the more you help these decision-makers by reducing confusion, ambiguity, or weak assumptions in your graphs, the better.

5.5.1.1 Scatter Plots

With that in mind, let's move to the first technique. Scatter plots are probably the most intuitive technique for problems with two objectives—it's no coincidence we chose this type of visualization to show some examples in the preceding pages. By plotting f_1 on one axis and f_2 on the other, the shape of the curve formed by the non-dominated points already shows us the trade-off between them: in some cases, we have a linear and constant trade-off. In others, we can identify that small gains in one objective require large sacrifices in the other in certain regions.

One way decision-makers can interact with this type of graph is by identifying the "knee point" of the frontier. This is the region where the curve's slope changes most drastically. This point often represents a balance point between both objectives. Another use is the application of visual limits: decision-makers might desire something like "only solutions with a cost below X are interesting" or "performance needs to be at least Y," and the graph readily shows which solutions on the frontier meet these criteria.

There's also another advantage to this technique, depending on the solution used to generate this type of graph. Unfortunately, we can't demonstrate it here due to the static nature of a book, but there are libraries and solutions (like Plotly, PowerBI, Tableau, and others) that allow for interactive manipulation of graphs, highlighting certain regions of the graph or clicking on specific solutions to access more details.

That said, we can also use this type of technique with three dimensions. First, it's worth mentioning the difficulty of generating a good visualization of this graph: depending on the distribution of points across the three dimensions, the viewing angle might not favor good decision-making—compare, for example, the two angles of the same Pareto front approximation in Fig. 5.14. In such cases, interactivity would be necessary to change the angle and better understand the distribution of solutions.

Another interesting alternative for scenarios with three or more objectives is the analysis of multiple 2D scatter plots, where we always show pairs of objectives. Consider Fig. 5.15, for example. Here, we have one visualization for each pair, such as f_1 and f_2, f_1 and f_3, and f_2 and f_3. This helps visualize the trade-offs for each pair individually, but we somewhat suffer from the lack of simultaneity in this analysis: we cannot see how a solution relates to all three objectives and the other solutions, considering all these objectives at the same time.

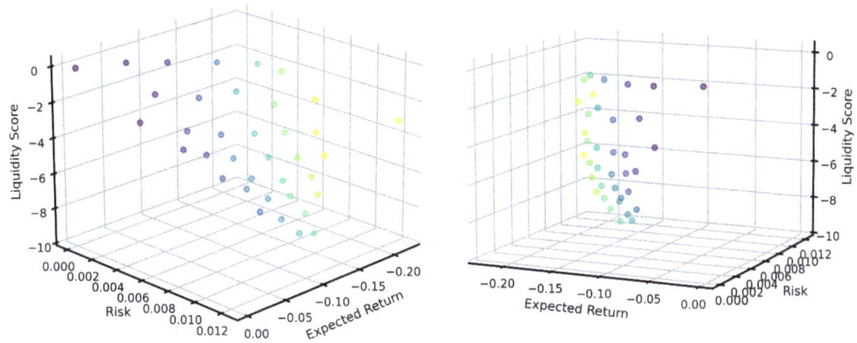

Fig. 5.14 Comparison of two viewing angles of the same three-dimensional Pareto front approximation, illustrating how the angle can affect clarity and decision-making. *Source* The author, 2025

5.5.1.2 Radar Plots

Another alternative is the radar plot. An example of this is shown in Fig. 5.16. These plots can be useful if the number of variables and alternatives is not too high. This figure, for example, displays four alternatives (i.e., solutions) in an illustrative MOP with five objectives. A line (and no longer a point, as was the case with scatter plots) represents each alternative. Since all objectives are considered minimization objectives, better solutions should be as close to the center as possible. As all alternatives are non-dominated, decision-makers can see where each alternative outperforms the others in a single visualization.

Although radar plots offer an intuitive way to compare the performance of multiple alternatives across multiple objectives, they also have some limitations. Perhaps the main one for our context is scalability: if we have many solutions (i.e., many lines) overlapping, comparing the trade-offs between them becomes a very difficult task. We also have an interpretation problem here: since the objectives are spread around the center, someone without prior knowledge of this type of graph might erroneously interpret that an objective is more associated with an objective immediately beside it than with other objectives on the opposite side of the graph. After all, we have this intuition of opposite sides in other circular visualizations like an analog clock (8 o'clock is close to 7 and 9, and much farther from 3) or a color wheel (red is close to orange and purple, and diametrically opposed to green). Therefore, it might be worthwhile to consider another type of visualization for multiple dimensions, such as parallel coordinates.

5.5.1.3 Parallel Coordinates

We have seen this type of visualization technique being employed, for example, to compare the performance among different ML models produced by a solution like

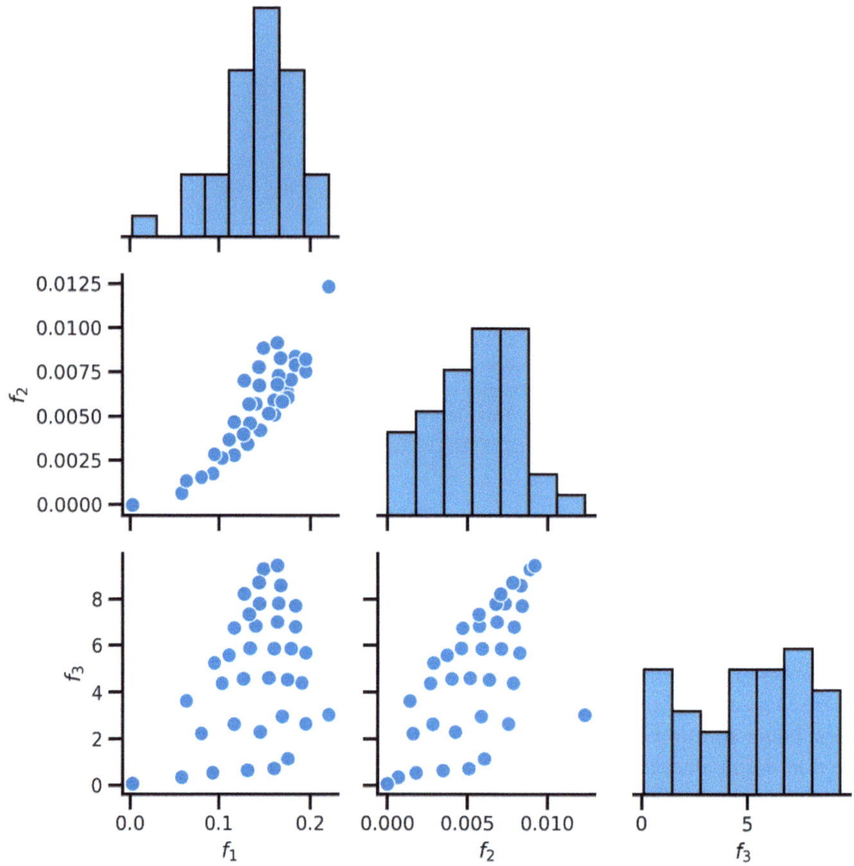

Fig. 5.15 Visualization of multiple 2D scatter plots for analysis of three objectives (f_1, f_2, f_3). Each graph shows the relationship between a pair of objectives for the same set of solutions. The difficulty mentioned in the text is that, although we see the pairwise trade-offs, a solution's overall performance across all three objectives simultaneously is not directly visualized in a single graph. *Source* The author, 2025

MLflow. In such cases, we need to simultaneously compare different conflicting metrics to choose which ML model would be most suitable according to the data scientists (i.e., the decision-makers) for that specific and subjective context of the company they are currently working for. This description sounds very much like a classic Multi-Objective Optimization problem, doesn't it?

Well, these techniques are quite useful for visualizing more than three dimensions at once. In them, each candidate solution is represented by a line that crosses multiple parallel vertical axes, as you can see in Fig. 5.17. Each of these axes represents an objective or criterion of our Optimization problem. The position where a solution's line crosses each axis corresponds to the value of that solution for that specific objective.

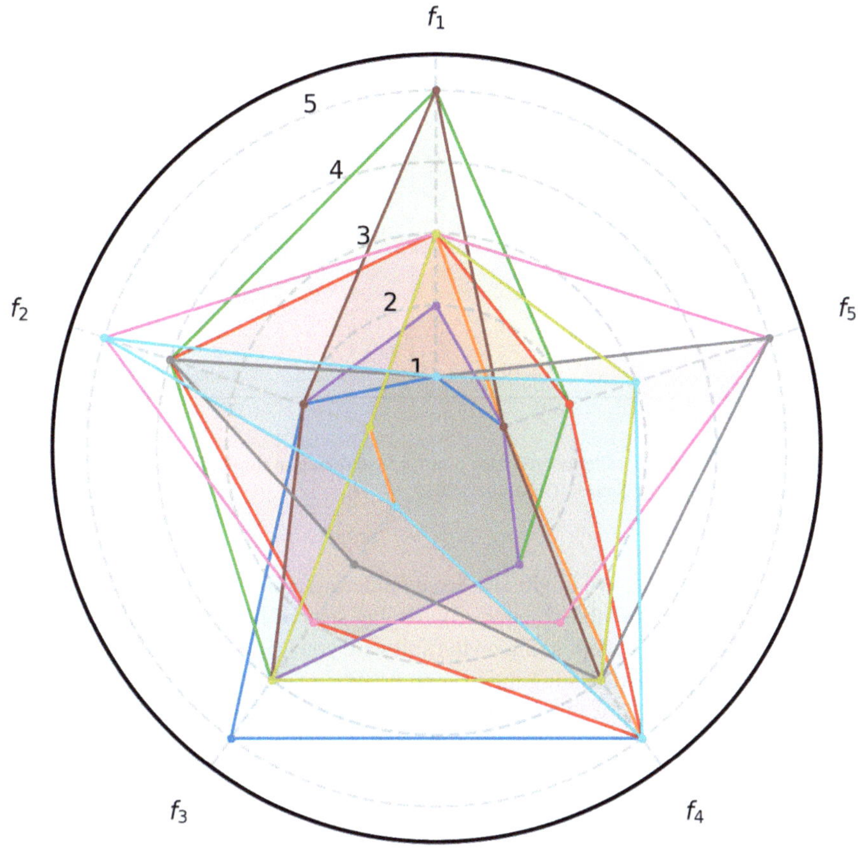

Fig. 5.16 Example of a Pareto front approximation with five objectives presented in a radar plot. *Source* The author, 2025

One of the reasons we like this visualization is the ability to identify relationships and trade-offs between objectives. For example, if many lines that show good values on one objective f_1 tend to show poor values on another objective f_2, this manifests visually as lines that "go down" on f_1 and "go up" on f_2. Or, perhaps, we might see crossing patterns between lines, indicating conflicts between objectives. On the other hand, if we see many lines following similar trajectories, we could interpret that we have objectives that behave in a correlated manner for that set of solutions.

The same idea of interactivity that we mentioned in the previous cases also applies here. In this instance, using a tool or library that allows selecting some lines or a certain range of values would help us explore a specific region of solutions, or see how solutions that meet a minimum requirement of one objective are distributed across other conflicting objectives.

Of course, here too there are some difficulties. The first is availability: although this technique is relatively well-known, not all visualization software has this tech-

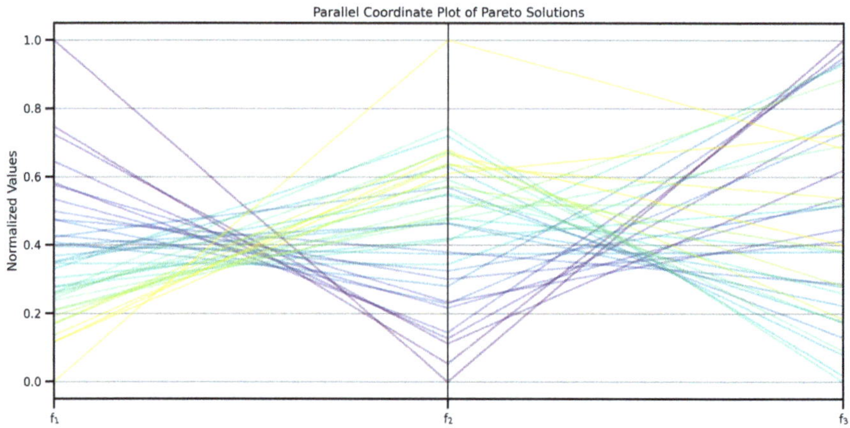

Fig. 5.17 Example of a parallel coordinates plot for visualizing multiple dimensions (objectives) and identifying trade-offs. *Source* The author, 2025

nique available for use. We might also have rendering and ordering problems: some computers might not be able to correctly display several hundreds or thousands of lines on their screen. Or, depending on the software you use, it might be worth inverting, reordering, and normalizing the axes to facilitate visualization and comparison of solutions. That said, if we have four or more objectives, it can indeed be interesting to use this technique for comparing results.

5.5.1.4 Heat Map

Another visual technique we can use is the heat map. If you've ever done any machine learning work, it's quite likely you've encountered this type of graph during exploratory data analysis. Here, we use the heat map representing solutions and objectives as rows and columns, respectively.

Thus, let's imagine a matrix where the rows represent the different candidate solutions from the Pareto front and the columns represent the different objectives we are considering. In this case, each cell in this matrix will contain the value of a specific objective for a particular solution. The heat map will also represent these numerical values on a predefined color scale. For example, in a problem where all objectives are to be minimized, lower values might be represented by cooler colors (like blue or green) and higher values by warmer colors (like yellow or red). Or, to facilitate visualization, we could use only a scale of shades, like from a lighter blue to a darker blue, or from a lighter red to a darker red.

The advantage of these maps lies in the possibility of providing an overview for multiple solutions and multiple dimensions. Therefore, we generally like this type of technique for visually detecting patterns among solutions or some interesting groupings of solutions. Consider, for example, a scenario where we have a column with many solutions displaying a color that represents poor values. This case might signal a particularly difficult objective to optimize or one where there is a strong conflict with other objectives.

That said, because it's a more visual and color-oriented technique, we need to pay some attention so that it doesn't end up biasing decision-making. Consider this: a poorly chosen color scheme could hide important patterns in the data or lead to incorrect interpretations—for some people, red represents something bad and green represents something good. Now, imagine if we had an Optimization problem where one objective is minimization and another is maximization. And, in this example, green represents low values and red, high values. If we don't process the data, for the minimization objective, green will represent something good, and for the maximization objective, green would represent something bad. Do you see how confusing this could be?

We could also have a problem with the numerical scale: if we don't normalize the objective values before producing the heat map, the colors wouldn't be represented correctly.

Similarly, we might have an issue if we don't aggregate the rows and columns so that similar items are close to each other. If we don't do this, we wouldn't be able to correctly show some interesting patterns in the data and, ultimately, our heat map, which should help us, might look like a disorganized grid with many mixed colors.

5.5.2 *Mathematical Techniques*

It's not a rule or requirement to use a visual technique (i.e., with graphs) for an MCDM process. In fact, there are even times when we don't want a visualization where all solutions from a Pareto front approximation carry the same weight. Sometimes, we want a more objective ranking of the frontier solutions, considering explicit preferences from decision-makers. In these cases, we can use several mathematical techniques.

5.5.2.1 TOPSIS (Technique for Order of Preference by Similarity to Ideal Solution)

It's quite likely that you've opened a spreadsheet program like Excel, Google Sheets, or LibreOffice Calc at some point, right? Now, imagine we've placed all the solutions from a Pareto front of a Multi-Objective Optimization problem with three objectives into this spreadsheet for analysis. If we have 100 solutions, our spreadsheet would contain at least 100 rows (one per solution) and 3 columns (one for each objective f_1, f_2, and f_3), wouldn't it?

Now, what if we wanted to sort these solutions? In these software programs, we could even establish an order among the columns: for example, we could first sort the solutions by f_3. As a tie-breaker among some solutions, we could then sort by f_1.

And, as another tie-breaker, we could then sort by f_2. However, notice that at no point did we sort by all three objectives *simultaneously*: we always need to explicitly state which objective is most important in these programs by default. And this is somewhat contradictory, considering that in Multi-Objective Optimization problems, we'd like to consider all objectives simultaneously, wouldn't we?

To address this, we have techniques like TOPSIS. It's a simple yet intuitive technique. The best solution on the frontier is the one that is, simultaneously, as close as possible to the ideal point (the best imaginable value for all objectives) and as far as possible from the nadir (the worst imaginable value for all objectives).[12]

We have its pseudocode in Algorithm 9. Here, we start with a set of m alternatives (i.e., solutions) and n criteria (i.e., objectives). Then, we construct an $m \times n$ decision matrix, where the rows represent the solutions from the Pareto front and the columns represent our objectives. It's important that this matrix is normalized so that all objectives are on the same numerical scale, thus avoiding distortions caused by different magnitudes among the objectives.

Also, observe that in the pseudocode, we need a weight vector W for each objective. This vector must be provided by the decision-makers and should contain the relative importance for each objective. For example, we would assign a weight of 1 to all objectives if they all have exactly the same importance in the decision-making process. Note that these values are subjective and will influence the ranking. After all, they are used to create a weighted matrix.

From the weighted basis, we determine two hypothetical points: the *Positive Ideal Solution* (or ideal point), which combines the best possible values for each weighted objective, and the *Negative Ideal Solution* (or nadir point), which combines the worst values. The next step is to calculate, for each solution on the frontier, its Euclidean distance to both the ideal point and the nadir point.

With these distances in hand, we calculate the relative closeness coefficient (P_i, or C_j^* in some books and articles) for each solution j. This metric, ranging from 0 to 1, indicates how close the solution is to the ideal point and how far from the nadir point. A higher score signifies a better solution. The solutions are then sorted according to this coefficient, providing us with a ranking that assists decision-makers in their choice (Fig. 5.18).

Attention

Some TOPSIS implementations assume all objectives are to be minimized. When using an existing implementation, verify this beforehand. Invert the sign of your maximization objectives' values to transform them into minimization objectives if this is your case. Furthermore, check if the implementation already normalizes all values or if it expects you to provide these values already normalized. After all, normalization affects the quality of the results.

---

[12] We previously discussed the ideal point and nadir point in Sect. 5.3.2.1.

Algorithm 9 TOPSIS Algorithm

Inputs:
  $A = A_1, A_2, \ldots, A_m$: Set of m alternatives.
  $C = C_1, C_2, \ldots, C_n$: Set of n criteria.
  $X = [x_{ij}]_{m\times n}$: Decision matrix, where x_{ij} is the value of alternative A_i with respect to criterion C_j.
  $W = [w_1, w_2, \ldots, w_n]$: Vector of criteria weights, where w_j is the weight of criterion C_j and $\sum_{j=1}^{n} w_j = 1$.
  I_j: Nature of criterion C_j ('benefit' to maximize, 'cost' to minimize).
Output:
  $Rank$: Ranking of alternatives, from best to worst.
Step 1: Construct the Normalized Decision Matrix (R)
for all criterion C_j from $j = 1$ to n **do**
  $\text{Denominator}_j \leftarrow \sqrt{\sum_{i=1}^{m} x_{ij}^2}$
  for all alternative A_i from $i = 1$ to m **do**
    $r_{ij} \leftarrow x_{ij}/\text{Denominator}_j$ ▷ Element of the normalized matrix R
  end for
end for
Step 2: Construct the Weighted Normalized Decision Matrix (V)
for all criterion C_j from $j = 1$ to n **do**
  for all alternative A_i from $i = 1$ to m **do**
    $v_{ij} \leftarrow w_j \times r_{ij}$ ▷ Element of the weighted matrix V
  end for
end for
Step 3: Determine the Positive Ideal Solution (A^+) **and the Negative Ideal Solution** (A^-)
for all criterion C_j from $j = 1$ to n **do**
  if criterion C_j is of 'benefit' type (or maximization) **then**
    $v_j^+ \leftarrow \max_i(v_{ij})$ ▷ Positive ideal value for criterion C_j
    $v_j^- \leftarrow \min_i(v_{ij})$ ▷ Negative ideal value for criterion C_j
  else if criterion C_j is of 'cost' type (or minimization) **then**
    $v_j^+ \leftarrow \min_i(v_{ij})$ ▷ Positive ideal value for criterion C_j
    $v_j^- \leftarrow \max_i(v_{ij})$ ▷ Negative ideal value for criterion C_j
  end if
end for
$A^+ \leftarrow (v_1^+, v_2^+, \ldots, v_n^+)$
$A^- \leftarrow (v_1^-, v_2^-, \ldots, v_n^-)$
Step 4: Calculate the Euclidean distances of each alternative to the Positive Ideal Solution (S_i^+) **and to the Negative Ideal Solution** (S_i^-)
for all alternative A_i from $i = 1$ to m **do**
  $S_i^+ \leftarrow \sqrt{\sum_{j=1}^{n}(v_{ij} - v_j^+)^2}$ ▷ Distance to the positive ideal solution
  $S_i^- \leftarrow \sqrt{\sum_{j=1}^{n}(v_{ij} - v_j^-)^2}$ ▷ Distance to the negative ideal solution
end for
Step 5: Calculate the Relative Closeness of each alternative to the Ideal Solution $(P_i$ **or** $C_i^*)$
for all alternative A_i from $i = 1$ to m **do**
  $P_i \leftarrow S_i^-/(S_i^+ + S_i^-)$ ▷ $P_i \in [0, 1]$
end for
Step 6: Rank the Alternatives
$Rank \leftarrow$ Sort alternatives A_i in descending order of P_i.
Return $Rank$

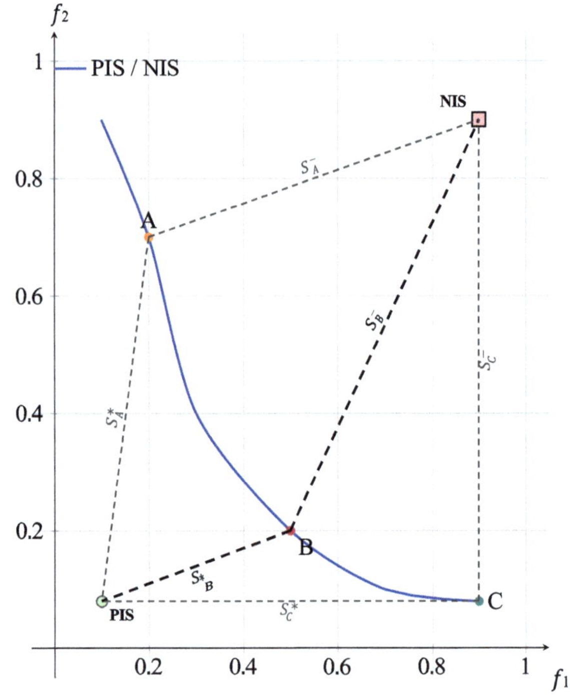

Fig. 5.18 Illustration of the distances of three candidate solutions (A, B, and C) to the Positive Ideal Solution (PIS) and the Negative Ideal Solution (NIS) for the TOPSIS method. *Source* The author, 2025

With that in mind, let's consider an example to understand how we could apply TOPSIS. Suppose a renewable energy company has obtained, through a Multi-Objective Optimization algorithm, a Pareto Front with 50 wind farm configurations, optimizing two objectives: maximizing annual energy generation (f_1) and minimizing visual impact (f_2, on a scale of 0 to 10, where 0 is better). The decision-makers assign a weight of 0.6 for generation and 0.4 for visual impact. Here, we would have:

1. First, we evaluate whether the TOPSIS algorithm the company uses already normalizes the data and if it allows us to specify whether each objective is for minimization or maximization.

   In our case, we find that it normalizes automatically (one less task for us).

   We also discover that the algorithm the company uses does not distinguish between which objectives are of the *benefit* type (maximization) or *cost* type (minimization) as we had seen in the pseudocode, but rather assumes all problems are for maximization. Therefore, we will need to invert the sign of f_2 so that TOPSIS considers a higher visual impact value as better.
2. Then, the values of f_1 and f_2 for the 50 solutions are normalized (for example, Min-Max results in a scale of [0.0, 1.0], where 1 is best).
3. The normalized matrix is weighted by 0.6 and 0.4.
4. We then identify the ideal point ([1.0, 1.0]) and the nadir point ([0.0, 0.0]).

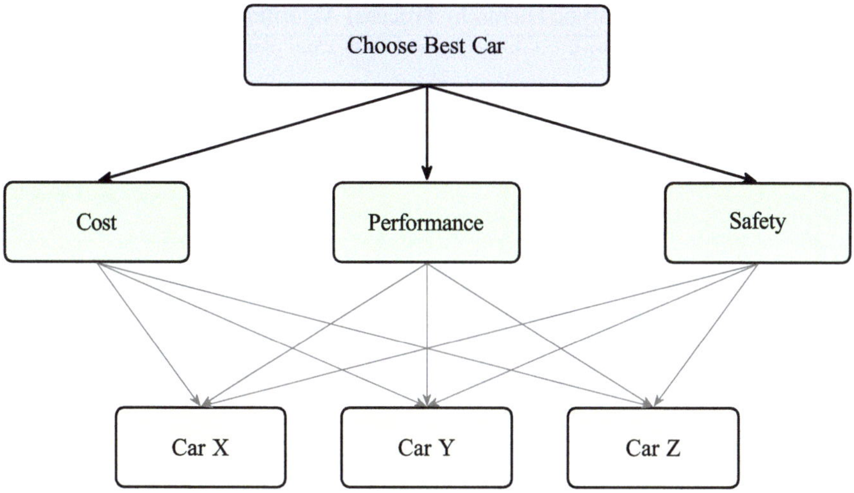

Fig. 5.19 Example of the hierarchical structure of AHP. *Source* The author, 2025

5. For each of the 50 solutions, we calculate the distances S_j^* (to the ideal point) and S_j^- (to the nadir point).
6. Afterward, we calculate the relative closeness $P_i \leftarrow S_i^- / (S_i^+ + S_i^-)$ for all 50 solutions.
7. Finally, we sort the solutions so that those with the highest P_i appear first. Then, the decision-makers can focus on the top n solutions for a final qualitative analysis.

5.5.2.2 AHP (Analytic Hierarchy Process)

AHP is another very interesting alternative—especially when decision-making involves qualitative criteria or when a hierarchical analysis of the problem is needed.

As you can see in the pseudocode of Algorithm 10, the first step in AHP is to structure the decision problem into a hierarchy. At the top, we have the overall objective of the problem (reusing the wind farm example from TOPSIS, it could be "select the best wind farm configuration"). Below this are the main criteria (i.e., our objectives $f_1, \ldots, f_k$). If necessary, we can decompose these criteria into sub-criteria. We also have a set of alternatives (i.e., the solutions from a Pareto front). One example of it is seen in Fig. 5.19, where we have three solutions to choose considering three main criteria: cost, performance, and safety.

Algorithm 10 AHP (Analytic Hierarchy Process) Algorithm

Require: Goal of the decision; Set of criteria $C = \{C_j\}$; Set of alternatives $A = \{A_i\}$; Saaty's Scale (typically 1-9); (Optional) Sub-criteria SC_{jk} for C_j.
Ensure: Ranked alternatives $Rank$; Global scores $GlobalScores$.
Step 1: Define Problem Hierarchy
Establish the hierarchy: Level 0 (Goal), Level 1 (Criteria C_j), then any Sub-criteria SC_{jk} under C_j, and finally the lowest Level (Alternatives A_i).
Step 2: Construct Pairwise Comparison Matrices
Initialize $\mathcal{M} \leftarrow \emptyset$ ▷ Set to store all comparison matrices
for all parent node P_k in hierarchy (Goal, or a Criterion/Sub-criterion with children $E_1, \ldots, E_p$) **do**
  $M_{P_k} \leftarrow$ CONSTRUCTPAIRWISEMATRIX($\{E_1..E_p\}$, w.r.t. P_k, Saaty's Scale) ▷ M_{P_k} elements $a_{ii} = 1, a_{ji} = 1/a_{ij}$
  Add M_{P_k} to $\mathcal{M}$.
end for
Step 3: Calculate Local Priority Vectors and Check Consistency
$LocalWeights \leftarrow \emptyset$ ▷ Map to store local priority vectors (weights)
for all matrix $M \in \mathcal{M}$ (derived for children of P_k) **do**
  $(W_M, CR_M) \leftarrow$ CALCULATEPRIORITIESANDCR(M, Random Index (RI) Table) ▷ W_M typically from principal eigenvector of M
  if $CR_M > 0.10$ **then** ▷ Or other acceptable consistency threshold
    Alert: Inconsistent judgments for P_k (CR = CR_M). **Judgments must be revised.**
    return Failure or Request Revision.
  end if
  $LocalWeights[P_k \rightarrow \text{children}] \leftarrow W_M$. ▷ Store validated local weights
end for
Step 4: Aggregate Weights for Global Alternative Scores
$W_{criteria} \leftarrow LocalWeights[\text{Goal} \rightarrow C]$. ▷ Weights of main criteria
for all alternative $A_l \in A$ **do**
  $GlobalScore_l \leftarrow$ CALCULATEGLOBALSCORE($A_l, C, W_{criteria}, LocalWeights$, hierarchy). ▷ Synthesizes weights across levels, considering sub-criteria if present
end for
Let $GlobalScores$ be the vector of all $GlobalScore_l$ values.
Step 5: Rank Alternatives
$Rank \leftarrow$ SORTALTERNATIVESBYSCORE(A, $GlobalScores$, descending).
return $Rank, GlobalScores$.

Within the algorithm, a key step of AHP is pairwise comparison. Typically, Saaty's scale, a numerical scale from 1 to 9, is used, where 1 indicates equal importance and 9 indicates extreme importance.[13] By making these comparisons, the algorithm populates a matrix, and from this matrix, the criteria (objectives) weights vector is derived.

Illustrating Saaty's Scale We believe an explanation of how these comparisons work might be helpful. Therefore, let's consider a simple example, like choosing your next vacation destination (which we hope happens soon and is an excellent experience for you).

[13] Saaty's scale assumes the following values: 1 for equal importance, 3 when one criterion is moderately more important than the other, 5 when one criterion is strongly more important than the other, 7 when one criterion is very strongly more important than the other, and 9 when one criterion is extremely more important than the other. We also use even numbers when a criterion falls between some of these values. For example: if the importance between one criterion and another is between equal importance and moderately more important, we would assign the weight 2.

Table 5.3 Comparison matrix

Criterion	Cost (C_1)	Attractions (C_2)	Duration (C_3)
Cost (C_1)	a_{11}	a_{12}	a_{13}
Attractions (C_2)	a_{21}	a_{22}	a_{23}
Duration (C_3)	a_{31}	a_{32}	a_{33}

Table 5.4 Filled comparison matrix

Criterion	Cost (C_1)	Attractions (C_2)	Duration (C_3)
Cost (C_1)	1	7	3
Attractions (C_2)	1/7	1	1/5
Duration (C_3)	1/3	5	1

So, in the first step, you defined the overall goal: to choose the best vacation destination. In the first level of the hierarchy, you mapped out three main criteria:

- C_1: minimize the cost of the trip;
- C_2: maximize the number of tourist attractions you can visit;
- C_3: maximize the duration of the trip, in days.

In the next step, we will perform pairwise comparisons between these criteria to determine how important they are in relation to your overall goal, using Saaty's scale. We will use a comparison matrix as a base, as you can see in Table 5.3.

Now, we need to replace these cells with values. First, the importance of a criterion compared to itself is always the same value. Therefore, we will replace a_{11}, a_{22}, and a_{33} with the value 1.

After that, the decision-makers (in this example, ourselves) must ask: "how much more important is cost compared to the duration of the trip, considering our overall goal is to choose the best vacation destination?" We will ask this question for all other combinations in the matrices.

Now, let's suppose you believe that cost C_1 is moderately more important than the trip duration C_3. This would correspond to the value 3. Therefore, a_{13} would have the value 3, or "cost is 3 times more important than the trip duration." Similarly, the inverse also applies: a_{31} has the value $\frac{1}{3}$, or "the trip duration is 3 times less important than cost."

We will fill in the values for all pairs, eventually arriving at the results in Table 5.4.

After producing the matrix for the criteria, we would repeat the same process for the alternatives with respect to each criterion individually. That is: we would compare the relative performance of each pair of solutions for each objective. During this process, we would also evaluate the consistency of human judgments—after all, sometimes due to confusion, doubt, or simply a lack of attention from insufficient coffee, we might provide inconsistent weights in some way (for example, someone might say C_1 is better than C_2, C_2 is better than C_3, and C_3 is better than C_1, but the math wouldn't add up that way).

Finally, the local weights calculated at each level of the hierarchy are combined to obtain the global score for each alternative. In the final ranking, alternatives with the highest global scores would be listed first.

5.5.2.3 ELECTRE (ELimination Et Choix Traduisant La RÉalité)

The ELECTRE algorithms (yes, because they are a family of algorithms—similar to the case of the NSGA variants we saw earlier) also produce a ranking of solutions, but they adopt a different methodology. They are based on the concept of outranking: that is, how much an alternative (solution) $\mathbf{x}_A$ is at least as good as, or preferable to, an alternative $\mathbf{x}_B$, considering all criteria simultaneously but without necessarily forcing a total trade-off between them.

A Pause to Compare Methods This is the third time we've discussed a ranking method, as we've already covered TOPSIS and AHP. And, depending on how you're feeling right now, the explanations might start to blend and become confusing. Allow us to help clarify things before we continue. For this, let's return to our analogy of choosing the best vacation destination that we mentioned earlier when discussing AHP. And, to help things along, imagine that each of these methods is represented by one of your friends.

Your friend who reasons like TOPSIS first establishes the "perfect vacation destination" (something utopian: the best food in the world, the best tourist attractions in the world, and the longest duration we can manage) and the "worst vacation destination in the world," which would be the exact opposite. Then, for each destination we're considering for a trip (like Buenos Aires, Manila, Nairobi, and Jaipur), they would measure how far each of these places is from the perfect (and fictitious) destination and how close they are to the nightmarish one. And, after that, they would opt for the destination closest to the dream.

Your friend who reasons like AHP would think: "for this perfect vacation, what's most important? Cost? Attractions? Duration?" They would make those pairwise comparisons (like cost vs. attractions using Saaty's scale) to define exact weights for each criterion. Then, for each possible destination (like Buenos Aires, Nairobi, Manila, and Jaipur), they compare them pairwise with respect to each criterion—something like "considering *only* cost, is Buenos Aires better than Manila? By how much?" In the end, they come up with a spreadsheet full of numbers, weights, and a clear, unique ranking: "based on my analysis, Manila is the best option with 85 points, followed by

Jaipur with 82, Buenos Aires with 78, and Nairobi with 70." Notice here that AHP provides an order based on judgments quantified by relative importance.

And then we come to ELECTRE. Instead of looking for a perfect ranking or the "absolute ideal," we want to eliminate clearly inferior options and identify a group of good options, even if we can't say which one is definitively the best in every sense. For example: for each pair of destinations, we would ask, "on what basis can I say that Buenos Aires outranks Jaipur (meaning, how much is Buenos Aires at least as good as, or preferable to, Jaipur)?" We make this decision based on concordance and discordance.

In concordance, we look at all criteria (i.e., the objectives "cost," "attractions," and "duration") and sum the weights of the criteria where one city is better than or equal to another. If this sum of weights exceeds a concordance threshold you've defined (for example, "at least 60% of the criteria's importance supports Buenos Aires"), then the concordance criterion is satisfied.

In discordance, we observe the criteria where Buenos Aires is worse than Jaipur. Here, we also have the concept of *veto*. Is there, for example, any criterion where Jaipur excels so much, so very much compared to Buenos Aires, that this factor alone would remove Buenos Aires from comparison? If so, Buenos Aires does not outrank Jaipur (i.e., we have a veto here), and the discussion ends. On the other hand, if there's no veto, we analyze what the biggest disadvantage of one city is over another for each criterion (objective), and we accept that Buenos Aires is worse than Jaipur in a given criterion as long as it's not a huge difference. This is the discordance criterion.

Ultimately, one city would outrank another only if both the concordance and discordance criteria are met simultaneously. We would do this for all solutions (in this case, all cities).

Algorithm 11 contains the pseudocode for the implementation of one of the simplest versions of this family of algorithms: ELECTRE I. This algorithm has some similarities to the other methods we've seen so far, such as having a list of all alternatives (or solutions), all criteria (or objectives), a weight vector for all criteria, and a decision matrix. That said, we also have some new things:

- We also have veto thresholds v_j establishing what the critical value would be for each objective to determine when one solution is "much better" than another.
- We also have the minimum proportion of criteria weights for which one solution must be at least as good as another solution for us to have concordance in the outranking. This deals with concordance.
- And, finally, we also have a maximum level for a single objective that is tolerated for one solution to still outrank another. This deals with discordance.

Of course, if the criteria have very different units or scales, we will need to normalize the objectives before proceeding. After that, we will calculate the concordance and discordance indices for all pairs of solutions. Once calculated, we can construct an outranking matrix for concordance and another for discordance. In the end, we will aggregate the information from both matrices into a single aggregated outranking matrix.

Algorithm 11 ELECTRE I Algorithm

Require: Set of alternatives $A = \{A_1, \ldots, A_m\}$; Set of criteria $C = \{C_1, \ldots, C_n\}$; Decision matrix $X = [x_{ij}]$ (value of A_i for C_j); Criteria weights $W = [w_j]$ (sum to 1); Concordance threshold c_{min}; Discordance threshold d_{max}; (Optional) Veto thresholds $V = [v_j]$ for criteria C_j.
Ensure: Kernel $K \subseteq A$ (set of non-outranked or preferred alternatives).
Step 1: Pre-processing (Optional)
(If criteria scales differ significantly) $X \leftarrow$ NORMALIZEDECISIONMATRIX(X). ▷ Hereafter, X refers to the matrix used (original or normalized).
Step 2: Calculate Concordance and Discordance Indices
for all ordered pairs of alternatives (A_k, A_l) where $k \neq l$ **do**
  $C(A_k, A_l) \leftarrow$ COMPUTECONCORDANCEINDEX(A_k, A_l, X, W). ▷ Sum of weights where A_k is at least as good as A_l.
  $D(A_k, A_l) \leftarrow$ COMPUTEDISCORDANCEINDEX(A_k, A_l, X). ▷ Max normalized performance difference where A_l is strictly better than A_k.
end for
Step 3: Determine Outranking Relations (S_{kl})
Initialize outranking matrix S (all elements to 0 or relation to false).
for all ordered pairs of alternatives (A_k, A_l) where $k \neq l$ **do**
  $isConcordant \leftarrow (C(A_k, A_l) \geq c_{min})$.
  $isNotDiscordant \leftarrow (D(A_k, A_l) \leq d_{max})$.
  $noVetoTriggered \leftarrow$ **true**.
  if Veto thresholds V are defined **then**
    $noVetoTriggered \leftarrow$ **not**CHECKVETO(A_k, A_l, X, V).
  end if
  if $isConcordant$ **and** $isNotDiscordant$ **and** $noVetoTriggered$ **then**
    $S_{kl} \leftarrow 1$. ▷ A_k outranks A_l (denoted $A_k\ S\ A_l$)
  else
    $S_{kl} \leftarrow 0$.
  end if
end for ▷ The matrix S defines the outranking graph $G = (A, S)$.
Step 4: Identify the Kernel from the Outranking Graph
$K \leftarrow$ IDENTIFYKERNEL(A, S). ▷ E.g., a common approach is to find alternatives not outranked by any other in A.
return K.

Then, the final result of ELECTRE is not necessarily a complete and linear ranking like in TOPSIS or AHP, but it can be a partial ordering (an outranking graph) or the identification of a "kernel" of best alternatives (this is the case for ELECTRE I). Each version of ELECTRE is adapted for different types of decision problems: in II, III, and IV, we have a complete ranking. In the case of ELECTRE TRI, we can have a classification into predefined categories (Fig. 5.20).

5.5.2.4 PROMETHEE (Preference Ranking Organization METHod for Enrichment of Evaluations)

Like ELECTRE, PROMETHEE is also a family of outranking methods. The difference here is that we explicitly use generalized preference functions for each objective. These functions describe the difference between two alternatives in a degree of preference between the interval [0, 1] and need to be defined by the decision-makers.

In this type of algorithm, decision-makers must first choose a type of preference function $P_i(d)$ that best represents how their preference for a solution $\mathbf{x}_A$ over a solution $\mathbf{x}_B$ increases as the difference $d = f_i(\mathbf{x}_A) - f_i(\mathbf{x}_B)$ (where i is a maximiza-

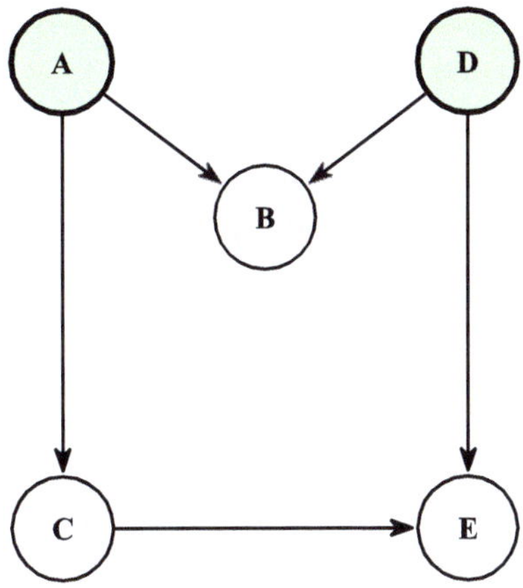

Fig. 5.20 Example of an outranking graph resulting from the ELECTRE method, where A outranks B and C, while C outranks E, and D outranks B and E. No solutions outrank A and D. *Source* The author, 2025

tion objective, or $f_i(\mathbf{x}_B) - f_i(\mathbf{x}_A)$ if f_i is a minimization objective) increases. There are different algorithms for the PROMETHEE family. Among these, Algorithm 12 shows PROMETHEE II, which produces a complete ranking of solutions.

Here, there's a good degree of freedom for decision-makers: the way the performance difference between two alternatives on a specific criterion translates into a degree of preference for one over the other can vary enormously. PROMETHEE recognizes this and offers six types of generalized preference functions.[14] Each models a different rationality, allowing for fine-tuning to the decision-makers' perception for each criterion:

- Usual Criterion (Type I): This is the simplest form. Any positive difference, however small, in favor of one alternative over another results in maximum preference. Otherwise, there is no preference. It's a function generally used when any advantage, even minimal, is considered decisive.
- Quasi-Criterion or U-Criterion (Type II): In this function, small differences are considered irrelevant. There is an indifference threshold used to define when a difference becomes significant for our analysis.
- V-Criterion (Type III): Here, preference grows linearly as differences also grow. Preference is null if differences are negative or zero. We generally use it when the intensity of preference is proportional to the magnitude of the difference, up to a certain limit.
- Level Criterion or Step-Criterion (Type IV): Here we have an indifference threshold and a strict preference threshold. There is no preference if the difference is less than or equal to the indifference threshold, and preference is maximal if it is greater than or equal to the strict preference threshold. Anything in this interval is understood as an "intermediate" preference.

[14] In this case, there are six classic types of functions. Naturally, new articles frequently extend and revisit this grouping.

Algorithm 12 PROMETHEE II Algorithm

Inputs:
$A = A_1, A_2, \ldots, A_m$: Set of m alternatives.
$C = C_1, C_2, \ldots, C_n$: Set of n criteria.
$X = [x_{ij}]_{m \times n}$: Decision matrix, where x_{ij} is the value of alternative A_i with respect to criterion C_j.
$W = [w_1, w_2, \ldots, w_n]$: Vector of criteria weights, where w_j is the weight of criterion C_j and $\sum_{j=1}^{n} w_j = 1$.
Type of Generalized Preference Function $P_j(d_j)$ for each criterion C_j.
(Optional) Parameters for each preference function (e.g., thresholds q_j, p_j, s_j).
Output:
Rank: Complete ranking of alternatives, from best to worst, with their net flows.
Step 1: Determine the Generalized Preference Function for each Criterion ▷ For each criterion C_j, the decision-maker chooses one of the six generalized preference functions proposed by Brans and Vincke, and defines its parameters if necessary.
Common types of functions $P_j(d_j)$, where $d_j(A_k, A_l) = x_{kj} - x_{lj}$ (for benefit criterion):
1. Usual (No strict preference until one is better): $P_j(d_j) = 0$ if $d_j \leq 0$, $P_j(d_j) = 1$ if $d_j > 0$.
2. Quasi-Criterion (with indifference threshold q_j): $P_j(d_j) = 0$ if $d_j \leq q_j$, $P_j(d_j) = 1$ if $d_j > q_j$.
3. Criterion with Linear Preference (with strict preference threshold p_j):
$P_j(d_j) = 0$ if $d_j \leq 0$, $P_j(d_j) = d_j/p_j$ if $0 < d_j \leq p_j$, $P_j(d_j) = 1$ if $d_j > p_j$.
4. Level Criterion (with thresholds q_j and p_j):
$P_j(d_j) = 0$ if $d_j \leq q_j$, $P_j(d_j) = 0.5$ if $q_j < d_j \leq p_j$, $P_j(d_j) = 1$ if $d_j > p_j$.
5. Criterion with Linear Preference and Indifference Zone (with q_j and p_j):
$P_j(d_j) = 0$ if $d_j \leq q_j$, $P_j(d_j) = (d_j - q_j)/(p_j - q_j)$ if $q_j < d_j \leq p_j$, $P_j(d_j) = 1$ if $d_j > p_j$.
6. Gaussian (with standard deviation s_j): $P_j(d_j) = 0$ if $d_j \leq 0$, $P_j(d_j) = 1 - \exp(-d_j^2/(2s_j^2))$ if $d_j > 0$. ▷ For cost criteria, $d_j(A_k, A_l) = x_{lj} - x_{kj}$. The function $P_j(d_j)$ still measures the degree of preference of A_k over A_l.
Step 2: Calculate Aggregate Preference Indices ($\Pi(A_k, A_l)$) ▷ For each pair of alternatives (A_k, A_l).
for all pair of alternatives (A_k, A_l) **do**
  $\Pi(A_k, A_l) \leftarrow 0$
  for all criterion C_j from $j = 1$ to n **do**
    $d_j(A_k, A_l) \leftarrow x_{kj} - x_{lj}$ ▷ Or $x_{lj} - x_{kj}$ if C_j is cost
    $\Pi(A_k, A_l) \leftarrow \Pi(A_k, A_l) + w_j \times P_j(d_j(A_k, A_l))$
  end for
end for ▷ $\Pi(A_k, A_l)$ represents the degree to which A_k is preferable to A_l considering all criteria.
Step 3: Calculate Positive ($\phi^+(A_k)$) and Negative ($\phi^-(A_k)$) Outranking Flows
for all alternative A_k from $k = 1$ to m **do**
  $\phi^+(A_k) \leftarrow \frac{1}{m-1} \sum_{l=1, l \neq k}^{m} \Pi(A_k, A_l)$ ▷ Positive flow: how much A_k dominates others
  $\phi^-(A_k) \leftarrow \frac{1}{m-1} \sum_{l=1, l \neq k}^{m} \Pi(A_l, A_k)$ ▷ Negative flow: how much A_k is dominated by others
end for
Step 4: Calculate Net Outranking Flow ($\phi(A_k)$) - For PROMETHEE II
for all alternative A_k from $k = 1$ to m **do**
  $\phi(A_k) \leftarrow \phi^+(A_k) - \phi^-(A_k)$
end for
Step 5: Rank Alternatives (PROMETHEE II)
Rank $\leftarrow$ Sort alternatives A_k in descending order of $\phi(A_k)$.
If $\phi(A_k) > \phi(A_l)$, then A_k is preferable to A_l $(A_k P A_l)$.
If $\phi(A_k) = \phi(A_l)$, then A_k is indifferent to A_l $(A_k I A_l)$.
Return *Rank*, $\phi(A_k)$

- V-Criterion with Indifference Zone (Type V): This function combines Types II and III. Small differences, up to an indifference threshold, are irrelevant. After that, preference begins to grow linearly as the difference increases. It reaches its maximum when the difference reaches the strict preference threshold. Therefore, we use it in cases where there is an initial range of indifference, followed by a gradual increase in the importance of the difference.

- Gaussian Criterion (Type VI): In this function, preference increases smoothly and continuously, following the shape of a Gaussian curve (or "bell curve"), as the positive difference increases. We use this function for criteria where the transition from indifference to strict preference is not abrupt nor strictly linear, but rather gradual, and with more sensitivity to change around intermediate differences.

The decision-maker analyzes each criterion and chooses one of the six preference functions that best represents how they perceive performance differences for that criterion. Then, we map this difference to a value between 0 (no preference of a given solution over another) and 1 (strict preference of a given solution over another).

Then, for each pair of solutions, we calculate an aggregate preference index. This index is the weighted sum of the preference degrees for all objectives, ranging from 0 to 1. With this, we calculate the outranking flows, where:

1. The *positive flow* ($\phi^+(A)$) measures how strongly solution $\mathbf{x}_A$ outranks (is preferable to) all other solutions, representing the "strength" of $\mathbf{x}_A$.
2. The *negative flow* ($\phi^-(A)$) measures how strongly solution $\mathbf{x}_A$ is outranked by (is less preferable than) all other solutions. It represents the "weakness" of $\mathbf{x}_A$.

With this, we calculate the difference between these two flows into a "net flow" to obtain a complete ranking: solutions with a higher net flow are ranked first, thus representing the most preferable alternatives.

5.5.3 Choosing and Combining MCDM Techniques

That said, when should you choose the correct technique to solve an Optimization problem? This might be a point of great discussion and controversy among professionals already working in the industry, but the fact is, after all, at the beginning of your career, you are not necessarily evaluated by your level of theoretical knowledge, but rather by how well you can solve the problems of teams composed of people.

Therefore, you might experience steeper career growth if you know how to use the most appropriate technique to solve each problem situation. Here, we are not stating that mastering the theoretical basis is optional: on the contrary, knowing when and where to apply them in the most suitable way is just as important as understanding how the different Optimization techniques work. That said, let's first divide them into visual techniques and mathematical techniques.

5.5.3.1 Choosing Visual Techniques

It is preferable to use a visual technique if we need, for example:

1. To group solutions along different objectives to better understand the nature of our Optimization problem;

2. To visually understand how different solutions across different objectives behave without a lengthy explanation being necessary. Here, imagine an ongoing meeting with several participants: sometimes, a table full of numbers doesn't convey this message as well as a graph full of colors and patterns in its shapes.
3. If we need to intuitively show trade-offs, behavioral patterns, or *outliers* among solutions.

The choice among visual techniques will depend on the amount of data, the number of objectives, and, mainly, the specific questions the decision-maker seeks to answer. Thus, we prefer scatter plots when we have only **two objectives** to compare. For example, we would use such a graph to see the trade-off between the cost of a material and its tensile strength for a structural component, or to compare energy consumption versus processing capacity of different microcontrollers.

On the other hand, we can use a combination of scatter plots (like a scatter plot matrix) when we have a slightly larger number of objectives (something like **between 3 and 7 objectives**). More than that would result in a graph that might be too large to facilitate analysis and comparison on your computer screen or during a presentation in a meeting room, for example. After all, remember: we want to get a general overview of all bivariate relationships between them at once, and a higher number of objectives makes this task difficult. Or, we can employ this technique if we want to quickly discover which pairs of criteria have the most evident correlations or conflicts from a visual standpoint.

That said, we might have a case where we have **few solutions** (up to about 10) and **between 3 and 10 objectives**. Here, a radar plot could be a good alternative. Since each solution overlaps the other on the same graph, and since each solution is represented by a line with a different color or dash pattern, it's not ideal to visually clutter the graph if we have a large number of solutions. Here, the order in which the objectives are organized affects the visual result and interpretation. Therefore, it's interesting to normalize the values in the objectives and also group the objectives so that those that are correlated can be close together on the graph.

Now, if we have **many objectives and many solutions**, it might be more interesting to use parallel coordinates or heat maps. Parallel coordinates work well when we have high dimensionality in the data. Furthermore, it is useful for identifying similar patterns among solutions or between some conflicting objectives. Just as it allows for the identification of patterns, it's also possible to identify solutions that completely deviate from these patterns: outliers will appear as lines with trajectories completely distinct from the vast majority of solutions.

Similarly, we can use heat maps if we are more interested in identifying "hotspots": combinations of solutions and objectives with distinct performance that may warrant further investigation. Or, even, to get a general overview of the structure of a decision matrix before applying mathematical ranking methods. However, just as was the case with radar plots, for heat maps, ordering also matters: to better identify patterns in the data, it's worthwhile to place similar solutions and objectives close to each other.

5.5.3.2 Choosing Mathematical Techniques

And, on the other hand, it's preferable to use a mathematical technique if we need, for example:

1. To justify why one decision was chosen over another. In visual methods, decision-makers analyze various solutions and choose one of them. In mathematical methods (specifically, those that generate a ranking), we choose a solution from among those that are best ranked.
2. Where we need to establish a formal ordering of the available alternatives based on performance across multiple criteria.
3. When it's necessary to explicitly incorporate decision-makers' preferences into the model, such as the relative importance of different objectives in the form of weights.

That said, the choice of a method will depend on how we want to organize the solutions to generate a classification. Thus, we prefer TOPSIS when we need to find a **solution that represents a good balance** among several objectives, and when these objectives are well-defined.

Now, there might be cases where subjectivity exists. That is, we might have situations where we need to somehow include the judgment and subjective preferences, as well as the experience of the decision-makers. In this case, AHP becomes a good option.

On the other hand, sometimes we don't necessarily want to find a single best alternative, but rather a subset of good solutions. Or, sometimes, we want to **penalize** poor performance in one or more objectives, or include thresholds of indifference and preference. Here, ELECTRE might be a better alternative. And, finally, sometimes we might want to work with **quantitative and qualitative data**, or want to model decision-makers' preferences for each objective. In these cases, we might prefer PROMETHEE.

5.5.3.3 Combining Techniques

Using just one technique can be somewhat limited for applying to a real problem. A visual technique, while excellent for exploration and intuitive analysis, may not be a good option for a justifiable final choice when we have multiple conflicting objectives: it's unlikely someone would look at a scatter plot and justify why a particular solution from the Pareto front was chosen over others if it doesn't have any visually distinguishing features.

A mathematical technique, on the other hand, even if it generates a good ranking, can be mistaken for a black box: after all, someone who doesn't understand how this technique works wouldn't be able to adequately critique its results.

Therefore, an alternative that might be interesting is the combination of these two types of techniques. First, we use a visual technique to show how the solutions are organized. Here, we want to get a quick overview of the solutions' performance and identify alternatives that are consistently good or bad across all criteria, or objectives where there is great performance variation among alternatives. After that, we can use a mathematical ranking technique, even though there isn't a universally best ranking method. At this point, we would already have an idea of how the solutions are distributed. We could even filter some solutions that are of interest to us for the ranking.

We could even extend our consideration about the non-existence of a universally good ranking method to any MCDM technique: that is, there isn't a single MCDM technique that is universally superior for all types of problems. The choice of one or more appropriate techniques, therefore, depends on the nature of the decision problem, the types of data, and the context of the decision-makers, among others. Perhaps that's why the MCDM process is rarely linear: an initial exploration in a graph will generate questions, suggestions, and adjustments that potentially lead to adjustments in the problem and the Optimization algorithm, producing then other MCDM solutions which, in turn, can lead to more questions, suggestions, and adjustments until we reach the point where people are comfortable with its operation. That said, don't understand this as something bad: it's the visualization of solutions that allows people to understand the Optimization problem and make good decisions based on it.

5.6 Conclusion

As explored in this chapter, MOO can handle real-world problems involving multiple, often conflicting, objectives. The concept of Pareto optimality was introduced to explain this paradigm. The chapter also addressed the difference between a priori methods, yielding a single solution based on predefined preferences, and a posteriori methods, which provide a range of solutions for later evaluation.

For a posteriori methods, we noted that generating a good Pareto front approximation is often an intermediate step. The subsequent phase involves MCDM, where various visual and mathematical techniques can translate the multiple solutions from the Pareto front approximation into one or a few actionable choices. Therefore, the point at which a decision-maker's preferences are incorporated guides the selection of the MOO method: if preferences are known beforehand and a single solution is required, an a priori method is suitable. Conversely, if preferences are subjective or depend on analyzing a Pareto front, an a posteriori method combined with MCDM techniques is appropriate for presenting the range of solutions.

With this foundation, the next chapter will explore applications of optimization techniques in various real-world scenarios, particularly those involving multiple conflicting objectives in complex systems.

Chapter 6
Applications of Optimization

Abstract The goal of this chapter is to connect optimization theory with actual practice by showcasing a wide range of real-world applications across different disciplines. Specifically, we present examples in the following Engineering areas: Industrial, Mechanical, Civil, Chemical, Electrical, Aerospace, and Computer. We also include examples specific to Data Science. The intention here is to demonstrate how both single-objective and multi-objective optimization techniques help solve complex challenges. These range from improving efficiency in traditional industrial processes, like supply chain management and plant operations, to designing advanced systems such as aerospace vehicles and integrated circuits. In Data Science, we also showcase how optimization is foundational to modern practices, covering machine learning model training, hyperparameter optimization, and generative AI.

Throughout the previous chapters, you've likely noticed our focus on analogies not necessarily tied to your profession, but rather on simpler examples. That is to say, while you might not be working with donuts or choosing cell phones, we've all undoubtedly gone through similar decision-making processes at some point in our lives, haven't we?

However, we haven't forgotten that your goal (or current reality) is to work as a data scientist or an engineer in the industry. Therefore, in this chapter, we aim to focus on how these optimization techniques are used in the real world, specifically within these professions.

After all, remember: understanding the concepts of the techniques we've covered so far is only half the battle. Typically, we're hired by companies and advance in our careers because we know how to apply these techniques to solve business problems. An academic assessment might inquire you on the differences between the NSGA-II and NSGA-III algorithms, but the real-world distinction lies in identifying optimization opportunities, differentiating between these and other techniques to address a business problem, and knowing how to choose the most suitable technique to deliver results for the company. Furthermore, it's about ensuring that people can interpret those results.

W. R. Monteiro and G. R. Meza, *A Practical Guide to Optimization in Engineering and Data Science*, https://doi.org/10.1007/978-3-032-04633-8_6

Tip
It's relatively common to see individuals in STEM (Science, Technology, Engineering, and Mathematics) programs place less importance on topics typically associated with the Humanities. For instance, at our University, we frequently encounter students who, if given the choice, would prefer to study more technical subjects involving equations and lines of code over those dealing with project and people management or interpersonal relationships. However, even if you prefer working more with systems and numbers than with people, you'll need to understand and be understood by people to grow in your career.

Thus, in this chapter, we will explore how optimization is practiced in organizations and by the teams working within them.

6.1 Types of Companies

So far, we've focused on the technical concepts of optimization, such as formulating an optimization problem, choosing between different optimization algorithms, and Multi-Criteria Decision-Making (MCDM) techniques. Notice that everything begins with defining a *problem*, and if we were to digress a bit on this word, we'd assume something is seeking a *solution*. We could even extrapolate this to say that we aim to use optimization to solve real-world problems, thereby improving the world by producing optimized products and processes.

Consider, for example, living in a house where the electricity never goes out because supply and demand are well balanced, allowing you to continue your activities without worrying about damaged appliances. Or, taking a flight without fear because the wing is designed for safety while also saving fuel. Or perhaps, discovering a great game, movie, song, or product recommended by a store because the recommendation algorithm truly understood your tastes.

You see, in each of these examples, optimization is happening in some form. And here, we want to ensure you understand that part of your role involves identifying where you can use optimization in your work and implement it correctly. Not everyone reading this book will design an airplane wing every day, but we can also design a logistics system that delivers all products on schedule, a production line that reduces idle time and increases financial returns, or perhaps calibrate a complex Machine Learning (ML) model.

That said, to apply these techniques in your work, there are still some issues we need to discuss.

6.2 Organizational Context

Depending on your field, you might at some point in your professional life hear terms like *Enterprise Engineering*, *Enterprise Architecture*, *Business Architecture*, and others. Fundamentally, the intent is similar: we want to create effective processes that align well with what the organization does and aims to do.

It's not uncommon to see someone in a technical field wanting to implement something they've just learned at their company, like a new programming language or framework. However, this thinking can be somewhat naive. Imagine, for instance, a small engine factory. This small factory is located in your hometown and has only one plant and one office. The Information Technology department consists of a team of just five people. This, as you might presume, is of a different magnitude than a tech unicorn where the Information Technology (IT) department has thousands of employees, or a multinational industry with billions of dollars available for investment.

Notice that the level of competition is different, and, consequently, so are the processes. If the team has five people, the departure or arrival of even one team member has a greater impact on this organization than it would in a multinational corporation. And, with a five-person team, the pool of technical knowledge is more limited. It's important for people to work with the same programming languages and code frameworks—after all, the job would be much more stressful if everyone had to learn different languages just because someone in a coordination or management role decided to reinvent the wheel. It also wouldn't make sense for the company to invest in management software that costs millions of dollars a year to maintain when something simpler could offer better cost-effectiveness.

We mention this so that you understand, before we move on to the next sections, that we don't necessarily always want the most complex optimization algorithm or the one that always achieves the best results from a technical standpoint. Sometimes, we might prefer a technique that is more robust, or yields results that are easier for laypeople to interpret, or is simply faster or lighter to run because the organization's computers aren't as powerful. Therefore, the way optimization is perceived, applied, and implemented varies drastically depending on the environment.

6.2.1 Small and Traditional Organizations

Let's start with those organizations that have been in the market for a long time, with well-established processes and a culture of "we've always done things this way, why change?". You know those places where processes, ideas, budgets, and objectives rarely change? Well then: we'll call these cases "traditional organizations."

In these organizations, we can apply optimization techniques, though it is rarely without any issue: significant challenges like resistance to change, potential data infrastructure limitations, and internal skepticism are often present and must be carefully considered rather than downplayed. We won't revolutionize these organizations with these techniques, but we can introduce small, **incremental** efficiency improvements. Imagine, for example, reducing the time for a process by 5%, or cutting waste by 5%. Although this might be a relatively small percentage, it can translate into significant economies of scale.

Here, our focus isn't to change the world overnight. Whether due to political issues or the relatively rigid processes of these organizations, don't expect to train an optimization algorithm that will cause a massive impact on the company. Firstly, there's the issue of resistance to change: it's natural for a certain skepticism to exist in these places regarding improvements suggested by our model, and the potential depth of this skepticism and resistance should not be underestimated.

Secondly, there's technology: it's common for these companies to view software not as an investment, but as a cost. Therefore, you'll likely encounter some obstacles to using software that would allow you to quickly implement an optimization algorithm. Some companies restrict software use beyond the basics, like spreadsheet software and an internet browser. Others limit which users can access specific optimization software the organization purchased years ago, still needing to justify that purchase by restricting the use of other tools. And still others simply see no reason to use software beyond what's already in daily use—that feeling of "*It works well with what we have. Why do it differently?*". On top of that, you won't always have all the necessary data available at that moment. Just as some data scientists may have struggled with in the past, there are places where data sources, like databases and spreadsheets, are disconnected, outdated, or simply inaccessible.

That said, for this type of organization, it might be worthwhile to start by proving value on a smaller scale. In these cases, you could use a simple tool like Excel Solver, or a free Python library, for example. In such scenarios, you wouldn't need to ask your organization to purchase or assign a paid license for you and your team, nor would you face significant hurdles with Information Security approvals. It might also be interesting to start with a simpler optimization problem, like a single-objective linear one.

Tip

Therefore, for these companies, try to focus on well-defined problems with clear gains, without promising the world. Aim to use the tools already available to you, and have the patience to explain your ideas thoroughly so they don't falter at the first hurdle.

6.2.2 Small and Non-traditional Organizations

We considered calling these "dynamic organizations," but opted for this less creative name. You see, rigid organizations can come in various sizes and ages. On the other hand, we can have organizations that defy the descriptions from the previous section: those with lean teams that embrace planned changes with fewer headaches. Note that *accepting change* doesn't equate to being disorganized; we're referring to cases where people decide on, expect, and even encourage changes due to market competition or company culture itself.

Here, we'd like you to think about companies that can't yet burn through much cash but have a strong desire and capacity for rapid growth. These are mostly small startups. In these companies, the goal is to produce things and make a positive impact quickly.

Consequently, we can't spend millions of dollars on expensive software or large sums on computational resources, like costly computers that would run our algorithm for many hours. To add to the challenge, many of these companies may have limited data due to being smaller and newer. It's also common for professional roles to be less defined: an Information Technology employee often performs the duties of a Data Engineer, Back-End Developer, Front-End Developer, DevOps Engineer, and even Project Manager. However, since we haven't yet invented a way for the day to last more than 24 hours, this means people generally won't have much time available to concentrate on a single activity like optimization. In these situations, it's important to produce something without compromising the delivery speed of other tasks.

Well, if speed and cost are important for these places, then it's worthwhile to focus on an implementation that is good enough. Just as a multi-objective optimization algorithm would *approximate* the true Pareto front, and sometimes it wouldn't be worth letting the algorithm run for additional hours in exchange for a tiny improvement on the front, the same applies here: sometimes a solution that is adequate (though not perfect) for daily operations is better than waiting hours or even days longer for a solution that is only slightly better.

It's also worth considering a solution that can be prototyped. It's common in these companies to test minimum viable products (MVPs): products that aren't perfect but prove that our idea works. And, considering the cost, it's also frequent to use a free and open-source solution for this.

Tip
Therefore, for these companies, consider MVPs where you favor good speed to prove the idea works, perhaps using a programming language and libraries that don't require paid licenses.

6.2.3 Large Organizations, Smaller Optimization Teams

Here, think of those gigantic organizations, like a multinational industry. These companies have tens or hundreds of thousands of employees and handle billions or trillions of dollars. However, this doesn't mean they have a large optimization team or a large Data Science team. Thus, it's an organization that is large on the outside—that is, to the market. But it doesn't mean it's a giant in the field of optimization. Consequently, it doesn't imply you'll have hundreds of other colleagues to support you with technical knowledge.

This type of scenario can be quite interesting. Consider this: we often have a lot of data and many opportunities, but little knowledge spread throughout the company. A single-objective optimization solution with a 0.5% profit improvement for a small company is, in fact, an almost imperceptible and hardly quantifiable gain. Sometimes this amount would barely cover the software execution costs and the time spent by the employee (in this case, you) to implement the problem. Now, for these large companies, the story is different: 0.5% can mean millions of dollars more or less.

Hence, there's high potential in any work you might develop. However, we face two main headaches: the first is related to systems. Whether due to contractual issues or Information Security, it's likely you'll encounter obstacles to using software that would allow you to quickly implement an optimization algorithm. Some companies restrict the use of previously unapproved software. Others limit which users can access specific optimization software the organization has already acquired, to manage costs and user license distribution. And in still others, you can only use software that the company has previously authorized.

The second headache is related to people. Often, teams work with a schedule of projects to be developed, and creating an optimization problem means, first and foremost, proposing time in the team's schedule for this project to be developed. And, from an accounting perspective, someone (or some other team) will have to cover your salary and any software expenses you might need. Therefore, the challenge is to convince other people and other departments that this optimization project is important.

That said, there's also a concern about ensuring the problem is implemented correctly. Let's return to our example of a 0.5% improvement in results. Although ideally a small 0.5% improvement can be beneficial, if our problem is poorly formulated, we could see a *worsening* of 0.5% or more in the results. Thus, it's also important to have a good understanding of the organization's processes to avoid losses that could culminate in a mysterious dismissal. Consequently, it can be interesting to start with an optimization project with a very reduced and well-defined scope to mitigate all these risks.

Tip

Therefore, for these companies, the challenge is more political than technical: we need to convince the right people that such a project is good and worthwhile. Once this first challenge is overcome, we also need to persuade them that we require access to the data available in the company's systems, as well as the necessary software to implement an optimization problem. Furthermore, we must thoroughly understand the business problems: large companies have complex rules that need to be correctly translated into a mathematical optimization formulation.

6.2.4 Large Organizations, Large Optimization Teams

We can also encounter scenarios of giant companies with large optimization (or Data Science) teams. This is the case for major banks, airlines, or multinational logistics companies. If you're looking for a career that uses these optimization concepts, this can be an excellent starting point. After all, you'll have a vast amount of data and a large team of professionals to show you how these things work in practice.

Here, the problems are complex and large scale. Thus, just as with the other large companies mentioned in the previous section, any small percentage improvement translates into savings of millions of dollars. However, the difference here lies in complexity and variety: we might have teams specialized in certain subtypes of optimization problems, or professionals with academic and professional experience in specific optimization subtypes. Consequently, there's a diversity of knowledge and problems, which is positive.

Perhaps paradoxically, the challenge can be one of change and innovation: in some of these companies, optimization processes are so consolidated and politically important that any suggestion might face significant resistance. And secondly, there's the organization itself: since many people work with optimization, silos exist. If you enjoy evolutionary algorithms and your team primarily works with linear programming, it will be difficult to practice the type of algorithm you prefer in this context. Therefore, the variety of problems you'll work on might be somewhat limited.

Tip
That said, for these companies, it's important to have a solid technical foundation, such as knowing the different families of optimization algorithms. Or, perhaps, being familiar with different libraries and programming languages that use optimization, as well as good software practices for accessing and refining data to an acceptable quality. With this, it's expected that you can propose improvements and research new things: a combination of autonomy in research, autonomy in executing algorithms, and autonomy in conversing and negotiating with people.

6.3 Applications

Just as we can subdivide the nature of work by company type, we can also subdivide optimization challenges by role and background. That is, each organization has specific interests and challenges related to optimization, and you can use your knowledge for this. Some companies are aware of these challenges, while others are not. Therefore, understand this section in two ways: first, to understand what kind of optimization challenges you might encounter in these places, and second, to propose challenges like those mentioned here if you are in one of these organizations and seek inspiration.

6.3.1 Industrial Engineering

This is a good starting point. After all, the objective here is to improve the productivity and efficiency of organizations. Problems dealing with cost reduction, quality maximization, and waste minimization fit here, for example. Furthermore, many of the examples we discuss in optimization studies are specific to Industrial Engineering: we maximize and minimize the operations of organizations, not necessarily of a single part or component. It's not that one is more or less important than the other; after all, no one would want to board an airplane with a poorly designed wing. We merely comment on the field's natural concern with optimization and the ease of visualizing optimization examples with a broader scope.

That said, let's begin with single-objective optimization problems. One of the most classic problems is the minimization of **production costs**. Here, the idea is simple: we have one or more factories where we need to produce as much as possible and, preferably, those products with the highest added value. However, we have constraints to obey: just as in the donut shop example at the beginning of the book, we need to respect constraints such as market demand, the production capacity of machines and workers, and the availability of raw materials.

Alternatively, still within the factory context, we might want to minimize the lead time, which is the total time required to produce one unit of a product. Minimizing this time is beneficial for making products available more quickly and thus increasing efficiency in the number of products manufactured per day. Here, the operational variables and constraints of the factory are similar to those in the production cost minimization problem, but the objective changes.

Still in production, sometimes we want to minimize the idleness of machines and people, or reduce the total time to complete all tasks in a given process. In these cases, we would have a job scheduling problem, which we've mentioned before. It is, indeed, an optimization problem whose objective is to find the best possible allocation of people, machines, and other resources for a predetermined number of tasks.

That said, we can also apply this logic to problems outside the production line. For example, we can use optimization algorithms to solve vehicle routing problems. Here, we want to distribute transit routes in a balanced way for a fleet of vehicles, aiming, for instance, to ensure all points the vehicles need to visit are covered, that we visit all points with the shortest possible distance traveled (or the shortest time), or that we are minimizing the total cost for the entire fleet to serve all points. In such cases, it's common to have constraints like vehicle load capacity, employee working hour limitations, or restrictions on vehicle loading and unloading times.

Similarly, we can also use multi-objective optimization algorithms. Consider, for example, a scenario where we're working in a multinational company and need to choose suppliers for our supply chain. Here, we'd like to simultaneously minimize acquisition cost and delivery time. That is, we want to acquire products at the lowest price, but we'd also like them to be delivered as quickly as possible. Or, if we're in a situation where these two objectives aren't necessarily conflicting, we could consider minimizing acquisition cost and maximizing product quality. Or even a mix of several objectives, such as cost, quality, delivery time, and supplier delivery consistency (after all, a supplier who promises fast deliveries but rarely fulfills them is of no use, right?), among others.

We could also consider a factory layout problem. As you know, when designing a factory, we need to think about where equipment, safety devices, workers, storage spaces, and other items will be located. We can't distribute things in a disorganized manner: we must respect the laws of Physics (two objects cannot occupy the same place at the same time), safety regulations (like the location of fire extinguishers and other safety devices, as well as spacing between objects and people), and management directives (maximizing profit). In this case, it would be beneficial to simultaneously reduce material movement between different production stages, maximize layout flexibility for future adjustments and expansions, minimize the time taken to transport products from one production stage to another, and minimize worker movement, among others. Thus, these are multiple layout objectives being optimized in parallel.

Or, imagine an integrated manufacturing project for a large industry. Here, we want to maximize the net operating return. I don't know if you've ever talked to someone in Accounting, but this isn't just a simple subtraction between what you sold and what you bought. We have to consider the purchase of raw materials, storage costs, transportation costs, processing, contractual penalties, and depreciation, among other factors. Therefore, this type of problem can be modeled as a complex multi-objective optimization problem.

That said, there are also new opportunities for optimization. Remember when we mentioned earlier that, depending on the type of organization, we could consider different kinds of influence that optimization could bring? Well, after the events of the early 2020s, such as the COVID-19 pandemic and isolationist trends in some countries, it has become a concern for some companies to model their optimization problems so that their supply chains not only seek to minimize cost and maximize efficiency, but also aim to maximize robustness (i.e., ensuring operations continue even if a new conflict occurs in a country affecting the production chain, or another global event like a pandemic), or consider fiscal and environmental impacts (like trade wars and natural disasters). These cases often require specific optimization techniques such as stochastic optimization or robust optimization, which involve distinct modeling approaches and algorithms beyond standard deterministic methods.

We might also be interested in optimizing a logistics chain from an environmental perspective (i.e., reducing greenhouse gas emissions) or considering fiscal and practical aspects. Consider, for example, a company with a fleet of trucks that is acquiring some electric vehicles. In addition to the environmental benefit, in some locations, there's a cost saving per kilometer driven. However, the range and recharging time of these vehicles must be taken into account in a routing problem. Thus, there's a conflict between the economy and environmental benefits of these vehicles, on one hand, and the practicality and availability time for their use, considering range and recharging time, on the other.

Or, sometimes, we might be interested in optimizing a production chain geared toward customized products. That is, some industries manufacture a large quantity of identical products (or, more directly, stock keeping units, or SKUs). Other industries, however, have a business model focused on product personalization in terms of size, color, prints, finishing, or other physical attributes. This type of demand needs to be considered for the business model to be sustainable or, in other words, to have an economically sustainable production chain despite the added complexity of personalization.

That being said, it's worth investing time to develop an optimization problem that can more adequately reproduce the real world. We know we previously mentioned the need to start with a small optimization problem with a reduced scope depending on the company, and this statement doesn't negate the other. Once the model is validated, some companies are interested in quickly scaling the problem. For example, a logistics company might be interested in optimizing vehicle routing, taking into account real-world complexities like product volume, vehicle capacity, and vehicle wear and tear, among others. If in this example we only considered weight and ignored dimensions, we could produce solutions that are optimal in theory but useless in practice. An Olympic four-person canoe is light enough to fit in a light commercial vehicle like a Ford Transit, but it wouldn't fit due to its length.

This being the case, an open question remains: many companies invest in more sensors, better equipment, and more data. In theory, having a wide variety and high velocity of data allows for better optimization models. We could even build more dynamic optimization problems that adapt to real-world changes. However, on the other hand, we also face an obstacle regarding the integration and processing of this data. Different equipment providers have their own rules and standards for accessing and processing this data, and not everyone who understands business rules well and has an engineering background also fully grasps the technical aspects (read: concepts of software engineering management and operation) to adequately operationalize this. This is where one of the conflicts begins between the ideal world of consultancies, sales engineers, and professional social media influencers, and the real world of employees in companies' Engineering departments: the discourse between the ease of product acquisition and integration versus the reality of technical hurdles for this to materialize.

6.3.2 Mechanical Engineering

Now, let's consider the field of Mechanical Engineering. Here, the scope varies widely: we could work on the scale of microelectromechanical devices or on the scale of vehicles and factories. We want to minimize or maximize concepts like strength, energy efficiency, and weight.

Following the pattern of Industrial Engineering, let's start by discussing opportunities for single-objective optimization. Probably the most elementary case lies in minimizing the volume or weight of structures. Think of components like beams, springs, or shafts. In these cases, we want to reduce the material used to produce these items, which, in practice, translates to lower costs and less weight or volume. These objectives are subject to safety constraints and operational limits, such as maximum deflection or maximum supported load.

Or, perhaps, we could focus on maximizing efficiency. In moving parts like bearings, we want to reduce energy losses due to friction or other reasons. Naturally, we'll need to respect constraints such as the equipment's operating ranges.

That said, we could work with both multi-objective and single-objective optimization for problems involving composite materials. In these cases, we might want to optimize the fiber orientation, or simultaneously maximize the natural frequency and the safety factor against material failure, for example.

Furthermore, we could absolutely consider a multi-objective optimization problem for suspension and engine designs. Consider an internal combustion engine, for example, in these cases, we might want to maximize torque and thermal efficiency at the same time. Or, perhaps, simultaneously minimize pollutant emissions and increase power. In the case of suspensions, we could aim to minimize the cost and the vibration transmitted from the road to the vehicle's body. Notice that these objectives are also conflicting: increasing suspension comfort can imply more expensive materials and a more complex system. Both scenarios negatively contribute to cost, and vice versa.

Lastly, we also mentioned heat exchangers at the beginning of the book. In these cases, we could go beyond a single-objective optimization problem that only reduces costs and convert it into a multi-objective optimization problem aiming for cost reduction and maximization of the heat transfer rate, as well as improving other variables equally relevant to the engine.

Considering these aspects, observe that optimization problems specific to this area shift focus from time to cost and, depending on the case, to strength, efficiency, performance, and sustainability. Additionally, it's possible to use more advanced optimization techniques to broaden the possibilities within a mechanical project. We could, for example, have projects employing computational fluid dynamics (CFD) or additive manufacturing (like 3D printing) to produce design and performance proposals for components that were previously unimaginable. In fact, additive manufacturing breaks many of the constraints imposed by traditional manufacturing processes, making the materialization of these complex shapes possible. These cases often rely on derivative-free optimization, surrogate modeling, or topology optimization.

6.3.3 *Civil Engineering*

Civil Engineering is also a very interesting field for optimization. Construction projects usually consume a large amount of financial resources. Pragmatically speaking, construction works are large projects where we coordinate people and materials. Some stages are prerequisites for others, and some need to be executed in parallel. In some cases, we need to order certain materials well in advance, and these aspects must be considered in a project. Large-scale works, such as buildings, bridges, ports, airports, and transportation routes, also affect society.

That said, we could have two keywords here: **safety** and **planning**. Structures need to stand for decades, but we also need to ensure that the work is well structured, planned, and executed.

With that, let's start with single-objective optimization problems. Here, we could begin with the problem of determining the optimal route for a road. We might want to minimize earthworks (cuts and fills) in the terrain for our road, which is nothing more than a cost minimization problem. We will also need to obey constraints such as slope, curves, and the minimum speed that the section would allow.

Or, perhaps, we could think about reducing the cost of a reinforced concrete structure. In these structures, we need to distribute elements like beams and columns to respect all technical standards, such as safety constraints and the use of the environment. On the other hand, these elements are expensive; therefore, finding the best distribution of these elements to reduce costs (and, sometimes, the construction time) is very important.

Speaking of construction time, we could also consider it as another single-objective optimization problem. We could think of optimizing the schedule to execute all stages in the shortest possible time while respecting constraints such as the order of tasks and the availability of schedules and weekdays for the teams.

Attention

Notice that the problem of task distribution for a large construction project is very similar in nature to distributing employees in a small factory or distributing teams in a small company. Although for the real world these problems have different orders of magnitude, sometimes for an optimization algorithm they have very similar structure. However, we can have very different algorithmic solutions (e.g., exact methods for smaller instances versus heuristics for larger ones) with critical differences in scale, data characteristics, and the potential presence of uncertainty between different companies.

Therefore, realize that sometimes a problem can be translated to another area of knowledge—interpret this as thinking outside the box. Also, always consider how these practical differences in scale and context impact algorithmic choice and actual complexity.

That said, it's also feasible to think about multi-objective optimization problems in this area. As an example, we could address an optimization problem for concrete slabs to simultaneously minimize cost and time.

Or, we could also consider a sustainability component in projects. Let's consider a scenario where we need to design a large bridge or viaduct. These places will be subject to a large volume of vehicle traffic, and therefore we need to consider safety as a primary prerequisite. Respecting safety, one might assume that the first objective is cost reduction, which in itself is true. On the other hand, we could also optimize for sustainability: between two equally safe and inexpensive bridges, it would be preferable to choose the one that generates less pollution during its construction and use, wouldn't it?

That said, we can also consider sustainability in other ways. Consider, for example, a water distribution company. Of course, we want to minimize the costs of water treatment and distribution, but we might also want to simultaneously minimize the environmental impact. Or, perhaps, maximize the resilience and service level of the network.

Alternatively, we could apply our knowledge to a project for the requalification of urban areas. In such projects, it's also in our interest to reduce the implementation and maintenance costs, but we also want to conceive a solution that is good for society. In other words, this means we would like to optimize, along with cost, the environmental impact by maintaining green areas and reducing waste generation and pollution, energy consumption, and maximizing the use of urban space.

This being the case, we can consider the future influence of optimization algorithms here in two ways. Firstly, Civil Engineering projects inherently deal with conflicting short-term and long-term objectives: sometimes, a current saving can turn into a large expense in the coming years or a reduction in the project's lifespan, for example. On the other hand, we can also consider applying our techniques to balance concepts of cost, performance, resilience, and sustainability in our projects. While we want inexpensive solutions, we also want solutions that are efficient and resistant to climate change. This statement can be applied to various contexts, such as Smart Cities projects, building projects in large cities suffering from climate change, or transportation network projects that need to withstand time and weather.

6.3.4 Chemical Engineering

Optimization can also be used here. In fact, we could consider using the techniques discussed in this book for various stages of plant operation or in the development of new products.

Following the pattern of other Engineering fields, let's begin with single-objective applications. One of the classic problems in this area is the optimization of a chemical reactor, considering variables such as temperature, pressure, reactants, and catalyst. Here, the goal is to maximize the production and efficiency of a specific product while adhering to constraints like the reactor's operational safety limits or the technical limitations of the equipment.

Similarly, we could adopt optimization concepts in production processes. For example, we could consider a problem of minimizing **resources** like steam, water, and electricity in a distillation or extraction process within a chemical plant. Likewise, the same plant might aim to maximize the efficiency of pollutant removal. As an example, we could consider an Advanced Oxidation Process (AOP) for degrading phenol into less toxic substances, taking into account process conditions like temperature, pH, and the initial phenol concentration.

If you have some familiarity with the field, you probably know that petrochemical plants usually involve many heating and cooling processes. Heat exchanger networks (HENs) are used for heating and cooling. Therefore, this presents a problem that can be either single objective or multi-objective. If it's single objective, we might be interested in designing a network that minimizes energy use while meeting minimum technical requirements. If it's multi-objective, we could also consider other objectives such as investment cost and heat recovery between hot and cold streams.

That said, we can also consider the process of choosing a solvent for a process in a chemical plant. The "best" solvent in this case may need to satisfy conflicting objectives such as cost, environmental impact, solvent performance, and safety in handling the solvent.

Furthermore, we can apply multi-objective optimization techniques to a problem of operating an oil refinery. In refineries, there are specific safety and operational constraints, and while respecting these, we need to optimize conflicting objectives such as minimizing storage tank changeovers, the time an atmospheric distillation unit (ADU) is not fed by tanks, and the number of hours the pipeline remains idle.

This being said, it's also common to think about multi-objective optimization problems considering environmental impact. We could, for example, continue producing a product with maximum efficiency in the production process or product quality, but also taking into account the minimization of environmental impacts in the production process.

Remember we mentioned at the beginning that one of the focuses here is efficiency, right? Well, optimizations in chemical processes at a molecular level can trigger a cascading effect. For instance, a more efficient chemical reaction implies less energy and fewer unwanted byproducts. In turn, we can design smaller reactors with fewer safety risks and lower infrastructure investments.

This oil case is interesting for highlighting the MCDM techniques we saw earlier. You see, complex chemical processes imply a high degree of uncertainty and variability. In these cases, we can prioritize applying robust optimization techniques and also access all solutions on a Pareto front so that plant operators have the flexibility to choose the best solution considering operational trends and future uncertainty.

6.3.5 Electrical Engineering

As in Chemical Engineering, the focus here is also on **efficiency** (now energy efficiency) and **safety**. The scale also varies: we can work on optimization problems in motors and signal processing up to large-scale power generation, transmission, and distribution.

Thus, let's begin with the scope of electric motors. Here, we have at least two classic optimization strategies. The first deals with maximizing the energy efficiency of a motor or, in other words, minimizing losses at a specific operating point or over a driving cycle. These losses can be caused by winding resistance or parasitic currents, for example.

Tips
Notice that a business problem dealing with "maximization" doesn't always translate into a maximization optimization problem. What a company might understand as "maximization of efficiency," an engineer might interpret as "minimization of losses." Here, maximization can be a consequence, not an end goal.

Another case concerns torque. In many applications, there are constraints on volume, weight, and cost, as is the case with some motors used in factories and vehicles. Respecting these constraints, we would need to design a motor that can provide the highest torque despite these limitations.

Naturally, it's also possible to apply optimization techniques in situations with a broader scope. As a first example, suppose we need to optimize an industrial plant to reduce electricity consumption. There are some space constraints and some equipment that really cannot be replaced. But what if we had the freedom to test repositioning some equipment or even replacing the equipment itself with more updated versions? In these cases, we could analyze the impact of changing the position or the equipment for refrigeration, compressed air, and lighting, for example.

Similarly, we can consider adopting multi-objective optimization algorithms for an energy transition considering sustainability, smarter, more flexible, or resilient grids. Or even a combination of all these factors. Consider, for example, an optimization algorithm used for developing an expansion plan for an energy distribution network with distributed generation. In these cases, it's common to use renewable energy sources that have variability in energy generation. Here, we want to minimize the costs of operating the network while simultaneously maximizing network reliability and reducing short-circuit levels and reverse power flow.

Alternatively, we could design an electronic power converter. Here, we traditionally have two conflicting objectives: on one hand, we want to minimize the total cost of the converter. On the other, we want to maximize power density and energy conversion efficiency. Similarly, an electric motor design can be conceived as a multi-objective optimization problem. In addition to minimizing losses or maximizing torque, we could simultaneously reduce the total mass of the motor and torque fluctuations.

This being the case, observe that the objectives in this area are interconnected and frequently conflicting. A motor or project that minimizes cost will rarely have the highest possible efficiency and reliability. And, on the other hand, a motor with maximum efficiency might be too heavy for the project. Or, a project with the highest power density might not always be the most reliable. Note, for example, the case of a distributed generation network. If we have low initial investment, we will have problems with reliability and maintaining good voltage levels in the future.

Furthermore, there is a trend in applying optimization in a more obscure way. You see, it's common to associate ML algorithms with smart grids and the Internet of Things (IoT). For example, we use ML models to predict energy generation trends, consumer demand, and equipment usage peaks. Here, we can use the predictions from these models to establish the variables and constraints for optimizing these networks. Or, we can even optimize the ML algorithms themselves, as we will see starting in Sect. 6.3.8.

6.3.6 Aerospace Engineering

In Aerospace Engineering, we deal with problems subject to wide variations in operating conditions such as extreme temperatures and large temperature ranges, high altitudes and vacuum, high speeds, and significant changes in atmospheric pressure. Here, there are many requirements for safety, reliability, performance, lightness, and efficiency.

Thus, let's start with an example mentioned throughout the book: one of the most common forms of optimization is in the design of an aircraft wing, as well as other parts of its surface. In these cases, we want to minimize drag assuming a specific flight condition. Minimizing drag translates to less fuel consumed and, in turn, greater range.

It's also possible to determine the optimal parameters for a multivariable Direct Thrust Control (DTC) of a turbofan engine. In these cases, we could optimize the controller so that it has the smallest possible error between the thrust command and the estimated result while respecting constraints that ensure the stability of the closed-loop system.

Or, we can employ optimization to find the ideal geometric shape of a wind tunnel contraction that minimizes the boundary layer height at the exit, while simultaneously ensuring that the flow remains attached to the contraction walls and respecting the limits of flow non-uniformity at the exit.

And, just as we saw in other Engineering fields, we can also apply multi-objective optimization here. In fact, perhaps the best known application is in *Multidisciplinary Design Optimization (MDO)*. Consider the following: any large project in any large company has political and technical conflicts. Sometimes, the management of one area never reaches a consensus with the management of another area. In other cases, the objectives of two departments are conflicting and they actively do not communicate with each other. And, sometimes, some areas have their own constraints. Let's imagine, for example, a food industry. The sales department might promise and sell products without necessarily worrying about delivery times and the production process, as it's in their interest to sell as much as possible considering the area's targets and commissions. The logistics department is concerned with ensuring everything is delivered completely and on time, but this requires prior planning. And the production department is concerned with ensuring that raw materials arrive in the correct quantity and are consumed in a timely manner, respecting the entire production plan. If the areas don't communicate, don't have shared goals, and aren't willing to concede some of their own objectives for the greater good of the organization, we would hardly have an optimized organization.

We mention this case to transplant this idea to the MDO problem. An aerospace vehicle is complex, and its performance depends on the interaction between different areas. The team responsible for aerodynamics would have the ideal vehicle design based on its own concepts, but this optimal design would not necessarily meet all the objectives of the structures team, which, in turn, would not necessarily meet all the objectives of the materials team, and none of these teams would likely make the cost management team very happy if they had complete freedom to do whatever they wanted.

Therefore, the idea of MDO is to optimize the vehicle's design by simultaneously considering all these teams and their conflicts. Sometimes we want a vehicle that, at the same time, has the lowest gross weight, the greatest range or endurance, and the lowest maintenance cost, for example.

Alternatively, it's possible to think of applying MDO to the integrated aero-structural design of a wing. That is, simultaneously optimizing the total weight and drag of the wing while respecting constraints such as structural elements, the aerodynamic shape of the wing, and limits on wing stresses and deformations.

Furthermore, with the advent of satellite constellations, we have the possibility of using multi-objective optimization techniques to find the smallest possible number of satellites that can simultaneously maximize service coverage, minimize communication latency, and minimize the total cost of launching and operating the satellites.

Moreover, it's possible to employ optimization techniques for modern trends in Engineering, such as using topological optimization techniques in the design of aerospace components with complex geometries. Here, the idea is to maximize performance and reduce weight. Another application is in optimizing the trajectories taken by shared drones in applications like precision agriculture and surveillance, for example.

This being said, there is a greater degree of complexity in optimization projects in this area—specifically, those of MDO. Firstly, due to organizational issues: we need to integrate specialists from various fields, with different levels of knowledge and understandings of the organization and interests that are often conflicting. And, even if all specialists want to cooperate, there are still challenges in integrating tools and mathematical models: teams use software that doesn't necessarily communicate with each other. Secondly, due to technical issues: modern MDO projects aim to reproduce simulations with maximum granularity and fidelity to real life, which implies a high computational cost. To try to resolve this complexity, some organizations certainly continue to practice optimization, but using more elegant approaches such as dividing into subproblems and using surrogate models.

6.3.7 Computer Engineering

Computer Engineering is an interesting case: while in other Engineering fields we saw the application of algorithms for producing physical products and works, here we extrapolate this. That is, there's an interest in projects like integrated circuits and microprocessors, but also in developing software such as operating systems, embedded systems, and large-scale software. The keyword here is **efficiency**.

In this way, let's think about a mobile device: notice that one of the concerns we had when discussing the example of choosing a cell phone in previous chapters was the battery. Most consumers would like a phone with a battery that lasts a long time, but they would also like a fast and lightweight phone. The problem here is that more powerful processors will consume more energy. A larger battery implies a larger device size and greater weight, which is not necessarily what consumers are looking for.

Thus, let's consider a slightly different and simpler single-objective optimization case: one of the main objectives when designing a cell phone or any other mobile device is to minimize total energy consumption. For this, we could manage which device components should always remain on and which components would stay in standby mode—we wouldn't need the GPS to be on all the time in a cell phone, but it's important for the antenna to function, for example. We could also modify the processor's frequency and voltage to have more performance (and, consequently, consume more energy) only when necessary.

Another scenario is in computer architecture: here, we want to design processors and memories while improving a performance metric like latency or throughput.

That said, we can also apply single-objective optimization concepts to a software project, such as minimizing the size of the code produced by a compiler, or the time needed to compile a complex project.

We could also consider a multi-objective optimization problem for designing an SoC (system-on-a-chip)-type integrated circuit. The challenge here is interesting: on one hand, we want to maximize the performance of this SoC by adding more memory, CPUs, GPUs, and other components. On the other, we want to minimize the chip area and energy consumption.

Or, we can consider a cloud resource optimization problem. Large companies spend a significant amount on operating their cloud-hosted systems. Furthermore, it's common for these companies to use different cloud solution providers in parallel, such as Microsoft Azure, Google Cloud, and Amazon Web Services. That said, for some companies, it might be interesting to optimize the distribution of applications and resources among different providers to simultaneously minimize cost and maximize availability. Or, in some cases, minimize cost and minimize latency.

Still in the context of latency, we could also apply a multi-objective optimization technique to packet routing in computer networks. Here, there are some interesting similarities with the logistics applications we mentioned on other occasions: we need to avoid traffic congestion at some points in the network, as well as avoid delays in packet delivery. Thus, we consider a problem that minimizes latency and, at the same time, maximizes data transmission.

That said, notice that the miniaturization of components implies the use of optimization since this process is not manual due to its complexity. Similarly, optimization is also used in designing software architectures, computer networks, and more complex algorithms. In most of these cases, we also note the presence of conflict between performance, power, area, and cost. More performance requires more energy or a larger chip area, and both translate into higher costs. In a software problem, area is replaced by a system reliability objective, for example.

6.3.8 Data Science

This is an interesting area. Firstly, let's consider traditional machine learning—that is, those algorithms we use to learn patterns from data, produce predictions, and make decisions based on new data. These algorithms are frequently the focus of Data Science teams in large companies, and this type of algorithm was one of the motivators for this book.

The reason is simple: although data scientists routinely work with data manipulation and the choice of ML models within their teams, in reality, optimization is at the heart of most of these models. After all, the "training" process of algorithms is, above all, about optimizing the hyperparameters of a predictive model to minimize a loss function that shows us how well a model is (or isn't) performing.

About Hyperparameters When we explored the different single-objective and multi-objective optimization techniques, we mentioned some parameters that each algorithm would have, didn't we? Similarly, you might recall that in linear regression, the coefficients β for each variable are the internal parameters that the algorithm optimizes to minimize the mean squared error. That is, they are the parameters used by the model itself to provide its predictions.

Now, what if we could go back to those optimization techniques and define parameters so that the computer finds the algorithm's own parameters? As if we could delegate to the computer the choice of the number of individuals per generation and the maximum number of generations for an evolutionary algorithm, and we would only control one or two parameters to direct this selection process. Or, perhaps, if we could control how the computer should find the values of β for a linear regression instead of defining it ourselves.

Well, this exists, and we call it *hyperparameters*. These are the values we define to guide how the training process itself will occur. Let's consider, for example, a very popular algorithm among data scientists: the *Random Forest*.

Imagine, for instance, that you're working at a financial institution and are tasked with creating an algorithm that must predict whether a customer should receive an increase in their credit limit. One (admittedly simplistic) way to do this is with a decision tree. It works like a flowchart of questions, such as

1. Has the client had an open account for more than 12 months? *(Yes/No)*
2. If yes, what is their age? *(Under 30/30 or older)*
3. ...
4. What was their average spending on the credit card in recent months? *(Less than USD 1000/USD 1000 or more).*

In the end, we'll have a decision: *Increase/Do not increase the limit.* Each question divides the data, and the tree training algorithm learns which are the best questions and in what order to ask them to properly separate those who should or should not have their credit limit increased.

Now, the idea of a random forest is of a greater magnitude. Instead of relying on a single decision tree, which can be very specific and not generalize well to new data, the random forest builds many different decision trees, each trained with a slightly different sample of your data and considering only a random subset of your variables at each split. For new predictions, the random forest consults all the trees in the forest and takes a "vote": the most voted decision or the average of the most voted decisions is the model's final prediction. Because there are many trees, the error tends to be smaller, and the robustness and predictive capability tend to be greater.

However, it's unfeasible to create all these questions manually: if it's already difficult for one decision tree, imagine for hundreds. Therefore, it's worth considering hyperparameters to assist us in this task. In the case of random forest, we have hyperparameters like the number of trees our model should have, the maximum depth of each tree, and the minimum number of samples required for each node in the tree. Note that these are not values that the random forest "learns" directly. We provide these settings, and the algorithm then builds the best forest given these constraints and guidelines.

Finding the "best" combination of hyperparameters for a model like linear regression, a random forest, a support vector machine (SVM), a decision tree, or a neural network is, fundamentally, an optimization problem. We want to find the set of hyperparameter values that minimizes a loss function, such as prediction error, or maximizes a performance metric like the F1-score.

That said, the ML algorithms we generally use in our implementations are imported from libraries like scikit-learn, LightGBM, Tensorflow, PyTorch, and

XGBoost. In these libraries, optimization is implemented in ways that, for data scientists choosing an algorithm, can seem almost invisible.

Frequently, the specific optimization mechanism an algorithm will use for training (i.e., to minimize the loss function and adjust its internal parameters) is itself a hyperparameter we can choose in the form of a string that defines which optimization algorithm will be employed.

For example, when training a logistic regression or a SVM in scikit-learn, you might find a hyperparameter called solver. The options for this solver can include strings like "`liblinear`," "`sgd`" (stochastic gradient descent), "`lbfgs`" (limited-memory Broyden-Fletcher-Goldfarb-Shanno, a quasi-Newton method), and "`newton-cg`" (Newton-conjugate gradient), among others. You see, each of these strings points to a different optimization technique that will be used internally by the library to find the best model parameters, but we don't have access to the implementation of these techniques themselves. While this abstraction is useful, this convenience can lead to a superficial understanding, and this is the reason why we recommend to emphasize that knowledge of optimizer characteristics, as it is highly beneficial for advanced ML applications. Naturally, the choice of solver can affect the quality of the solution found (how good the model is), as well as the training speed and the amount of memory used.

In other cases, especially in more complex algorithms like neural networks trained with Tensorflow or PyTorch, the optimization technique is already embedded in the architecture itself and in the training process, often selectable via simple string parameters (e.g., "`adam`," "`sgd`"). However, for data scientists and engineers, particularly those aiming for advanced practice or research, a deeper understanding of the characteristics of these underlying optimizers (e.g., their convergence properties, sensitivity to hyperparameters like learning rates, behavior on different types of loss landscapes) is important for effective model training, debugging, and innovation, and downplaying the need for this understanding could be a disservice.

That said, allow us a moment to look into these optimizers. Training a neural network is essentially an optimization problem where we seek the millions (or billions) of weights and biases of the network that minimize a loss function, which is highly non-convex and of extremely high dimension. For this task, gradient-based methods, calculated via backpropagation, are almost universally employed.

In Deep Learning frameworks like Tensorflow (with Keras) and PyTorch, we need to choose which optimization technique to use in addition to defining the architecture of the neural network itself.

Let's start with stochastic gradient descent (SGD), which we've already seen in Sect. 3.1.2.3. It updates the model's parameters in the opposite direction of the loss function's gradient, calculated over a small batch (mini-batch) of data. By working with these mini-batches, we can escape shallow local minima. However, it can be slower to converge and is quite sensitive to the choice of learning rate: a rate that's too high can cause it to "jump" over the optimum, while a rate that's too low makes training excessively time-consuming.

That said, using SGD can be a good option if you have a solid understanding of the ML problem, knowledge (which you have), and patience (which you might not have) to adjust the learning rate and its decay. Alternatively, we can use it as a baseline to compare our model with other optimizers.

Some SGD implementations also offer an additional parameter for momentum, which adds a fraction of the previous update to the current update. SGD with momentum, shown in Algorithm 1, helps to smooth out the oscillations that can occur with pure SGD and generally accelerates convergence because it allows the optimizer to "gain speed" in consistent gradient directions. Because it's more stable, it might be interesting to opt for this method over "pure" SGD if you notice it's oscillating too much or seems "stuck" in its training process. It's also an interesting alternative for more complex loss functions.

Algorithm 1 SGD with Momentum

Require: Dataset D composed of N instances
Require: Objective function $f(\mathbf{x}; D)$
Require: Initial parameters $\mathbf{x}_0$
Require: Learning rate η
Require: Number of epochs E
Require: Momentum term γ (e.g., 0.9)
Ensure: Optimized parameters $\mathbf{x}_{final}$
Initialize parameters: $\mathbf{x} \leftarrow \mathbf{x}_0$
Initialize velocity vector: $v \leftarrow \mathbf{0}$ (same dimension as $\mathbf{x}$)
for $e = 1$ to E **do** ▷ Loop over epochs
    Shuffle the dataset D randomly
    for each training example in D **do** ▷ Or use mini-batches
        Compute gradient: $g_t \leftarrow \nabla_{\mathbf{x}} f(\mathbf{x}; D)$
        Update velocity: $v \leftarrow \gamma v + \eta \cdot g_t$
        Update parameters: $\mathbf{x} \leftarrow \mathbf{x} - v$
    end for
end for
return $\mathbf{x}$ as $\mathbf{x}_{final}$

Besides SGD and SGD with momentum, we have methods with adaptive learning rates, such as AdaGrad, RMSProp, and Adam. These optimizers adjust the learning rate for each parameter individually during training. The idea here is to decrease the learning rate for parameters that receive frequent or large-magnitude updates, and increase (or keep higher) for parameters that are updated less frequently or with smaller magnitude.

Of these, AdaGrad is a gradient-based method (in fact, it stands for "Adaptive Gradient Algorithm") that accumulates the squares of past gradients for each parameter and divides the global learning rate by this accumulated sum, plus a small term to avoid division by zero. Its pseudocode is presented in Algorithm 2. It's good for problems with sparse data, like recommendation systems and natural language processing, because it better optimizes rarely seen features. However, it has a tendency to stop learning too early. This might be interesting if you explicitly want that, but generally, we want to explore a bit more in current deep learning problems.

Algorithm 2 Adagrad

Require: Dataset D composed of N instances
Require: Objective function $f(\mathbf{x}; D)$
Require: Initial parameters $\mathbf{x}$
Require: Global learning rate η (e.g., 0.01)
Require: Number of epochs E
Require: Smoothing term ϵ (e.g., 10^{-8}, to prevent division by zero)
Ensure: Optimized parameters $\mathbf{x}_{final}$

```
Initialize parameters: x ← x_0
Initialize accumulator for squared gradients: G ← 0 (vector of same dimension as x, stores G_{j,j})
for e = 1 to E do                                              ▷ Loop over epochs
    Shuffle the dataset D randomly
    for each training example in D do                          ▷ Or use mini-batches
        Compute gradient: g_t ← ∇_x f(x; D)
        for each parameter component j from 1 to dim(x) do
            Accumulate squared gradient: G_j ← G_j + g_{t,j}^2
            Update parameter component: x_j ← x_j − η / sqrt(G_j + ϵ) · g_{t,j}
        end for
    end for
end for
return x as x_final
```

We also have RMSProp (or root mean square propagation). Both it and Adam are commonly used as default methods in some deep learning frameworks. It's also a GD-based method and similar to AdaGrad. The difference here is that it modifies the accumulation of squared gradients to use an exponential moving average to prevent a large decay in the learning rate. With this, it converges faster and more stably than AdaGrad in non-convex problems. Furthermore, RMSprop can be more stable than Adam in non-stationary problems.

Then we have Adam (or adaptive moment estimation), shown in Algorithm 3. It's a combination of AdaGrad and RMSProp. From AdaGrad, Adam adopts the ability to maintain adaptive learning rates per parameter, which is useful for sparse gradients. From RMSProp, it adopts the idea of using a moving average of squared gradients to normalize updates, making it useful for non-stationary problems. It's one of the most popular optimizers because it converges quickly and performs well on various deep learning problems without needing fine-tuning of initial hyperparameters.

That said, there's also a variant of Adam, called AdamW. L2 regularization is a common technique to prevent overfitting by penalizing large weights. In traditional Adam, the L2 effect for some weights can be reduced depending on how this regularization is implemented. In AdamW, however, L2 is applied directly to the weights. In practice, this means AdamW can often achieve better generalization and performance compared to Adam with L2, especially for complex models like those using Transformers.

Algorithm 3 Adam

Require: Dataset D composed of N instances
Require: Objective function $f(\mathbf{x}; D)$
Require: Initial parameters $\mathbf{x}_0$
Require: Step size (learning rate) η (e.g., 0.001)
Require: Number of epochs E
Require: Exponential decay rates for moment estimates: β_1, β_2 (e.g., 0.9, 0.999)
Require: Smoothing term ϵ (e.g., 10^{-8})
Ensure: Optimized parameters $\mathbf{x}_{final}$

Initialize parameters: $\mathbf{x} \leftarrow \mathbf{x}_0$
Initialize 1st moment vector: $m \leftarrow \mathbf{0}$
Initialize 2nd moment vector: $v \leftarrow \mathbf{0}$
Initialize timestep: $t_{iter} \leftarrow 0$
for $e = 1$ to E **do** ▷ Loop over epochs
  Shuffle the dataset D randomly
  for each training example in D **do** ▷ Or use mini-batches
    $t_{iter} \leftarrow t_{iter} + 1$
    Compute gradient: $g_t \leftarrow \nabla_{\mathbf{x}} f(\mathbf{x}; D)$
    Update biased first moment estimate: $m \leftarrow \beta_1 m + (1 - \beta_1) g_t$
    Update biased second raw moment estimate: $v \leftarrow \beta_2 v + (1 - \beta_2) g_t^2$
    Compute bias-corrected first moment estimate: $\hat{m} \leftarrow m/(1 - \beta_1^{t_{iter}})$
    Compute bias-corrected second raw moment estimate: $\hat{v} \leftarrow v/(1 - \beta_2^{t_{iter}})$
    Update parameters: $\mathbf{x} \leftarrow \mathbf{x} - \eta \frac{\hat{m}}{\sqrt{\hat{v}} + \epsilon}$
  end for
end for
return $\mathbf{x}$ as $\mathbf{x}_{final}$

Practical Strategies There's no universally superior optimizer for all tasks, dataset types, and model architectures. Each problem will have a specific loss surface complexity, even if the problems are similar. Therefore, it's important to test and experiment to find the optimal combination for each specific scenario.

If you don't know where to start, try Adam or AdamW with their default parameters. They are robust and generally perform well. If the model is large scale (like transformers), AdamW might work better.

For Recurrent Neural Networks (RNNs), RMSProp might be a viable alternative if Adam or AdamW do not yield expected performance. Similarly, for sparse data, Adagrad could be interesting if Adam hasn't yielded good performance.

6.3.8.1 Hyperparameter Optimization

Hyperparameter optimization for a model is what can potentially bring significant improvement in results. Let's use AutoML (automated machine learning) as an example. There are market solutions that allow data scientists with limited time and individuals without theoretical experience in Data Science (whom we call *citizen data scientists*) to train multiple ML algorithms and, among them, choose the best one. With AutoML, we can automate the entire ML model development pipeline, from data preprocessing and feature engineering to model type selection, architecture optimization, and hyperparameter tuning. As an example, sometimes a random forest might perform worse than other models not because it's intrinsically uninteresting for that problem, but simply because it wasn't optimized. Sometimes, changing hyperparameters like the rules for splitting tree nodes and the number of estimators could already lead to much superior performance.

Of course, for this, we could perform a brute-force search, like *Grid Search*. Or, we can use some random search technique when we have a very large combination of hyperparameters, like *Random Search*. However, these two techniques have problems: *Grid Search* can take a very long time, and *Random Search* might randomly ignore some very good solutions.

To solve this, there are some Python libraries like Optuna and Ray Tune for optimizing hyperparameters using more intelligent search algorithms. Optuna, for example, uses sampling and pruning algorithms to efficiently find promising combinations of hyperparameters. This library supports various optimization algorithms, including Bayesian optimization and evolution-based algorithms.

Important

Bayesian optimization is an efficient approach for finding the optimum of black-box functions, such as the function that maps hyperparameters to a model's performance. It's useful in situations where evaluating a function is expensive, as is the case with training an ML model. After all, to evaluate each combination of hyperparameters, we have to invest several minutes or hours just to train that model, don't we?

Bayesian optimization works by building a probabilistic model to approximate the objective function (i.e., the model's performance as a function of hyperparameters). This model is updated with each new evaluation (i.e., tested hyperparameters and their result). Then, an acquisition function is used to decide which next set of hyperparameters to evaluate. Here, we balance exploration (evaluating hyperparameters in regions of high uncertainty) and exploitation (evaluating hyperparameters in regions where the surrogate model predicts good performance).

In each iteration, we evaluate the objective function with the hyperparameters suggested by the acquisition function, update the surrogate model with the new result, and repeat the cycle. Thus, we use information from past evaluations to make smarter decisions about where to sample next, and we find a good solution using few evaluations.

Ray Tune is another popular library that facilitates the parallelization and distributed execution of optimization tasks. It integrates with various other machine learning libraries and offers several search algorithms, in addition to allowing the implementation of advanced pruning logic and resource scheduling. It's common to use Ray Tune for large-scale optimizations, where multiple trainings can be executed concurrently on different CPU cores or even on clusters of machines.

That said, allow us to return briefly to the specific discussion of neural network optimizers: one of the main parameters we use in optimizers is the learning rate η. There's a technique called *learning rate scheduling* that allows us to reduce the learning rate over time. The intention here is simple: at the beginning of training, a higher learning rate can help the model converge more quickly to a good region of the parameter space. However, as the model approaches the optimum, a smaller learning rate is preferable to allow for finer adjustment and prevent the optimizer from overlooking the minimum. Implementing a learning rate scheduler generally leads to more stable convergence and better final model performance, without the need to manually find the "perfect learning rate."

6.3.8.2 Multi-objective Optimization

It's true that single-objective optimization is the basis for training many ML models, but this doesn't mean it's not interesting to use some multi-objective optimization technique. Let's go back to the AutoML case we mentioned earlier: more recent AutoML software (and research on explainability in ML) allow us to balance two conflicting objectives: accuracy and interpretability. Simpler models, like decision trees or a linear regression with few variables, tend to be more interpretable but may have more errors. A neural network model, on the other hand, might have fewer errors but is difficult (or impossible) to explain to regulatory bodies, laypeople, or stakeholders in general. With MOO, we could find a model that balances these two conflicting objectives.

Similarly, there are other conflicting objectives that can be optimized together using MOO techniques. The first case is balancing accuracy and robustness, where a model with high accuracy might be fragile to small perturbations like adversarial attacks. The second case is between accuracy and *fairness*: a model might have few errors, but it may have been trained on data with socioeconomic biases and could perpetuate these biases. The third case is between model performance (accuracy) and computational performance (latency or memory usage): in cases where we need to use this model on devices with limited computational resources, sometimes it's not in our interest to have a model with the best possible performance, but rather a lean model that balances accuracy and efficiency.

6.3.8.3 Generative AI

The focus of this area is on creating models capable of generating new, original content that *resembles* the data they were trained on. While in traditional ML we worked mainly with predictions and inferences from values in tables, images, and texts, here we produce content in the form of articles, texts, messages, code in programming languages, images that look like photographs, synthesized music, videos, and more. Naturally, optimization is also relevant for training these models. Many generative AI models, for example, were trained with frameworks like Tensorflow and PyTorch, which means, in practice, that the concepts we saw earlier about optimizers also apply here.

Therefore, the first classic case of using single-objective optimization here is in the loss function for training these models. Many LLMs are trained using a language modeling objective such as, for example, maximizing the probability of the next word (or token) in a sequence, given the preceding words. Considering the size of modern LLMs (with billions or trillions of parameters at the time of writing this book) and the large volume of data required for their training, optimizing the efficiency of this process is another challenge. Here, we also want to minimize the effort to train a model, whether in GPU hours, energy consumption, or total cost. For this to happen, we return once again to the optimization mechanisms we saw earlier, like gradient-based methods.

That said, the biggest challenges in generative AI arise when we start treating it as a multi-objective optimization problem.[1] You see, the output of a generative model should be high-quality content, but it should also be diverse. Consider, for example, a request to create an image of a garden with some people talking. It wouldn't be good if the people were floating or if they were in positions and actions that wouldn't make sense in real life. Similarly, it wouldn't be interesting to have many images within a small set of variations, much like the notion of diversity we introduced when we talked about the concept of a Pareto front. Notice that these two objectives can be conflicting: many diverse images among themselves don't necessarily maintain the same level of quality, and very high quality might discourage diversity. Furthermore, we also have the following challenges:

- **Quality and Fidelity to Instruction:** It's no use if the result is good but doesn't meet what was originally requested, right? A model excessively focused on the instruction might produce rigid results, while a model focused on overall quality might ignore nuances of the request.
- **Quality and Fairness:** Just as in traditional ML algorithms, we also want to generate content that is fair and doesn't reinforce harmful stereotypes, such as automatically associating certain professions with a specific gender. However, this reduction of biases might be perceived by some metrics as a reduction in fidelity or quality compared to the training data patterns if the latter are biased.
- **Efficiency and Quality:** Users of these models would like to have fast responses, but a high-quality model can be slower and more expensive to run. Therefore, for real-time applications or those that will run on limited computational resources, it may be necessary to optimize for a balance between output quality and content generation speed.

[1] In these cases, the Pareto front would represent the models or model hyperparameter configurations that are optimal considering two or more objectives.

6.4 Conclusion

A recurrent theme in this chapter is the multidimensional impact of optimization within organizations. Whether in a small startup or a large, established corporation, the ability to identify optimization opportunities and tailor solutions accordingly is a skill of an effective engineer or data scientist. To demonstrate this, we showed how engineers and data scientists in different organizations can apply optimization for innovation and problem-solving. These examples, ranging from minimizing manufacturing costs to maximizing energy system efficiency, and from refining machine learning model accuracy to navigating ethical considerations in generative AI, illustrate how theoretical algorithms and mathematical formulations from earlier chapters connect to industrial realities.

Furthermore, this chapter points out that successful optimization extends beyond mere algorithmic application. It involves a thorough understanding of specific business needs, the organizational context, and the ability to communicate complex results effectively to various stakeholders. We also wanted to show the close relationship between optimization and data science, particularly in training and tuning ML models and addressing the multi-objective challenges of generative AI. Consequently, we envision a future where these fields will become even more interconnected.

Chapter 7
Practical Optimization Examples with Python

Abstract This chapter offers a hands-on experience, demonstrating how to implement multiple optimization techniques using Python, a popular programming language for data scientists and engineers. The goal is to demonstrate how to use existing implementations of techniques seen in previous chapters, such as Linear Programming for production planning, Mixed-Integer Linear Programming for the knapsack problem, and Non-Linear Programming for engineering design and data-driven tasks. Solutions for Multi-Objective Optimization problems, such as truss and heat exchanger design, classifier tuning, and portfolio optimization, are also implemented. Additionally, this chapter covers Hyperparameter Optimization for machine learning models and multi-criteria decision-making using data visualization tools and mathematical techniques like TOPSIS. The goal is to show how to apply the technical concepts from previous chapters to real-world implementations using tools common in academia and industry.

7.1 Python for Optimization

We will use Python to develop the exercises in this chapter. It's a programming language widely used by universities to teach software development, and it's simpler to learn than C and C++, for example. Furthermore, it's the standard language for Data Science tasks: many existing artificial intelligence models today were built in this language.

On the other hand, it's also important to say that it's not the only language you'll encounter for working on Optimization. MATLAB, for example, is a cornerstone in engineering courses and academic research in these fields for simulating real-world engineering problems and numerical computation. Among the toolboxes it offers, there is one exclusively for Optimization, for instance. R is another language used for data analysis and statistical operations. It's a language especially used by academics outside of Computer Science courses.[1] We also have traditional languages, like C++

[1] As a curious note, complaints are not infrequent from people who have moved from Academia to Industry about the need to abandon R to develop algorithms with Python. Anecdotally, we notice this especially with our friends in Economics and Econometrics.

W. R. Monteiro and G. R. Meza, *A Practical Guide to Optimization in Engineering and Data Science*, https://doi.org/10.1007/978-3-032-04633-8_7

and Java. These languages allow for greater control over data types and enable the construction of custom solvers. However, they have a steeper learning curve: we perform the same operation with more lines of code in these languages compared to Python. Finally, there are also newer languages optimized for problems of high numerical complexity, such as Julia. That said, it still doesn't have the same popularity as the other languages.

7.1.1 But Why Python?

Continuing our discussion on programming languages, it's important to highlight one point: each programming language has its advantages and disadvantages, as well as its specific use cases. Python isn't superior to C++, Java, and other languages, and the reverse is also true. We understand this point is important since the choice and adoption of a programming language is usually a topic that generates many (good) discussions.

Important
Knowing Python before reading this chapter isn't a prerequisite, but it's highly recommended so you can better understand what we're doing. If you have the availability and time, we recommend learning the basics of Python before we move on. In addition to the resources traditionally offered by universities, you can consult a Python Crash Course: a quick course in Python. There are various types of these free courses available online with a high degree of quality and autonomy.

In recent years, Python has become a popular programming language due to the ease of learning how it works compared to other languages. That said, there are also other languages as simple as, or simpler than, Python. The second factor is the popularity of libraries. A library is a collection of algorithms produced by other people or companies to perform a specific job. We have, for example, specific libraries for producing graphics, for creating Machine Learning (ML) algorithms, for scientific computing, and for Optimization. And finally, we have versatility and community: it's not as if we have a set of libraries developed some years ago, and things are in slow decline. In fact, we have a large number of companies and users producing more libraries and improving existing ones across various fields. Therefore, using Python wouldn't restrict you solely to Data Science and Optimization but would also enable you to work in other Information Technology (IT) areas.

Tips
Think of libraries as additional items in a residence. When you buy or rent a home, it usually comes with the basics: walls, doors, windows, and sometimes some large furniture. This is like a programming language: it, by itself, has enough for you to start.

After you move into this house, you can add appliances and furniture so you can do your tasks and activities according to your needs. These items would be like libraries: specific things you connect (the technical term is `import`) to your code. You probably wouldn't build your own refrigerator: it's more worthwhile to buy a ready-made one, which has been evaluated and tested by many other people before. Similarly, it's sometimes worthwhile to use a ready-made library that has been validated by many other people and already has more reliable logic.

7.1.2 Setting Up

When you want to browse a website, you'd open a browser like Safari, Chrome, Firefox, or Opera, right? And if you wanted to create a spreadsheet, you'd open Excel, Google Sheets, or similar, correct? Similarly, to develop algorithms in Python, we'll need specific software for this.

To run the Python code examples in this chapter, you'll need a Python environment. We recommend starting with a cloud-based solution for its simplicity and zero-setup nature *unless* you already have experience in setting up your local environment or if you have your local environment already up and running.

7.1.2.1 Recommended: Google Colaboratory (Colab)

Google Colab is a free, cloud-based Jupyter notebook environment that requires no setup and runs entirely in your browser. It provides access to computing resources, including GPUs and TPUs, and comes pre-installed with many common data science libraries. We will use Colab to run all examples in this chapter.

Using Google Colab Using Google Colab is straightforward. First, you'll need to access Colab by navigating to https://colab.research.google.com/ using a Google account. Once there, you can create a new Python 3 notebook by selecting `File` → `New notebook`. Notebooks are composed of cells, and you'll primarily use "code cells" to write and execute Python code.

To run your code that you've typed into a cell, you can either click the "play" button to the left of the cell, press `Shift + Enter`, which runs the cell and moves to the next one (or creates a new one if you're at the end), or press `Ctrl + Enter`, which runs the cell and keeps the focus on it.

We strongly recommend spending a few minutes exploring Google Colab documentation, user guides, and tutorials before working with it.

While many libraries are pre-installed, if you need a specific one (like `Pymoo` or `PuLP`), you can install packages directly in a code cell using pip, Python's package installer. The command is prefixed with an exclamation mark (`!`) to indicate it's a shell command, like so:

```
!pip install numpy scipy matplotlib seaborn pandas
pymoo scikit-learn pulp ortools optuna pyDecision
```

Listing 7.1 Installing packages in Google Colab

You should run this cell once at the beginning of your notebook session if any of these packages are not already available or if you need a specific version. For this chapter, the listed packages are key. Your work should be automatically synchronized with your Google Drive.

7.1.2.2 Local Alternatives

If you prefer to work on your local machine, there are a couple of popular options. One is the **Anaconda Distribution**, available at https://www.anaconda.com/download. It's a free and open-source distribution of Python and R tailored for scientific computing and data science, simplifying package management and deployment. Anaconda comes bundled with Python, JupyterLab, Spyder (an integrated development environment, or IDE), and many pre-installed scientific packages. After installing Anaconda, you can launch JupyterLab from the Anaconda Navigator UI or your terminal.

Alternatively, you can set up **JupyterLab/Jupyter Notebook** directly. These are web-based interactive development environments for notebooks, code, and data. If you have Python installed on your system, you can install JupyterLab using pip with the command `pip install jupyterlab`, and then launch it by typing `jupyter lab` in your terminal. This will open JupyterLab in your web browser.

For these local setups, ensure you have Python (typically, a version no more than four years old) installed. You'll then use `pip` in your terminal (without the `!` prefix used in Colab) to install the necessary libraries for this chapter:

```
pip install numpy scipy matplotlib seaborn pandas
pymoo scikit-learn pulp ortools optuna pyDecision
```

Listing 7.2 Shell commands for installing libraries locally

7.1.2.3 Libraries

Secondly, Python has an extensive collection of open-source libraries for numerical computing (`NumPy`), scientific computing (`SciPy`), data analysis (`Pandas`), machine learning (`Scikit-learn`, `TensorFlow`, `PyTorch`), and, for our purposes in this chapter, Optimization (e.g., `SciPy.optimize`, `Pymoo`, `Pyomo`, `CVXPY`, `PuLP`, `OR-Tools`). Having these tools available means we can spend less time implementing Optimization techniques from scratch and more time formulating our Optimization problems.

Now, before moving forward, let's understand what these libraries have to offer in Table 7.1. Notice there are different libraries intended for different use cases. Therefore, they are not necessarily competitors with each other, and we can use multiple libraries in our daily activities in industry.

7.1.3 Running Code

The Python code examples in this chapter are provided as self-contained snippets. You can **execute** them by copying and pasting each snippet into a code cell in your chosen notebook environment (Colab, JupyterLab) and then running the cell to see the output.

In fact, data science notebooks, such as Jupyter or Colab, are very useful for cases like ours. They allow you to interactively run code in small chunks, inspect variables, and see results immediately, which is great for learning and debugging. They also enable you to mix code, text, and visuals by combining executable code, rich text explanations, mathematical equations (using LaTeX!), images, and plots all in one place, creating a detailed and shareable document. This is also one of the reasons why data scientists often like working with this kind of document.

Also, this structure promotes reproducibility, as a notebook captures the entire analysis workflow, making it easier for others (or your future self) to understand and reproduce your work. Therefore, we strongly recommend you to actively run the code, experiment with parameters, and try the exercises to get the most out of this chapter.

So, to start, we'd like you to create a new notebook in Colab from scratch. From there, in the first cell, we will install all required libraries before moving ahead. We've told you which libraries we needed in Sect. 7.1.2.1, but to make things easier for you, please copy and paste the command below into your first cell. The command below should be in a **single** line:

```
!pip install numpy scipy matplotlib seaborn pandas
pymoo scikit-learn pulp ortools optuna pyDecision
```

Listing 7.3 Installing packages in Google Colab

Then, run this cell. After waiting for some seconds, you should see a similar result to the one in Fig. 7.1. The overall layout may be a bit different on your computer,

Table 7.1 Comparison of Python libraries for optimization

Library	Main optimization type	Typical real-world applications	Main pros	Main cons
SciPy.optimize	General optimization (Non-Linear, constrained or unconstrained), Global	Curve fitting, model calibration, small to medium-scale engineering problems, parameter optimization in simulations	Easy to start for those already using NumPy/SciPy. Variety of solvers (gradient-based, gradient-free, global). Ideal for objective functions and constraints defined as Python functions. Good for rapid prototyping	Less efficient for large-scale or highly complex problems. Some solvers can be sensitive to initial parameters. Modeling can be less intuitive for structured Programming Optimization (PO) problems
PuLP	Linear Programming (LP), Mixed-Integer Programming (MIP)	Resource allocation, production planning, transportation problems, simple scheduling, diet problems	Very intuitive syntax, close to mathematical notation. Easy to learn and use for classic LP/MIP problems. Integrates with solvers like GLPK, CPLEX, Gurobi, and MOSEK	Limited to LP/MIP. Less flexible for very complex or non-linear modeling. Performance depends on the backend solver
Google OR-Tools	LP, MIP, Constraint Programming (CP), Vehicle Routing Problems (VRP), Graph problems	Logistics and supply chain, fleet routing, complex task scheduling, network design, portfolio optimization	High-performance solvers, especially for combinatorial and large-scale problems. Supports multiple programming languages	The syntax can be less algebraic and more programmatic for some models. The learning curve might be slightly steeper for some techniques
CVXPY	Convex programming	Optimal control, filter design, machine learning (e.g., support vector machines (SVMs), Lasso regression), quantitative finance, signal processing	Guarantees global optimality if the problem is convex. Automatic problem transformation for solvers. Supports various convex solvers, such as Clarabel, OSQP, and SCS. Powerful for problems with specific mathematical structures	Restricted to convex problems. May be less intuitive if you're unfamiliar with convex programming

(continued)

Table 7.1 (continued)

Library	Main optimization type	Typical real-world applications	Main pros	Main cons
Pyomo	Advanced Algebraic Modeling (e.g., LP, MIP, Non-linear Programming (NLP), Stochastic Optimization)	Planning and operation of energy systems, chemical process design, complex supply chain optimization, advanced financial modeling, optimization under uncertainty	Powerful and flexible for a range of problems. Allows for complex and structured algebraic modeling. Connects to many open-source and commercial solvers, like Gurobi, IPOPT, and CPLEX. Good for research and advanced industrial problems	Steeper learning curve than PuLP. Configuration and debugging can be more complex. The syntax might be verbose for simple problems
Pymoo	Multi-Objective Optimization (MOO)	Engineering design with trade-offs, portfolio optimization (risk vs. return), model calibration with multiple errors, feature selection in ML	Reference library for MOO. Implements multi-objective evolutionary algorithms (MOEAs) like NSGA-II/III, and MOEA/D. Includes performance metrics, visualizations, and tools for multi-criteria decision-making (MCDM)	Focused on MOO, primarily with heuristic/evolutionary methods. Might not be the first choice for single-objective problems (although it supports them) or where the exact mathematical structure is exploitable by other methods
Optuna	Hyperparameter Optimization (HPO) in Machine Learning	Tuning learning rates, number of layers/neurons in neural networks, regularization parameters, number of trees in Random Forests or Gradient Boosting Trees (GBTs). Optimization of ML pipelines	Easy to use and integrate with ML frameworks. Efficient search algorithms (including MOO-based). Good visualization and result analysis tools. ML framework agnostic	Primarily focused on HPO. Less suitable for general optimization problems that don't fit the 'evaluate a parameter configuration' paradigm
Hyperopt	Bayesian HPO	Similar to Optuna: tuning hyperparameters in ML models, especially when objective function evaluations (model training) are expensive	Popular for Bayesian optimization. Can be efficient when the number of evaluations is limited	Might have a slightly steeper learning curve than Optuna for beginners. Development appears less active compared to Optuna. More restricted focus on Bayesian HPO

(continued)

Table 7.1 (continued)

Library	Main optimization type	Typical real-world applications	Main pros	Main cons
Ray tune	Distributed and scalable HPO	Tuning hyperparameters for Deep Learning models on clusters, optimizing Reinforcement Learning systems, large-scale neural architecture search	Excellent for distributed HPO across multiple CPUs/GPUs and machines. Supports advanced HPO algorithms and schedulers. Highly scalable. Integrates with other HPO libraries, like Optuna and Hyperopt	Overhead might be greater for small problems or those not requiring distribution. Initial Ray environment setup can be more complex than libraries solely focused on HPO

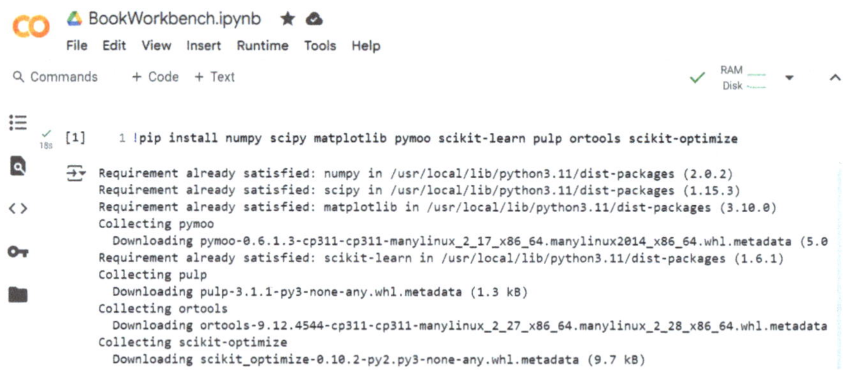

Fig. 7.1 The starting point in our notebook before running all exercises. *Source* The author, 2025

but notice the cell has been successfully executed, as seen in the small green tick on the left of the cell. This means that no errors occurred during the installation of our libraries, and we are ready to move on.

> **Important**
> For the next exercises, you can create a new cell and paste the code from there. You will only need to run this cell again in Colab if you go back to this notebook some hours after being idle. If you have installed these packages in your local environment (e.g., Anaconda and JupyterLab), you will not need to install them again in the future.

Best Practices Here, we're always starting from the premise that you'd come into contact with code development at some point in your life. Developing code, like any creative work (i.e., creating a document, a presentation, or a spreadsheet), is an art. It's important that your message isn't clear only in your head, but also to other people.

Imagine if on your very first day of work, you encounter code that was written years ago by someone who no longer works at the company. The code is poorly organized, poorly structured, and barely has any documentation. You can't even get a sense of where to start because of the disorganization. Of course, that wouldn't be good, would it?

Similarly, assume that your code should be understandable to other people with different ages and levels of experience. Never assume that everyone has the same life experience and worldview as you do. Therefore, not everyone would understand an algorithm you developed without much thought about who would maintain it in the future.

Therefore, we suggest you always keep your code organized. To begin with, try to structure it well: separate the problem definition in one part of the code, then place the logic of your Optimization algorithm, and finally, the logic to display the results. In some specialized Computer Science courses, it's common to study things like code organization, software design patterns, and concepts like functions and classes. It would be beneficial if you could learn a bit more about this and apply it to your problems.

Also, it's also good to use a code versioning tool. Currently, the most used versioning standard is Git, implemented by solutions like GitHub, GitLab, Bitbucket, and others. Every established company already has a solution in use, and it might be interesting for you to understand how this works as well.

Now, considering the Optimization problem itself: it's interesting to try to start simply. If we use few constraints and variables, it becomes easier to validate the code we're producing. Imagine, for example, if the solutions to this simplified problem make no sense at all. In that case, it would be easier to investigate whether the issue is in the code's structure or in the problem's formulation. We notice that it's common for less experienced people to turn a problem involving code into a snowball: in these cases, they implement code after code, without validating anything. In the end, it becomes so difficult to investigate what happened that significant time can be wasted, leading to frustration. It's in these situations that we can also find and solve small problems in the code, such as an incorrect mathematical sign or a misplaced decimal point.

Finally, it's also worthwhile to familiarize ourselves with the documentation of what we're using. Optimization libraries, programming languages, and solvers have documentation available on the internet for us to understand what different error messages and parameters mean. Look: it's even good to learn from examples to understand how to test and start, but what will really make a difference is knowing the documentation of what we use every day.

7.2 Single-Objective Optimization Examples

Let's start with some single-objective examples to demonstrate constrained and unconstrained optimization, as well as linear programming, mixed-integer linear programming, and non-linear programming. From now on, we strongly recommend you to copy and paste the code snippets below into the same notebook created in the previous section. Before running the code, try understanding what we did by considering comments, variable names, and the expected results from the code given the context of each example, as well as their mathematical definitions.

Practice is paramount: after having a rough idea of what the code does and how it behaves, try re-implementing the code yourself. For some people, manually re-typing each line of code and seeing how it works is a strategy. For others, modifying an example is another strategy. And for others, commenting each individual line of code with your own understanding is a viable strategy. Use whichever strategy is best for you.

7.2.1 Example 1: Optimal Production Planning

Imagine a company that produces two items, Product A and Product B. Each unit (such as kilogram, pound, or gallon) of Product A has a profit of \$20, while each unit of Product B has a profit of \$30. The production process has constraints: one unit of Product A requires 1 h of labor and 2 units of raw material, whereas one unit of Product B requires 1 h of labor and 1 unit of raw material.

On any given day, the company has a maximum of 40 h of labor and 60 units of raw material available. Therefore, our goal is to determine how many units of Product A and Product B should be produced daily to maximize the total profit.

7.2.1.1 Mathematical Formulation

This is a Linear Programming problem. Let x_A be the number of units of Product A produced, and x_B be the number of units of Product B produced.

Our main objective is to maximize the total profit P, which is given by the function:

$$\max P = 20x_A + 30x_B$$

However, you might remember we said earlier that several algorithms are designed to minimize functions. Therefore, let's convert that problem to minimization by optimizing the negative of P:

$$\min \quad -P = -20x_A - 30x_B$$

This objective is subject to a few constraints. The labor constraint dictates that the total labor hours used for both products cannot exceed 40 h:

$$1x_A + 1x_B \leq 40$$

Also, the raw material constraint states that the total raw material consumed must not exceed 60 units, or:

$$2x_A + 1x_B \leq 60$$

However, we must also respect the laws of physics. That is, no negative number of products can be produced:

$$x_A \geq 0$$

$$x_B \geq 0$$

Best Coding Practices Not everyone reading the code examples we will provide after this point will have extensive experience with coding. Therefore, we believe that it is important to give some tips to write code that solves your problems and is maintainable. We know the decision-makers will be happy if you solve their problems, but we also want you to make your peers who will maintain and extend your code happy, as well.

Therefore, let's start with the notations: mathematical notations like x_A or x_B are helpful for formulas, but we cannot use subscript or superscript when writing our variable names. With that, x_A could become `xA`, `x_a`, or a more descriptive name, such as `product_A_units`.

Also, specific programming languages have their own best practices. For Python, there is a style guide named *PEP 8*. It recommends using `snake_case` for variable and function names (e.g., `product_a_units`, `calculate_profit`). PEP 8 also strongly recommends that we prefer names that clearly reflect the variable's meaning within the optimization context.

So, while we are using `xA` just to map the mathematical notation into working code, in larger, more complex projects, we would rather use a more descriptive name like `units_of_product_a`. As a tip, imagine that after finishing your code you will live your life and forget about it for several months. After going back to your code you might be feeling confused with the code that you developed yourself. So, notice that what seems perfectly obvious to you during development can become quite obscure months later without good documentation and clear coding style. To avoid that, we recommend using clear and descriptive variable names, and adding comments to explain complex logic or assumptions that are not necessarily explicit just by reading our code.

Furthermore, please notice that an algorithm does not have the common sense as you do. You might have a clear understanding that time does not go backwards (at least in our Optimization problems), and that concepts such as negative height or negative costs would not exist. But if you don't include such rules in your code with constraints and bounds you might get unrealistic solutions for your Optimization problems.

Last, but not least: we will show you how we implement Optimization problems with the same programming language (Python), but with different libraries. You will see that they have different ways to operate and instantiate our variables, constraints, bounds, and objectives. Having that said, do not assume that you can just swap one library with another and expect that everything will work from the get-go. Often you will need to modify some extra lines of code and, to better understand how to operate each library, they always have their own documentation readily available on the internet. Please read the documentation of whichever library you will use to understand its own requirements, pros and cons, and best practices.

7.2.1.2 Python Implementation with SciPy

We can use `scipy.optimize.linprog` for this linear programming problem. Notice that in the first lines we are importing the NumPy library (so that we can create our coefficients) and the `linprog` interface from the SciPy library.

You will also notice that we are running the linear programming optimizer with the `highs` method. It will choose between simplex and interior-point method solvers depending on the problem, according to the documentation.

```
import numpy as np
from scipy.optimize import linprog

# Coefficients of the objective function (minimize -P)
c = np.array([-20, -30])

# Coefficients of the inequality constraints (A_ub @ x <= b_ub)
# "ub" means "upper bound"
A_ub = np.array([[1, 1], [2, 1]])
b_ub = np.array([40, 60])

# Bounds for variables (xA >= 0, xB >= 0)
x_A_bounds = (0, None)
x_B_bounds = (0, None)
bounds = [x_A_bounds, x_B_bounds]
```

```
16
17 print("Solving with SciPy linprog...")
18 result_scipy = linprog(c, A_ub=A_ub, b_ub=b_ub, bounds=bounds,
     method='highs')
19
20 if result_scipy.success:
21     optimal_xA = result_scipy.x[0]
22     optimal_xB = result_scipy.x[1]
23     max_profit = -result_scipy.fun
24     print(f"SciPy - Optimal production plan: Product A: {
       optimal_xA:.2f}, Product B: {optimal_xB:.2f}, Max Profit: ${
       max_profit:.2f}")
25     labor_used = np.dot(A_ub[0, :], result_scipy.x)
26     material_used = np.dot(A_ub[1, :], result_scipy.x)
27     print(f"SciPy - Resource Utilization: Labor: {labor_used:.2f
       }/{b_ub[0]}, Material: {material_used:.2f}/{b_ub[1]}")
28 else:
29     print(f"SciPy - Optimization failed: {result_scipy.message}")
```

Listing 7.4 Python code for optimal production planning using SciPy

Important
Notice that on the left of the code snippet we have the line numbers. It means that the pasted code in your notebook should follow the same structure. Due to the book margins, a few lines are split (e.g., lines 18, 24, and 27). When pasting into your own notebook, make sure that they are following the same order as shown in the book.

For example, check Fig. 7.2. This is part of the code in Listing 7.4. Note the top cell respects the line count as seen in the code snippet (the if-else block in the snippet has 10 lines). The top cell also respects the same code indentation from the snippet. The bottom cell, on the other hand, does not match the line count nor the indentation. Remember: the line order and the indentation matter.

7.2.1.3 Python Implementation with PuLP

Now, let's try implementing the same problem using another library: `PuLP`. It is a Python library for linear and integer programming. Notice that we are instantiating our problem in a more algebraic way. We also can name our variables, problems, and constraints, such as `UnitsProductA_PuLP`, but make sure that each variable has its own unique name.

Also, note a few things:

1. PuLP supports maximization problems (`LpMaximize`).
2. We can type the lower and upper bounds (`lowBound` and `upBound`, respectively) for each variable, as well as their type (`Continuous`, `Binary` or `Integer`).

```
if result_scipy.success:
    optimal_xA = result_scipy.x[0]; optimal_xB = result_scipy.x[1]
    max_profit = -result_scipy.fun
    print(f"SciPy - Optimal production plan: Product A: {optimal_xA:.2f}, Product B: {optimal_xB:.2f}, Max Profit: ${max_profit:.2f}")
    labor_used = np.dot(A_ub[0, :], result_scipy.x)
    material_used = np.dot(A_ub[1, :], result_scipy.x)
    print(f"SciPy - Resource Utilization: Labor: {labor_used:.2f}/{b_ub[0]}, Material: {material_used:.2f}/{b_ub[1]}")
else:
    print(f"SciPy - Optimization failed: {result_scipy.message}")
```

```
if result_scipy.success:
optimal_xA = result_scipy.x[0]; optimal_xB = result_scipy.x
[1]
max_profit =-result_scipy.fun
print(f"SciPy-Optimal production plan: Product A: {
optimal_xA:.2f}, Product B: {optimal_xB:.2f}, Max Profit: ${
max_profit:.2f}")
labor_used = np.dot(A_ub[0, :], result_scipy.x);
material_used = np.dot(A_ub[1, :], result_scipy.x)
print(f"SciPy-Resource Utilization: Labor: {labor_used:.2f
}/{b_ub[0]}, Material: {material_used:.2f}/{b_ub[1]}")
else:
print(f"SciPy-Optimization failed: {result_scipy.message}")
```

Fig. 7.2 A sample of a good copy (top) and a bad copy (bottom) of a code snippet. *Source* The author, 2025

```
import pulp

prob = pulp.LpProblem("OptimalProductionPlan_PuLP", pulp.
    LpMaximize)
xA = pulp.LpVariable("UnitsProductA_PuLP", lowBound=0, cat='
    Continuous')
xB = pulp.LpVariable("UnitsProductB_PuLP", lowBound=0, cat='
    Continuous')
prob += 20 * xA + 30 * xB, "TotalProfit" # Objective
prob += xA + xB <= 40, "LaborConstraint" # Constraint 1
prob += 2 * xA + 1 * xB <= 60, "RawMaterialConstraint" #
    Constraint 2

print("Solving with PuLP...")
prob.solve()

print(f"PuLP - Status: {pulp.LpStatus[prob.status]}")
if pulp.LpStatus[prob.status] == "Optimal":
    optimal_xA_pulp = pulp.value(xA)
    optimal_xB_pulp = pulp.value(xB)
    max_profit_pulp = pulp.value(prob.objective)
    print(f"PuLP - Optimal production plan: Product A: {
    optimal_xA_pulp:.2f}, Product B: {optimal_xB_pulp:.2f}, Max
    Profit: ${max_profit_pulp:.2f}")
    labor_used_pulp = optimal_xA_pulp * 1 + optimal_xB_pulp * 1
    material_used_pulp = optimal_xA_pulp * 2 + optimal_xB_pulp * 1
    print(f"PuLP - Resource Utilization: Labor: {labor_used_pulp
    :.2f}/40, Material: {material_used_pulp:.2f}/60")
else:
    print("PuLP - Optimal solution not found.")
```

Listing 7.5 Python code for optimal production planning using PuLP

7.2.1.4 Python Implementation with OR-Tools

Finally, let's test Google's `OR-Tools`, which is a powerful suite for combinatorial optimization, such as linear programming and mixed-integer linear programming.

Notice that at the start we need to create the solver: without it, we cannot move forward. In this case, we are using GLOP, which was developed by Google themselves and uses a dual simplex approach under the hood.

```python
from ortools.linear_solver import pywraplp

solver = pywraplp.Solver.CreateSolver('GLOP') # GLOP is Google's
    LP solver
if not solver:
    print("OR-Tools: GLOP solver not available.")
else:
    # Variables: xA and xB, non-negative
    xA = solver.NumVar(0, solver.infinity(), 'xA_OR')
    xB = solver.NumVar(0, solver.infinity(), 'xB_OR')

    # Constraints
    # Labor: xA + xB <= 40
    solver.Add(xA + xB <= 40)
    # Raw Material: 2*xA + xB <= 60
    solver.Add(2 * xA + xB <= 60)

    # Objective function: 20*xA + 30*xB
    solver.Maximize(20 * xA + 30 * xB)

    print("\nSolving with OR-Tools (GLOP)...")
    status = solver.Solve()

    if status == pywraplp.Solver.OPTIMAL:
        print(f"OR-Tools - Optimal production plan: Product A: {xA
    .solution_value():.2f}, Product B: {xB.solution_value():.2f},
    Max Profit: ${solver.Objective().Value():.2f}")
        print(f"OR-Tools - Resource Utilization: Labor: {xA.
    solution_value() + xB.solution_value():.2f}/40, Material: {2*
    xA.solution_value() + xB.solution_value():.2f}/60")
    else:
        print("OR-Tools - The problem does not have an optimal
    solution.")
```

Listing 7.6 Python code for optimal production planning using OR-Tools

7.2.1.5 Comparison

Now, let's compare the results for all techniques. If you made sure you have copied the code correctly, respecting the indentation and lines of code organization, you should have ended up with:

```
Solving with SciPy linprog...
SciPy - Optimal production plan: Product A: 0.00, Product B:
    40.00, Max Profit: $1200.00
SciPy - Resource Utilization: Labor: 40.00/40, Material: 40.00/60
-----------------------------------------------
```

```
Solving with PuLP...
PuLP - Status: Optimal
PuLP - Optimal production plan: Product A: 0.00, Product B: 40.00,
     Max Profit: $1200.00
PuLP - Resource Utilization: Labor: 40.00/40, Material: 40.00/60
-----------------------------------------------
Solving with OR-Tools (GLOP)...
OR-Tools - Optimal production plan: Product A: 0.00, Product B:
    40.00, Max Profit: $1200.00
OR-Tools - Resource Utilization: Labor: 40.00/40, Material:
    40.00/60
```

Listing 7.7 Results for the Example 1

The three techniques converged to the same solution, although each library has a different way to model and solve LP problems. SciPy's `linprog` is matrix-based and efficient for numerically defined problems. It is good to use if your data is already defined in NumPy arrays.

On the other hand, PuLP is more readable by non-software engineers. It supports various solvers and is good to build complex models from scratch, as working with dozens or hundreds of coefficients with SciPy can be quite confusing. OR-Tools often has very good performance with very complex problems. It is also algebraic and supports different solvers.

7.2.2 *Example 2: Linear Programming, Extended*

Now, what if we extend the previous problem by adding a third product? Product C will have a profit of \$25 per unit, and producing one unit of Product C requires 2 h of labor and 1.5 units of raw material.

The constraints are the same: we still have the same 40 h and 60 units of raw material available. Therefore, our goal here is to investigate how the introduction of Product C will affect the optimal production plan and maximum profit.

7.2.2.1 Mathematical Formulation

Let x_A, x_B, x_C be the units of Product A, B, and C, respectively. Modifying our formulation from the previous example we would have:

$$\begin{aligned}
\min \; -P &= -20x_A - 30x_B - 25x_C \\
\text{subject to} & \\
& 1x_A + 1x_B + 2x_C \leq 40 \\
& 2x_A + 1x_B + 1.5x_C \leq 60 \\
& x_A, x_B, x_C \geq 0
\end{aligned} \tag{7.1}$$

where the first constraint relates to the labor constraint, the second relates to the raw material constraint, and the third relates to the non-negativity.

7.2.2.2 Python Implementation with SciPy

Now, let's implement this problem using SciPy for sampling purposes. Notice it's pretty much the same code used for the SciPy implementation in Example 1, but with a few lines changed to account for the inclusion of x_C. We have included comments to help you understand which lines have been modified.

```python
import numpy as np
from scipy.optimize import linprog

# Adding a third value for xC
c = np.array([-20, -30, -25])

# Coefficients of the inequality constraints (A_ub @ x <= b_ub)
# "ub" means "upper bound"

# Recapping what the lines below should mean:
# xA + xB + 2xC <= 40  (Labor)
# 2xA + xB + 1.5xC <= 60 (Raw Material)
A_ub = np.array([[1, 1, 2], [2, 1, 1.5]])
b_ub = np.array([40, 60])

# Bounds for variables (xA >= 0, xB >= 0, and adding xC >= 0)
x_A_bounds = (0, None)
x_B_bounds = (0, None)
x_C_bounds = (0, None)
bounds = [x_A_bounds, x_B_bounds, x_C_bounds]

# simplex and interior-point method solvers
print("Solving with SciPy linprog...")
result_scipy = linprog(c, A_ub=A_ub, b_ub=b_ub, bounds=bounds,
    method='highs')

if result_scipy.success:
    optimal_xA = result_scipy.x[0]
    optimal_xB = result_scipy.x[1]
    optimal_xC = result_scipy.x[2] # also including xC
    max_profit = -result_scipy.fun
    print(f"SciPy - Optimal production plan: Product A: {optimal_xA:.2f}, Product B: {optimal_xB:.2f}, Product C: {optimal_xC:.2f}, Max Profit: ${max_profit:.2f}")
    labor_used = np.dot(A_ub[0, :], result_scipy.x)
    material_used = np.dot(A_ub[1, :], result_scipy.x)
    print(f"SciPy - Resource Utilization: Labor: {labor_used:.2f}/{b_ub[0]}, Material: {material_used:.2f}/{b_ub[1]}")
else:
    print(f"SciPy - Optimization failed: {result_scipy.message}")
```

Listing 7.8 Python code for extended production planning (SciPy)

7.2.2.3 Python Implementation with PuLP

To demonstrate how we often adapt code to address changes in our Optimization problems, let's also modify our code implementing PuLP to account for x_C.

```
import pulp

prob = pulp.LpProblem("OptimalProductionPlan_PuLP", pulp.
    LpMaximize)
xA = pulp.LpVariable("UnitsProductA_PuLP", lowBound=0, cat='
    Continuous')
xB = pulp.LpVariable("UnitsProductB_PuLP", lowBound=0, cat='
    Continuous')
xC = pulp.LpVariable("UnitsProductC_PuLP", lowBound=0, cat='
    Continuous') # Adding xC
prob += 20 * xA + 30 * xB + 25 * xC, "TotalProfit" # Objective,
    modified to include xC
prob += xA + xB + 2 * xC <= 40, "LaborConstraint" # Constraint 1,
    modified to include xC
prob += 2 * xA + 1 * xB + 1.5 * xC <= 60, "RawMaterialConstraint"
    # Constraint 2, modified to include xC

print("Solving with PuLP...")
prob.solve()

print(f"PuLP - Status: {pulp.LpStatus[prob.status]}")
if pulp.LpStatus[prob.status] == "Optimal":
    optimal_xA_pulp = pulp.value(xA)
    optimal_xB_pulp = pulp.value(xB)
    optimal_xC_pulp = pulp.value(xC) # Including xC
    max_profit_pulp = pulp.value(prob.objective)

    # Modifying the three lines below to also account for xC
    print(f"PuLP - Optimal production plan: Product A: {
    optimal_xA_pulp:.2f}, Product B: {optimal_xB_pulp:.2f},
    Product C: {optimal_xC_pulp:.2f}, Max Profit: ${
    max_profit_pulp:.2f}")
    labor_used_pulp = optimal_xA_pulp * 1 + optimal_xB_pulp * 1 +
    optimal_xC_pulp * 2
    material_used_pulp = optimal_xA_pulp * 2 + optimal_xB_pulp * 1
     + optimal_xC_pulp * 1.5
    print(f"PuLP - Resource Utilization: Labor: {labor_used_pulp
    :.2f}/40, Material: {material_used_pulp:.2f}/60")
else:
    print("PuLP - Optimal solution not found.")
```

Listing 7.9 Python code for extended production planning (PuLP)

7.2.2.4 Comparison

Now, let's compare the results from both techniques. You can see that both techniques found the same solutions. Having that said, notice the optimal solution is not suggesting the production of Products A and C, and we are not using all the raw material available.

```
Solving with SciPy linprog...
SciPy - Optimal production plan: Product A: 0.00, Product B:
    40.00, Product C: 0.00, Max Profit: $1200.00
SciPy - Resource Utilization: Labor: 40.00/40, Material: 40.00/60
--------------------------------------------
Solving with PuLP...
PuLP - Status: Optimal
PuLP - Optimal production plan: Product A: 0.00, Product B: 40.00,
     Product C: 0.00, Max Profit: $1200.00
PuLP - Resource Utilization: Labor: 40.00/40, Material: 40.00/60
```

Listing 7.10 Results for the Example 1

Tips
As an extension of this algorithm, what if you modify the amounts of time and raw materials to see the solutions these libraries yield for you? Or, how about you introduce some constraints to ensure we at least produce a minimum amount of each one of these products? Or what would happen if we introduce additional products with their own coefficients?

7.2.3 Example 3: The 0/1 Knapsack Problem

The third example is a little bit different, and a traditional problem commonly addressed in Optimization classes. Imagine a hiker preparing for a trip. She has a knapsack that can carry a maximum weight of W_{max}. She has a list of N items, each with a specific weight w_i and an associated value (or utility) v_i.

The goal in our scenario is to choose which items to pack to maximize the total value of items in the knapsack without exceeding the maximum weight capacity. Each item can either be fully included (where we assign the value 1) or not at all (where we assign the value 0)—hence the name 0/1 knapsack problem.

For illustration purposes, let's consider a concrete example where the knapsack weight is $W_{max} = 15$ kg. Here, we have the following items:

- Item 1: $w_1 = 12$ kg, $v_1 = 4$
- Item 2: $w_2 = 2$ kg, $v_2 = 2$
- Item 3: $w_3 = 1$ kg, $v_3 = 1$
- Item 4: $w_4 = 1$ kg, $v_4 = 2$
- Item 5: $w_5 = 4$ kg, $v_5 = 10$

7.2.3.1 Mathematical Formulation

This is a Mixed-Integer Linear Programming (MILP) problem. Let x_i be a binary decision variable for each item i: $x_i = 1$ if item i is included in the knapsack, and $x_i = 0$ if item i is not included.

The objective, as you remember, is to maximize total value V:

$$\max V = \sum_{i=1}^{N} v_i x_i$$

Transforming it into a minimization problem results in:

$$\min V = -\sum_{i=1}^{N} v_i x_i$$

Also, remember that our total weight must not exceed the total capacity W_{max}:

$$\sum_{i=1}^{N} w_i x_i \leq W_{max}$$

And, of course, we should allow each variable to be either 0 (not included in the knapsack) or 1 (included in the knapsack), with no other values allowed:

$$x_i \in 0, 1 \quad \forall i = 1, \ldots, N$$

7.2.3.2 Python Implementation with PuLP

For this example, let's try an implementation with PuLP, which is well-suited for MILP problems. One of the advantages of this library is that it enables us to set the variable types: since we are only interested in having each value set as 0 or 1, we could set all of them as being the `Binary` type.

```
import pulp

# Problem Data
items_data = {
    "item1": {"weight": 12, "value": 4},
    "item2": {"weight": 2, "value": 2},
    "item3": {"weight": 1, "value": 1},
    "item4": {"weight": 1, "value": 2},
    "item5": {"weight": 4, "value": 10},
}
item_names = list(items_data.keys())
max_weight = 15

```

```
# Creating the MILP problem (and we can set it as a maximization problem)
prob_knapsack = pulp.LpProblem("KnapsackProblem", pulp.LpMaximize)

# Decision Variables: x_i = 1 if item_i is chosen, 0 otherwise (and we are using the "Binary" type for all our variables)
x = pulp.LpVariable.dicts("x", item_names, cat="Binary")

# Objective Function: Maximize total value
prob_knapsack += pulp.lpSum([items_data[i]["value"] * x[i] for i in item_names]), "TotalValue"

# Constraint: Total weight <= max_weight
prob_knapsack += pulp.lpSum([items_data[i]["weight"] * x[i] for i in item_names]) <= max_weight, "WeightConstraint"

# Solve the problem
print("\nSolving Knapsack Problem with PuLP...")
prob_knapsack.solve()

# Print the results
print(f"Knapsack - Status: {pulp.LpStatus[prob_knapsack.status]}")
if pulp.LpStatus[prob_knapsack.status] == "Optimal":
    print(f"Knapsack - Maximum Value: {pulp.value(prob_knapsack.objective):.2f}")
    total_packed_weight = 0
    print("Items to pack:")
    for i in item_names:
        if pulp.value(x[i]) == 1:
            print(f"  - {i} (Weight: {items_data[i]['weight']}, Value: {items_data[i]['value']})")
            total_packed_weight += items_data[i]['weight']
    print(f"Total packed weight: {total_packed_weight:.2f} kg (Max allowed: {max_weight} kg)")
else:
    print("Knapsack - Optimal solution not found.")
```

Listing 7.11 Python code for the 0/1 Knapsack problem using PuLP

This implementation introduced a few new things. First, we have created a data dictionary named `items_data`. Each item (e.g., "item1") is a key that lets us quickly look up its details (like its weight and value). The goal is to dynamically modify our repository of items in a single place instead of modifying several lines of code.

This `items_data` is reused a little bit later in the `pulp.LpVariable.dicts`. The idea here is to create several variables with a single line of code. Imagine if we had 10,000 items: it would be pretty tiresome to manually create each single `pulp.LpVariable`, right?

Then, you can see that some lines containing `for i in item_names`. It means, basically, *"for each item in our list (item_names), do something."* By "something" it could be getting its value from `items_data` and multiplying it by its decision variable. With that, we are repeating a task for every item in a list without writing the code out for each variable separately.

Although the knapsack problem can be computationally very expensive for large problems, we can use PuLP's standard solver in problems like ours. Also, notice we are showing the results of the Optimization. In fact, we are displaying only the items selected (`pulp.value(x[i]) == 1`). With that, we have the following results:

```
Solving Knapsack Problem with PuLP...
Knapsack - Status: Optimal
Knapsack - Maximum Value: 15.00
Items to pack:
- item2 (Weight: 2, Value: 2)
- item3 (Weight: 1, Value: 1)
- item4 (Weight: 1, Value: 2)
- item5 (Weight: 4, Value: 10)
Total packed weight: 8.00 kg (Max allowed: 15 kg)
```

Listing 7.12 Results for the Example 1

7.2.4 Example 4: Parameter Estimation for a Non-Linear Model (Curve Fitting)

Now, instead of an example resembling a real-world situation, let's consider the scenario where we are fitting a non-linear model $y(t) = Ae^{-kt} + C$ to noisy data (such as the kind of data we expect in real-world situations) by minimizing the sum of squared errors (SSE). This is a non-linear least squares problem.

7.2.4.1 Mathematical Formulation

Remember: our goal is to minimize the sum of squared errors (SSE). To do that, we need to measure the error between the model prediction and the real value for each one of the N observed data points, where each i-th data point is defined as (t_i, y_i^{obs}), and each prediction is defined as $y(t_i; A, k, C) = Ae^{-kt_i} + C$.

Having that said, our objective is:

$$\min SSE(A, k, C) = \sum_{i=1}^{N} (y(t_i; A, k, C) - y_i^{\text{obs}})^2$$

7.2.4.2 Python Implementation with `scipy.optimize.curve_fit`

`scipy.optimize.curve_fit` is specifically designed for curve fitting using non-linear least squares.

```
import numpy as np
from scipy.optimize import curve_fit
import matplotlib.pyplot as plt

# 1. Define the model function that we want to fit to the data.
def model_func(t, A, k, C):
```

```
7       return A * np.exp(-k * t) + C
8
9   # 2. Generating synthetic data with noise.
10  np.random.seed(42) # For reproducibility
11  t_data = np.linspace(0, 10, 50)
12  A_true, k_true, C_true = 5.0, 0.5, 1.0
13  y_true = model_func(t_data, A_true, k_true, C_true)
14  noise_level = 0.5
15  y_noisy = y_true + noise_level * np.random.normal(size=len(t_data)
        )
16
17  # 3. Provide initial guesses for the parameters [A, k, C].
18  initial_guesses_cf = [1.0, 0.1, 0.5]
19
20  # 4. Perform curve fitting
21  print("\nSolving with SciPy curve_fit...")
22  try:
23      params_opt_cf, params_cov_cf = curve_fit(model_func, t_data,
        y_noisy, p0=initial_guesses_cf, method='lm')
24      A_opt_cf, k_opt_cf, C_opt_cf = params_opt_cf
25      print(f"curve_fit - Optimal parameters found:")
26      print(f"  A_opt = {A_opt_cf:.3f} (True: {A_true})")
27      print(f"  k_opt = {k_opt_cf:.3f} (True: {k_true})")
28      print(f"  C_opt = {C_opt_cf:.3f} (True: {C_true})")
29      std_dev_errors_cf = np.sqrt(np.diag(params_cov_cf))
30      print(f"curve_fit - Standard deviation errors:")
31      print(f"  A_err = {std_dev_errors_cf[0]:.3f}, k_err = {
        std_dev_errors_cf[1]:.3f}, C_err = {std_dev_errors_cf[2]:.3f}"
        )
32
33      # Creating a nice scatterplot to make things easier for us
34      plt.figure(figsize=(12, 7))
35      plt.scatter(t_data, y_noisy, label='Noisy Data', color='blue',
         s=20, alpha=0.6)
36      plt.plot(t_data, y_true, label='True Model', color='black',
        linestyle=':', linewidth=2)
37      plt.plot(t_data, model_func(t_data, A_opt_cf, k_opt_cf,
        C_opt_cf), label=f'curve_fit (A={A_opt_cf:.2f}, k={k_opt_cf:.2
        f}, C={C_opt_cf:.2f})', color='red', linewidth=2)
38  except RuntimeError:
39      print("curve_fit failed. Could not find optimal parameters.")
```

Listing 7.13 Python code for non-linear parameter estimation using `curve_fit`

Now, let's spend some time explaining this initial data setup. First, to create our noisy dataset of 50 data points, we generate `t_data` values in the [0, 10] range (line 11). We then define the "true" model parameters A, k, and C (line 12). In real-world problems, these true values are unknown, and the challenge is for an optimizer to discover them. However, since we are creating synthetic data, we set these true values ourselves. Real-world data is also usually noisy. Therefore, we add noise (lines 14 and 15) to the "clean" `y_true` values, which were initially produced by our `model_func` (defined in lines 6–7) using these true parameters.

So, basically, lines 10–15 served the purpose of generating synthetic (and noisy) data. In true applications, this data would be swapped by sensor data, financial data, or other real-world noisy data.

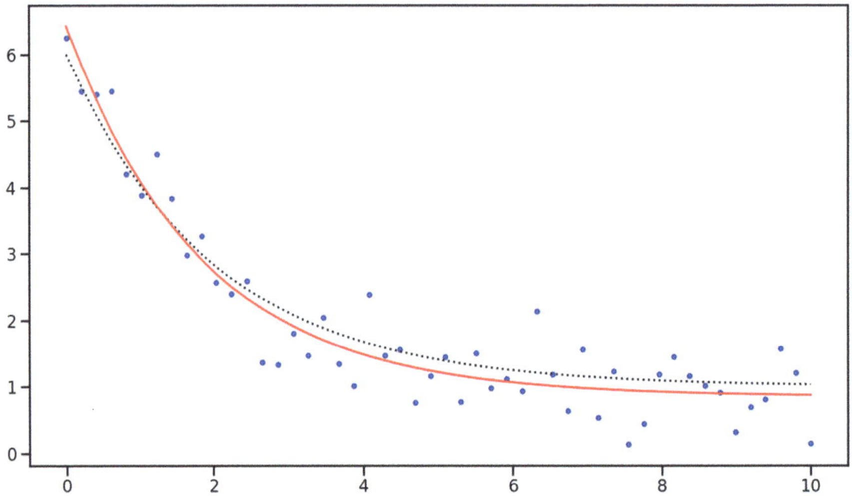

Fig. 7.3 The output from the SciPy `curve_fit` (red solid line) against the original noisy data (blue dots) and the source curve before adding noise (black dotted line). *Source* The author, 2025

Now, our challenge is to have a model that calculates the best values for A, k, and C given our noisy data. Our Optimization algorithm (`curve_fit` in line 23) requires the mathematical representation we believe represents our data (`model_func`), the noisy data input (`t_data`) and output (`y_noisy`), and our initial guess of what are the right values for A, k, and C.

The core idea behind `curve_fit` is to find the set of parameters for our `model_func` that makes the function's output match our `y_noisy` data as closely as possible. Or, in other words, the algorithm tries to minimize the sum of the squares of the errors between the predicted values (using a combination of values for A, k, and C that it found) and the values in `y_noisy` through several iterations.

The results are the parameters it found (lines 23 and 24). To evaluate it, we also generate a scatterplot with the original data, a lineplot containing the predicted curve with the parameters found by our model (red line), the curve with the original parameters used to generate the noisy data (black dotted line), and the noisy data (blue dots). This plot is shown in Fig. 7.3.

7.2.4.3 Python Implementation with `scipy.optimize.differential_evolution`

The implementation above used the `"lm"` value for the *method* parameter of `curve_fit` (line 23). According to SciPy's documentation, it means that we were using the Levenberg-Marquardt algorithm, which sometimes behaves similarly to a gradient descent method (when far from the solution), and sometimes closer to a Gauss-Newton method (when closer to the solution).

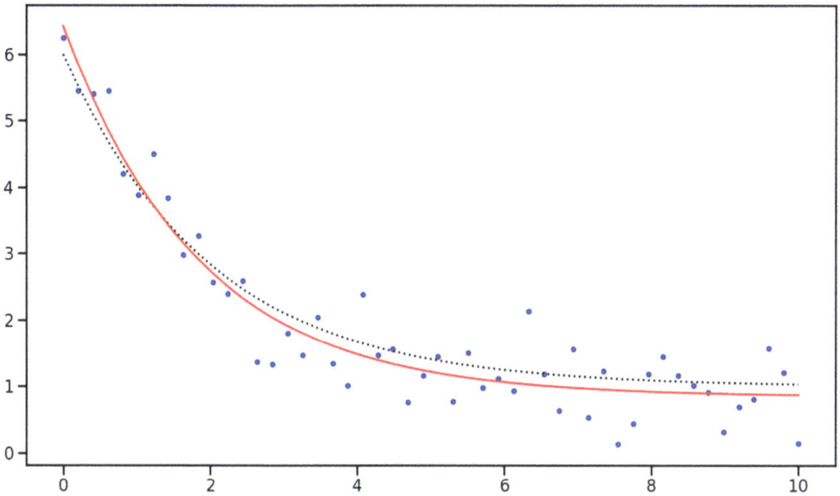

Fig. 7.4 The output from the SciPy `differential_evolution` (red solid line) against the original noisy data (blue dots) and the source curve before adding noise (black dotted line). *Source* The author, 2025

However, we could also try something different. Differential Evolution is a population-based global optimization algorithm. We will use the `curve_fit` code as a baseline, but replace it with SciPy's `differential_evolution`. You can notice there are a few differences here, as noted in the comments in green. First, we need to create an evaluation function to calculate the SSE—something that was not required when using `curve_fit`. However, the initial guesses are not required anymore. Instead, we inform the bounds we want the algorithm to explore. Also, you can see the output of this algorithm in Fig. 7.4.

```
import numpy as np
from scipy.optimize import differential_evolution # Changing the import
import matplotlib.pyplot as plt

# 1. Adding an extra function to include the objective function we want to minimize
def model_func(t, A, k, C):
    return A * np.exp(-k * t) + C

def objective_func(params, t_data_arg, y_noisy_arg): # Objective for DE
    A, k, C = params
    y_predicted = model_func(t_data_arg, A, k, C)
    return np.sum((y_noisy_arg - y_predicted)**2)

# 2. Generating synthetic data with noise.
np.random.seed(42) # For reproducibility
t_data = np.linspace(0, 10, 50)
A_true, k_true, C_true = 5.0, 0.5, 1.0
y_true = model_func(t_data, A_true, k_true, C_true)
```

```python
noise_level = 0.5
y_noisy = y_true + noise_level * np.random.normal(size=len(t_data))

# 3. Provide initial bounds for the parameters [A, k, C].
param_bounds = [(0, 10), (0, 2), (0, 5)]

# 4. Perform DE
print("\nSolving with SciPy differential_evolution...")
try:
    # Modifying the function call
    params_opt_de = differential_evolution(objective_func, bounds=param_bounds, args=(t_data, y_noisy), seed=42)
    A_opt_de, k_opt_de, C_opt_de = params_opt_de.x
    print(params_opt_de)

    print(f"differential_evolution - Optimal parameters found:")
    print(f"  A_opt = {A_opt_de:.3f} (True: {A_true})")
    print(f"  k_opt = {k_opt_de:.3f} (True: {k_true})")
    print(f"  C_opt = {C_opt_de:.3f} (True: {C_true})")

    # Creating a nice scatterplot to make things easier for us
    plt.figure(figsize=(12, 7))
    plt.scatter(t_data, y_noisy, label='Noisy Data', color='blue', s=20, alpha=0.6)
    plt.plot(t_data, y_true, label='True Model', color='black', linestyle=':', linewidth=2)
    plt.plot(t_data, model_func(t_data, A_opt_de, k_opt_de, C_opt_de), label=f'differential_evolution (A={A_opt_de:.2f}, k={k_opt_de:.2f}, C={C_opt_de:.2f})', color='red', linewidth=2)
except RuntimeError:
    print("differential_evolution failed. Could not find optimal parameters.")
```

Listing 7.14 Python code for parameter estimation using `differential_evolution`

7.2.4.4 Discussion

You might have noticed that, for this example, we were able to fit our curve using both strategies. `curve_fit` is a specialized function designed for non-linear least squares minimization and is able to quickly converge to a precise solution if our initial parameter guesses are reasonably close to the optimal values and the problem landscape is relatively smooth. Therefore, misguided guesses can negatively affect its performance. Also, it is susceptible to giving us a suboptimal fit by getting trapped in local minima if the initial guesses are poor or if the parameter space is complex and multi-modal.

On the other hand, `differential_evolution` is a population-based approach that is more robust in finding the global minimum. We were not required to inform our initial guesses, but we had to define the cost function so that it was able to minimize. Generally, this method is more computationally intensive than `curve_fit` as it evaluates the model function many more times (as it is population-based, for each iteration we are evaluating all individuals from that population).

7.2.5 Example 5: Alternative Model Fitting

But, what if instead of using the non-linear model in the previous example, we use a quadratic function of the form $y(t) = at^2 + bt + c$ for the same data? It means that we want to find the parameters a, b, c that best fit the data and compare its goodness of fit (e.g., SSE) to this new model.

7.2.5.1 Mathematical Formulation

The objective here is to minimize SSE for the quadratic model. The definitions from the previous example remain mostly the same. We still need to measure the error between the model prediction and the real value for each one of the N observed data points, but now each prediction is defined as $y(t_i; a, b, c) = at_i^2 + bt_i + c$.

Having that said, our objective now is:

$$\min SSE(a, b, c) = \sum_{i=1}^{N} (y(t_i; a, b, c) - y_i^{\text{obs}})^2$$

7.2.5.2 Python Implementation with `scipy.optimize.curve_fit` and `scipy.optimize.differential_evolution`

We'll use the same `t_data` and `y_noisy` from Sect. 7.2.4.

```
import numpy as np
from scipy.optimize import curve_fit, differential_evolution
import matplotlib.pyplot as plt

# 1. Define the model function that we want to fit to the data.
def model_func(t, a, b, c):
    return a * t**2 + b * t + c    # Equation changed

def objective_func(params, t_data_arg, y_noisy_arg): # Objective
     for DE
    a, b, c = params
    y_predicted = model_func(t_data_arg, b, b, b) # this is a typo
      on purpose, but we'll talk about it
    return np.sum((y_noisy_arg - y_predicted)**2)

# 2. Generating synthetic data with noise.
np.random.seed(42) # For reproducibility
t_data = np.linspace(0, 10, 50)
a_true, b_true, c_true = 5.0, 0.5, 1.0 # Renaming from A_true,
     k_true and C_true
y_true = model_func(t_data, a_true, b_true, c_true)
noise_level = 0.5
# changing the noise to account for the exponential scale
y_noisy = y_true + noise_level * np.random.normal(size=len(t_data)
     , scale=(t_data+2)**2)

```

```python
# 3. Provide initial guesses for the parameters [a, b, c].
initial_guesses_cf = [5.0, 0.5, 1.0]

# 4. Preparing the figure to include our visualizations
plt.figure(figsize=(12, 7))
plt.scatter(t_data, y_noisy, label='Noisy Data', color='blue', s=20, alpha=0.6)
plt.plot(t_data, y_true, label='True Model', color='black', linestyle=':', linewidth=2)

# 5. Perform curve fitting
print("\nSolving with SciPy curve_fit...")
try:
    params_opt_cf, params_cov_cf = curve_fit(model_func, t_data, y_noisy, p0=initial_guesses_cf, method='lm')
    a_opt_cf, b_opt_cf, c_opt_cf = params_opt_cf # Renaming from A_opt_cf, k_opt_cf, C_opt_cf
    print(f"curve_fit - Optimal parameters found:")
    print(f"  a_opt = {a_opt_cf:.3f} (True: {a_true})")
    print(f"  b_opt = {b_opt_cf:.3f} (True: {b_true})")
    print(f"  c_opt = {c_opt_cf:.3f} (True: {c_true})")
    std_dev_errors_cf = np.sqrt(np.diag(params_cov_cf))
    print(f"curve_fit - Standard deviation errors:")
    print(f"  a_err = {std_dev_errors_cf[0]:.3f}, b_err = {std_dev_errors_cf[1]:.3f}, c_err = {std_dev_errors_cf[2]:.3f}")

    # Adding a line for this model
    plt.plot(t_data, model_func(t_data, a_opt_cf, b_opt_cf, c_opt_cf), label=f'curve_fit (a={a_opt_cf:.2f}, b={b_opt_cf:.2f}, c={c_opt_cf:.2f})', color='red', linestyle='--', linewidth=2)
except RuntimeError:
    print("curve_fit failed. Could not find optimal parameters.")

# 6. Provide initial bounds for the parameters [a, b, c].
param_bounds = [(0, 10), (0, 2), (0, 5)]

# 7. Perform DE
print("\nSolving with SciPy differential_evolution...")
try:
    # Modifying the function call
    params_opt_de = differential_evolution(objective_func, bounds=param_bounds, args=(t_data, y_noisy), seed=42)
    a_opt_de, b_opt_de, c_opt_de = params_opt_de.x
    print(params_opt_de)

    print(f"differential_evolution - Optimal parameters found:")
    print(f"  a_opt = {a_opt_de:.3f} (True: {a_true})")
    print(f"  b_opt = {b_opt_de:.3f} (True: {b_true})")
    print(f"  c_opt = {c_opt_de:.3f} (True: {c_true})")

    # Adding a line for this model
    plt.plot(t_data, model_func(t_data, a_opt_de, b_opt_de, c_opt_de), label=f'differential_evolution (a={a_opt_de:.2f}, b={b_opt_de:.2f}, c={c_opt_de:.2f})', color='purple', linestyle='-.', linewidth=2)
except RuntimeError:
```

```
68      print("differential_evolution failed. Could not find optimal
        parameters.")
```

Listing 7.15 Python code for quadratic model fitting

7.2.5.3 Comparison

In Fig. 7.5 we can see both models fitted against the noisy data. However, the simpler model (`curve_fit`) found a better curve.

However, there was a problem in the implementation of the problem optimized by `differential_evolution`: in line 11, we have a typo in our code. Instead of `model_func(t_data_arg, a, b, c)`, we are calling `model_func (t_data_arg, b, b, b)` as a typo (on purpose, mind you). Therefore, even with powerful optimization, fitting the "wrong" model structure will generally lead to a poorer fit compared to the right implementation or a simpler model. How about fixing that line back to `model_func(t_data_arg, a, b, c)` and comparing the results?

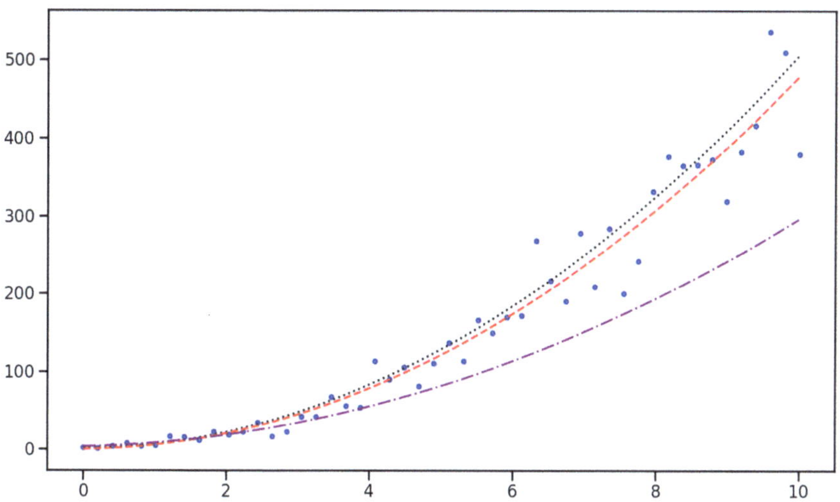

Fig. 7.5 The quadratic function curve fitting using the SciPy `curve_fit` (red dashed line) and `differential_evolution` (purple dashed and dotted line) methods against the original noisy data (blue dots) and the source curve before adding noise (black dotted line). *Source* The author, 2025

7.2.6 *Example 6: Designing a Cylindrical Water Tank*

Now, let's go back to a real-world example. Here, we want to design a cylindrical water tank that holds a specific volume of water while minimizing the amount of material used or, in other words, minimizing its surface area. As is the case with many water tanks worldwide, this tank should be closed at both ends.

7.2.6.1 Mathematical Formulation

So, let's start with our variables. The amount of material (that is, the surface area S) is calculated after the radius of the tank (r) and its height (h). So, these are our main variables.

Therefore, our objective can be defined as:

$$\min S(r, h) = 2\pi r^2 + 2\pi r h$$

Remember we've mentioned a "specific volume of water," right? It means we also need to define an equality constraint on which the volume $V = \pi r^2 h$ must be equal to a fixed value $V_0 = 10$ m^3. It also means that this is a constrained NLP.

Usually, optimization libraries expect that an equality constraint must always be equal to zero. Therefore, we can rewrite our constraint g as $g(r, h) = \pi r^2 h - V_0 = 0$. Last, but not least: the radius and height must respect the laws of physics. Therefore, $r > 0$ and $h > 0$.

7.2.6.2 Python Implementation with SciPy

In this problem, let's keep using SciPy. We use `scipy.optimize.minimize` with the Sequential Least Squares Programming (SLSQP) method. In this case, we have an equality (`'type': 'eq'`) constraint and a very small value as the lower bound for the variables r and h.

```
import numpy as np
from scipy.optimize import minimize

V0_base = 10.0
def surface_area_base(x):
    r, h = x
    return 2*np.pi*r**2 + 2*np.pi*r*h

def volume_constraint_base(x):
    r, h = x
    return np.pi*r**2*h - V0_base

cons_base = ({
    'type': 'eq',
    'fun': volume_constraint_base
})
```

```
bounds_base = [(1e-6, None), (1e-6, None)]
x0_base = [1.0, V0_base/(np.pi*1.0**2)]

print(f"\nSolving Base Water Tank Problem (V0={V0_base}m^3)...")
result_tank_base = minimize(surface_area_base, x0_base, method='
    SLSQP', bounds=bounds_base, constraints=cons_base)

if result_tank_base.success:
    r_opt, h_opt = result_tank_base.x
    print(f"Base Tank - Optimal: r={r_opt:.3f}m, h={h_opt:.3f}m,
    Area={result_tank_base.fun:.3f}m^2, h/(2r)={h_opt/(2*r_opt):.3
    f}")
else:
    print(f"Base Tank - Optimization failed: {result_tank_base.
    message}")
```

Listing 7.16 Python code for water tank design

7.2.7 *Example 7: Tank Design with Height Constraint*

It is perfectly possible to fine-tune an optimization problem after finding an optimal solution. Therefore, let's consider that the decision-makers want to introduce a new height constraint to the water tank problem. Being more specific, the height h of the tank must be no more than 1.5 times its radius r. All other constraints remain as is.

7.2.7.1 Mathematical Formulation

For this example, we will keep the equality constraint of $V_0 = 10\ \text{m}^3$ and the variables (i.e., the height h and the radius r). The objective is also the same.

So, what changes in our case is the introduction of a new inequality constraint: $h \leq 1.5r$. However, as it happened with the equality constraint, often the implementation of the inequality constraints in Optimization libraries must be standardized in a way to always be less than, or equal to zero. Therefore, we can modify this constraint to be $h - 1.5r \leq 0$.

Therefore, our Optimization problem is defined as:

$$\begin{aligned} &\min\ S(r,h) = 2\pi r^2 + 2\pi r h \\ &\text{subject to}\quad \pi r^2 h - V_0 = 0 \\ &\qquad\qquad\quad h - 1.5r \leq 0 \\ &\qquad\qquad\quad r > 0, h > 0 \end{aligned} \tag{7.2}$$

7.2.7.2 Python Implementation with SciPy

The code in this example is based on the code in Example 6. The new inequality constraint $h \leq 1.5r$ forced the solution away from the unconstrained optimum, increasing the surface area in the process.

```
import numpy as np
from scipy.optimize import minimize

def surface_area_base(x):
    r, h = x
    return 2*np.pi*r**2 + 2*np.pi*r*h

def volume_constraint_base(x):
    r, h = x
    return np.pi*r**2*h - V0_base

# New function
def height_radius_constraint(x):
    r, h = x
    return 1.5*r - h     # Constraint: h <= 1.5r, or 1.5r - h >= 0

cons_base = ({
    'type': 'eq',
    'fun': volume_constraint_base
}, { # New inequality constraint
    'type': 'ineq',
    'fun': height_radius_constraint
})

bounds_base = [(1e-6, None), (1e-6, None)]
x0_base = [1.0, V0_base/(np.pi*1.0**2)]

# Adjusted the print message below
print(f"\nSolving Constrained Water Tank Problem (V0={V0_base}m^3,
     h <= 1.5r)...")
result_tank_base = minimize(surface_area_base, x0_base, method='
    SLSQP', bounds=bounds_base, constraints=cons_base)

if result_tank_base.success:
    r_opt, h_opt = result_tank_base.x
    print(f"Base Tank - Optimal: r={r_opt:.3f}m, h={h_opt:.3f}m,
    Area={result_tank_base.fun:.3f}m^2, h/(1.5r)={h_opt/(1.5*r_opt
    ):.3f}")
else:
    print(f"Base Tank - Optimization failed: {result_tank_base.
    message}")
```

Listing 7.17 Python code for water tank design with height constraint

For comparison purposes, check below the outcomes from the two examples: first, the base (unconstrained) problem. Second, the constrained problem.

```
    Solving Base Water Tank Problem (V0=10.0m^3)...
    Base Tank - Optimal: r=1.168m, h=2.335m, Area=25.695m^2, h/(2r
    )=1.000
    ---------------------------------------------
    Solving Constrained Water Tank Problem (V0=10.0m^3, h <= 1.5r)
    ...
```

```
Base Tank - Optimal: r=1.285m, h=1.928m, Area=25.939m^2, h
/(1.5r)=1.000
```

Listing 7.18 Results comparing the Examples 6 and 7

7.2.8 Example 8: Constrained Truss Design

Now, let's think about one of the classical optimization problems: the two-bar truss structure. The goal is to minimize the total volume of material used (f_1) subject to the maximum allowable stress in the bars defined as:

$$\sigma_{\text{allowable}} = 100 \text{ MPa } (100 \times 10^6 \text{ Pa})$$

We also may have a few values that we cannot change (i.e., constants): the load is $F = 20$ kN, and the half-span is $B = 1$ m. These values are non-negotiable. On the other hand, we can modify the height H of the truss structure (between 0.1 m and 2 m) and the cross-sectional area A (between 10^{-5}m^2 and 10^{-3}m^2).

7.2.8.1 Mathematical Formulation

So, let's organize our problem into an equation. Let's start with the volume f_1: it depends on the area A and the height H:

$$f_1(H, A) = 2\,A\sqrt{B^2 + H^2}$$

The stress should respect the maximum allowable stress, or $\sigma(H, A) \leq \sigma_{\text{allowable}}$. This stress is defined as $\sigma(H, A) = \frac{F\sqrt{B^2+H^2}}{2HA}$.

Also, we should not forget the bounds of our variables H and A. Also, we can replace our constants with their numerical values (i.e., F with 20 and B with 1). Therefore, the complete definition of our Optimization problem is:

$$\begin{aligned} \min\ & f_1(H, A) = 2\,A\sqrt{1 + H^2} \\ \text{subject to}\ & \frac{20\sqrt{1 + H^2}}{2HA} - \sigma_{\text{allowable}} \leq 0 \\ & 0.1 \leq H \leq 2 \\ & 10^{-5} \leq A \leq 10^{-3} \end{aligned} \tag{7.3}$$

7.2.8.2 Python Implementation with SciPy

Now, let's implement this problem with SciPy. We will keep using the SLSQP technique, but notice that we are using a slightly different approach. First, notice that we have created the functions with the responsibility of validating the objective and the constraint (lines 10–23). Then, we declare the inequality (lines 25–28), as well as the bounds (line 29) and our initial guess (line 30). The optimizer itself is declared in line 33, while in the subsequent lines we present these results to the user.

```
1 import numpy as np
2 from scipy.optimize import minimize
3
4 # Problem constants
5 F_load_truss = 20e3   # N
6 B_half_span_truss = 1.0   # m
7 sigma_allowable_truss = 100e6   # Pa (100 MPa)
8
9 # Objective function: we need to minimize volume
10 def truss_volume(x): # x = [H, A]
11     H, A = x
12     L_bar = np.sqrt(B_half_span_truss**2 + H**2)
13     return 2 * A * L_bar
14
15 # Constraint function: the stress should be less than or equal to sigma_allowable
16 def stress_constraint_truss(x): # x = [H, A]
17     H, A = x
18     if H <= 1e-6 or A <= 1e-9: # avoiding division by zero if bounds are not strict enough
19     return np.inf   # a large number to indicate infeasibility
20
21     L_bar = np.sqrt(B_half_span_truss**2 + H**2)
22     stress = (F_load_truss * L_bar) / (2 * H * A)
23     return sigma_allowable_truss - stress
24
25 truss_cons = ({
26     'type': 'ineq',
27     'fun': stress_constraint_truss
28 })
29 truss_bounds = [(0.1, 2.0), (1e-5, 1e-3)]    # Bounds for [H, A] (as seen in the mathematical definition)
30 truss_x0 = [1.0, 5e-5]    # Initial guess for H, A
31
32 print("\nSolving Mono-Objective Constrained Truss Design with SciPy minimize (SLSQP)...")
33 result_truss_mono = minimize(truss_volume, truss_x0, method='SLSQP', bounds=truss_bounds, constraints=truss_cons)
34
35 if result_truss_mono.success:
36     H_opt_mono, A_opt_mono = result_truss_mono.x
37     vol_opt_mono = result_truss_mono.fun
38     stress_at_opt = (F_load_truss * np.sqrt(B_half_span_truss**2 + H_opt_mono**2)) / (2 * H_opt_mono * A_opt_mono)
39     print(f"Mono Truss - Optimal: H={H_opt_mono:.3f}m, A={A_opt_mono:.4e}m^2")
```

```
    print(f"  Min Volume={vol_opt_mono:.4e}m^3, Stress={
    stress_at_opt/1e6:.2f}MPa (Allowable: {sigma_allowable_truss/1
    e6:.2f}MPa)")
    print(f"  Constraint func val: {stress_constraint_truss(
    result_truss_mono.x):.2e} (should be >=0 and small if active)"
    )
else:
    print(f"Mono Truss - Optimization failed: {result_truss_mono.
    message}")
```

Listing 7.19 Python code for mono-objective constrained truss design

From this execution, we got the following results:

```
    Solving Mono-Objective Constrained Truss Design with SciPy
    minimize (SLSQP)...
    Mono Truss - Optimal: H=1.000m, A=1.4141e-04m^2
    Min Volume=4.0000e-04m^3, Stress=100.00MPa (Allowable: 100.00
    MPa)
    Constraint func val: 8.11e-03 (should be >=0 and small if
    active)
```

Listing 7.20 Result of the Example 8

7.2.9 Example 9: SVM Hyperparameter Tuning

Let's go back to an ML model example. One of the most known ML models is an SVM, and one of its most known implementations uses the scikit-learn library. In our context, let's consider we want to train an SVM classifier (known in scikit-learn as "SVC") using the RBF kernel. Here, our challenge is to find the best values for the hyperparameters C (regularization parameter) and γ (kernel coefficient) with the goal of maximizing the cross-validated accuracy based on a given dataset. In our example, we will use a toy dataset[2] known as the "Breast Cancer Wisconsin Dataset."

7.2.9.1 Mathematical Formulation

So, our goal is to maximize the average cross-validation accuracy of an SVM trained with parameters C and γ. We can define it as $f(C, \gamma)$.

However, remember that most Optimization algorithms are developed with minimization problems in mind. Therefore, we can modify this objective as being the minimization of the error rate, or:

$$1 - f(C, \gamma)$$

[2] A small and simple dataset used for teaching, learning, and testing purposes.

The parameters C and γ usually have a range of acceptable values, and often on a log scale. Let's consider that C can be a value between the $[10^{-3}, 10^{3}]$ range, and γ can be a value between the $[10^{-4}, 10^{1}]$ range. Therefore:

$$10^{-3} \leq C \leq 10^{3}$$

$$10^{-4} \leq \gamma \leq 10^{1}$$

7.2.9.2 Python Implementation with Optuna

In this case, let's try a real library specialized in hyperparameter optimization: Optuna. Here, we are using optimization to automate our search for the best hyperparameters (`C` and `gamma`) for SVC. To do that, we are creating an objective function (`svm_objective_optuna`) to be minimized. For any given set of hyperparameters suggested by Optuna, we will train an ML model, evaluate its performance using stratified cross-validation, and then we will return a single number: the error rate to be minimized.

Also, notice both parameters we want to optimize vary over orders of magnitude. In this case, to make things easier we are using the `log=True` parameter in the `trial.suggest_float` functions to sample values in a logarithmic scale.

```
import numpy as np
from sklearn.svm import SVC
from sklearn.model_selection import cross_val_score,
    StratifiedKFold
from sklearn.preprocessing import StandardScaler
from sklearn.datasets import load_breast_cancer
import optuna

# Loading a toy dataset
X, y = load_breast_cancer(return_X_y=True)

# Scaling all independent variables
scaler = StandardScaler()
X_scaled = scaler.fit_transform(X)

# Objective function for Optuna
def svm_objective_optuna(trial):
    # Define the hyperparameter search space
    # Optuna's suggest_float with log=True is suitable for C and
    gamma, which are often tuned on a logarithmic scale.
    C = trial.suggest_float('C', 1e-3, 1e3, log=True)
    gamma = trial.suggest_float('gamma', 1e-4, 1e1, log=True)
    model = SVC(C=C, gamma=gamma, random_state=42) # Using a fixed
     random_state for SVC

    # Using StratifiedKFold for balanced class representation in
    folds
    cv = StratifiedKFold(n_splits=3, shuffle=True, random_state
    =42) # Reduced splits for speed
    accuracy = cross_val_score(model, X_scaled, y, cv=cv, scoring=
    'accuracy').mean()
```

```
    # Optuna minimizes the returned value, so we return the error rate.
    error_rate = 1.0 - accuracy
    return error_rate

print("Optimizing SVM Hyperparameters with Optuna...")

# Create a study object and specify the direction as 'minimize'.
# Using TPESampler with a seed for reproducibility, similar to the random_state parameter.
sampler = optuna.samplers.TPESampler(seed=42)
study = optuna.create_study(direction='minimize', sampler=sampler)

# Start the optimization. n_trials is the number of times the objective function will be called.
study.optimize(svm_objective_optuna, n_trials=100)

# Get the results
best_params_optuna = study.best_params
best_C_optuna = best_params_optuna['C']
best_gamma_optuna = best_params_optuna['gamma']
min_error_rate_optuna = study.best_value
max_accuracy_optuna = 1.0 - min_error_rate_optuna

print(f"\nOptuna SVM Tuning - Optimal Parameters Found:")
print(f"  Best C: {best_C_optuna:.4f}")
print(f"  Best gamma: {best_gamma_optuna:.4f}")
print(f"  Max Cross-Validated Accuracy: {max_accuracy_optuna:.4f} (Min Error Rate: {min_error_rate_optuna:.4f})")
```

Listing 7.21 Python code for SVM hyperparameter tuning

This code optimizes with the `optuna.samplers.TPESampler` technique. It is a sophisticated Bayesian optimization algorithm which, instead of randomly guessing or exhaustively trying all combinations (like the grid search we have mentioned earlier), balances exploration and exploitation to find the best results. From this execution, we got the following result:

```
    Optuna SVM Tuning - Optimal Parameters Found:
    Best C: 7.5139
    Best gamma: 0.0148
    Max Cross-Validated Accuracy: 0.9807 (Min Error Rate: 0.0193)
```

Listing 7.22 Result of the Example 9

7.3 Multi-Objective Optimization (MOO) Examples

7.3.1 Example 10: Two-Bar Truss Design

In this example, we will revisit the symmetric two-bar truss problem seen in Sect. 7.2.8. The constants are the same: we still need to support a load $F = 20$ kN with a half-span $B = 1$ m. The variables are also the same: the height H and the bar cross-sectional area A.

However, we will change our objectives: in addition to minimizing the total volume of material used (f_1), we will also minimize the stress (f_2) at the same time.

7.3.1.1 Mathematical Formulation

We will reuse the same variables, constraints, and the objective f_1 seen in that example. Now, we will also include the objective f_2 to minimize stress. We can define it as:

$$> f_2(H, A) = \frac{F\sqrt{B^2 + H^2}}{2HA}$$

And then we would remove the stress from the constraints. Therefore, adapting it as a Multi-Objective Optimization problem results in:

$$\begin{aligned} \min \mathbf{f}(H, A) &= \begin{bmatrix} f_1(H, A) = 2\,A\sqrt{1 + H^2} \\ f_2(H, A) = \frac{20\sqrt{1+H^2}}{2HA} \end{bmatrix} \\ \text{subject to} \quad & 0.1 \leq H \leq 2 \\ & 10^{-5} \leq A \leq 10^{-3} \end{aligned} \tag{7.4}$$

7.3.1.2 Python Implementation with Pymoo (NSGA-II, SPEA2, and MOEA/D)

Now, let's use the `pymoo` library to implement this problem and compare three popular MOO algorithms we saw earlier: NSGA-II, SPEA2, and MOEA/D. You will notice that we are able to define our Multi-Objective Optimization Problem differently from the other libraries, with a definition closer to Software Engineering and farther from Mathematics.

In this example, we wanted you to pay attention to the following points:

1. First, we define the class `TwoBarTrussMOOProblem` containing the problem with its variables (2), objectives (2), lower and upper bounds for all variables, and constraints (none).
2. Then, we define the optimization algorithm limits, such as the maximum number of generations, the number of individuals in the population, and the seed number to allow our reproducibility.
3. Then, we define the three optimization algorithms (NSGA-II, SPEA2, and MOEA/D).
4. The last 30 rows are dedicated to show figures integrating the Pareto front approximations and Pareto sets from the three algorithms for comparison purposes.

```python
import numpy as np
from pymoo.core.problem import Problem
from pymoo.algorithms.moo.nsga2 import NSGA2
from pymoo.algorithms.moo.spea2 import SPEA2
```

```
from pymoo.algorithms.moo.moead import MOEAD
from pymoo.util.ref_dirs import get_reference_directions # For MOEA/D
from pymoo.optimize import minimize as pymoo_minimize
import matplotlib.pyplot as plt

# --- Problem Constants ---
F_load_moo = 20e3       # N (20 kN)
B_half_span_moo = 1.0   # m

# --- Define the MOO Problem Class ---
class TwoBarTrussMOOProblem(Problem): # Renamed to avoid clash
    def __init__(self):
        super().__init__(n_var=2, n_obj=2, n_constr=0, xl=np.array([0.1, 1e-5]), xu=np.array([2.0, 1e-3]))

    def _evaluate(self, x, out, *args, **kwargs):
        H_vals = x[:, 0]; A_vals = x[:, 1]
        L_bar = np.sqrt(B_half_span_moo**2 + H_vals**2)
        obj1_volume = 2 * A_vals * L_bar

        # Add small epsilon to denominator to prevent division by zero if H or A are at their absolute minimum bound
        # though problem bounds should ideally prevent H=0 or A=0.
        epsilon_denom = 1e-9
        obj2_stress = (F_load_moo * L_bar) / (2 * H_vals * A_vals + epsilon_denom)
        out["F"] = np.column_stack([obj1_volume, obj2_stress])

# --- Instantiate the Problem ---
problem_moo = TwoBarTrussMOOProblem()

# --- Algorithm Settings ---
pop_size_moo = 100
n_gen_moo = 100 # Number of generations
random_seed_moo = 42

# --- Solve with NSGA-II ---
algorithm_nsga2 = NSGA2(pop_size=pop_size_moo, eliminate_duplicates=True)
print("\nStarting NSGA-II optimization...")

res_nsga2 = pymoo_minimize(problem_moo, algorithm_nsga2, termination=('n_gen', n_gen_moo),
seed=random_seed_moo, verbose=False)
print("NSGA-II optimization complete.")

# --- Solve with SPEA2 ---
algorithm_spea2 = SPEA2(pop_size=pop_size_moo)
print("\nStarting SPEA2 optimization...")

res_spea2 = pymoo_minimize(problem_moo, algorithm_spea2, termination=('n_gen', n_gen_moo), seed=random_seed_moo, verbose=False)
print("SPEA2 optimization complete.")

# --- Solve with MOEA/D ---
# MOEA/D requires reference directions. For 2 objectives, this is straightforward.
```

```
ref_dirs = get_reference_directions("das-dennis", 2, n_partitions=99) # n_partitions controls number of subproblems/solutions
algorithm_moead = MOEAD(ref_dirs=ref_dirs, n_neighbors=15, prob_neighbor_mating=0.7)
print("\nStarting MOEA/D optimization...")

res_moead = pymoo_minimize(problem_moo, algorithm_moead, termination=('n_gen', n_gen_moo), seed=random_seed_moo, verbose=False)
print("MOEA/D optimization complete.")

# --- Process and Visualize Results ---
print("\n--- Truss MOO Results ---")

# Plotting Pareto Fronts (Objective Space)
plt.figure(figsize=(12, 8))
results_dict = {"NSGA-II": res_nsga2, "SPEA2": res_spea2, "MOEA/D": res_moead}
colors = {"NSGA-II": "blue", "SPEA2": "green", "MOEA/D": "purple"}
markers = {"NSGA-II": "o", "SPEA2": "s", "MOEA/D": "^"}

for name, res in results_dict.items():
    if res.F is not None and len(res.F) > 0:
        print(f"{name} found {len(res.F)} solutions on its Pareto front.")
        plt.scatter(res.F[:,0], res.F[:,1]/1e6, s=40, facecolors='none', edgecolors=colors[name], marker=markers[name], label=f'{name} Front')
    else:
        print(f"{name} did not yield results.")

plt.title('Pareto Front Comparison: NSGA-II vs SPEA2 vs MOEA/D')
plt.xlabel('Volume ($m^3$)')
plt.ylabel('Stress (MPa)')
plt.legend()
plt.grid(True)
plt.savefig('truss_pareto_front.png', dpi=300)
plt.show()

# Plotting Pareto Sets (Decision Variable Space)
fig_sets, axs_sets = plt.subplots(1, 3, figsize=(21, 6), sharex=True, sharey=True)
fig_sets.suptitle('Pareto Set Comparison (Decision Variables H vs A)', fontsize=16)

for i, (name, res) in enumerate(results_dict.items()):
    ax = axs_sets[i]
    if res.X is not None and len(res.X) > 0:
        ax.scatter(res.X[:,0], res.X[:,1], s=30, facecolors='none', edgecolors=colors[name], marker=markers[name])
        ax.set_title(f'Pareto Set ({name})')
        ax.set_xlabel('Height H (m)')
        ax.grid(True)

        if i == 0:
            ax.set_ylabel('Area A ($m^2$)') # Only set y-label for the first plot
            ax.ticklabel_format(axis='y', style='sci', scilimits=(0,0))
```

```
    else:
        ax.set_title(f'Pareto Set ({name}) - No results')

plt.savefig('truss_pareto_set.png', dpi=300)
plt.show()

print("\nTruss MOO plotting complete.")
```

Listing 7.23 Python code for two-bar truss MOO design with NSGA-II, SPEA2, and MOEA/D

The Pareto front approximations found by the three models are in Fig. 7.6, while the corresponding Pareto sets are in Fig. 7.7.

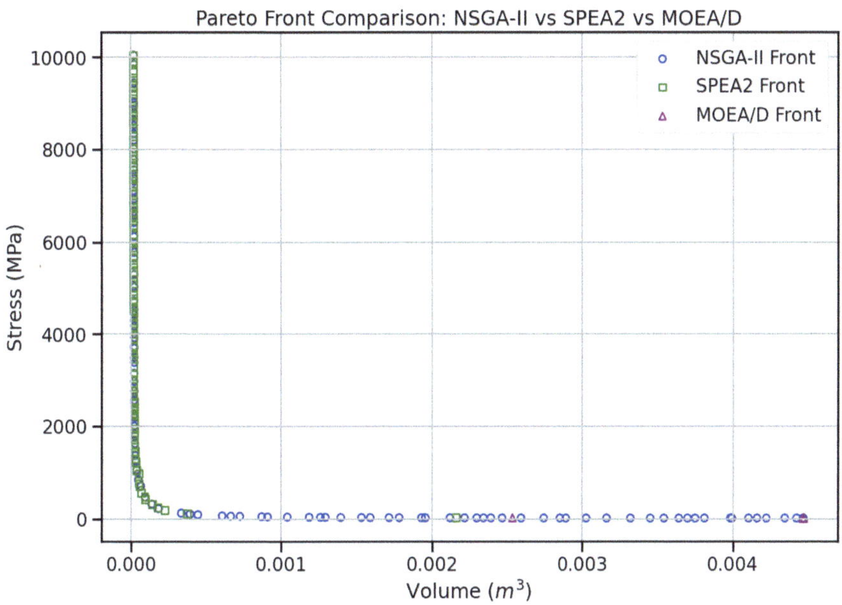

Fig. 7.6 The Pareto front approximations for the two-bar truss design problem through three different techniques. *Source* The author, 2025

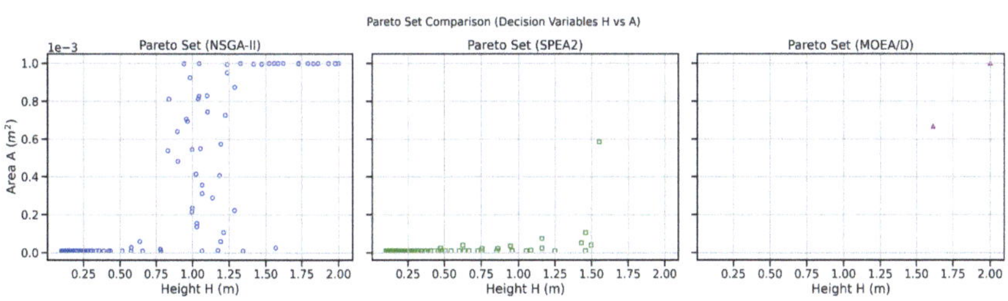

Fig. 7.7 The Pareto sets for the two-bar truss design problem through three different techniques. *Source* The author, 2025

The MOO algorithms each provide an approximation of the Pareto front. You might have noticed that, for this specific problem, we have differences in convergence, diversity, and extreme points.

Also, remember the pros and cons of each model? You can see here in practice: NSGA-II has good crowding distance and spreads solutions across its own front approximation. SPEA2 uses a fine-grained fitness assignment and density estimation, resulting in more diverse solutions. MOEA/D decomposes the multi-objective problem into a number of scalar optimization subproblems and solves them simultaneously, and as a result, we have a result sensitive to the choice of reference directions and aggregation functions.

7.3.2 *Example 11: Classifier Tuning*

We have mentioned in Sect. 6.3.8 that, in some situations, data scientists are faced with the challenge of balancing accuracy and other conflicting objectives, such as interpretability, fairness, and computing resources. Let's consider a random forest model, for example. In this model, increasing the number of trees (`n_estimators`) or the maximum depth of trees (`max_depth`) will improve its chances of having increased accuracy at the expense of having a more complex model, which takes more time to train and uses more memory to serve predictions.

Therefore, in this example, we want to find the best hyperparameters for a random forest classifier that, at the same time, minimizes the model complexity (measured by the `n_estimators` parameter multiplied by `max_depth`) and maximizes the cross-validated model accuracy. We will also use a toy dataset in this example—in this case, the wine dataset.

7.3.2.1 Mathematical Formulation

First, let's rename a few things before moving forward. Starting with the variables, let $x_1 =$ `n_estimators` and $x_2 =$ `max_depth`.

Also, let f_1 be the first objective (accuracy) and f_2 be the second objective (complexity, or the number of estimators). Also, remember that f_1 should be converted into a minimization objective since several Optimization libraries expect to work only with minimization objectives.

We also need to define the bounds of hyperparameters. In the case of a random forest for this example, we may test between 10 and 200 estimators, and between 2 and 20 maximum depth per tree. With these points considered, we can write our problem as:

$$\min \mathbf{f}(x_1, x_2) = \begin{bmatrix} f_1(x_1, x_2) = 1 - \text{Accuracy}(x_1, x_2) \\ f_2(x_1) = x_1 * x_2 \end{bmatrix}$$
$$\text{subject to} \quad 10 \leq x_1 \leq 200$$
$$2 \leq x_2 \leq 20 \tag{7.5}$$

7.3.2.2 Python Implementation with Pymoo

```
import numpy as np
from sklearn.ensemble import RandomForestClassifier
from sklearn.model_selection import cross_val_score,
    StratifiedKFold
from sklearn.datasets import load_wine
from pymoo.core.problem import Problem
from pymoo.algorithms.moo.nsga2 import NSGA2
from pymoo.optimize import minimize as pymoo_minimize
from pymoo.core.variable import Real, Integer # For variable types
     if needed, though NSGA2 handles it
import matplotlib.pyplot as plt

# Load wine dataset
X_rf, y_rf = load_wine(return_X_y=True)

class RandomForestTuningProblem(Problem):
    def __init__(self):
        super().__init__(n_var=2, n_obj=2, n_constr=0, xl=np.array
    ([10, 2]), xu=np.array([100, 15]))

    def _evaluate(self, x, out, *args, **kwargs):
        n_solutions = x.shape[0]
        obj_values = np.full((n_solutions, 2), np.nan)

        for i in range(n_solutions):
            n_estimators = int(round(x[i, 0])) # Ensure integer
            max_depth = int(round(x[i, 1]))    # Ensure integer
            if max_depth == 0:
            max_depth = None # Treat 0 as None for max_depth if
    desired, or ensure min bound > 0

            model = RandomForestClassifier(n_estimators=
    n_estimators, max_depth=max_depth, random_state=42, n_jobs=1)
    # Use n_jobs=1 for safety in some parallel contexts

            # Using StratifiedKFold for balanced class
    representation in folds
            cv = StratifiedKFold(n_splits=3, shuffle=True,
    random_state=42) # Reduced splits for speed
            try:
            accuracy = cross_val_score(model, X_rf, y_rf, cv=cv,
    scoring='accuracy').mean()
            except ValueError: # Can happen if max_depth is too
    small or other invalid params
            accuracy = 0.0

            obj_values[i, 0] = 1.0 - accuracy  # Objective 1:
    Minimize (1 - accuracy)
```

```
            obj_values[i, 1] = n_estimators * max_depth  # Objective 2: Minimize complexity

        out["F"] = obj_values

# Instantiate and solve
problem_rf = RandomForestTuningProblem()
algorithm_rf = NSGA2(pop_size=20, eliminate_duplicates=True)
print("\nOptimizing Random Forest Hyperparameters (MOO)...")
res_rf = pymoo_minimize(problem_rf, algorithm_rf, termination=('n_gen', 50), seed=42, verbose=False)

# Plotting results
if res_rf.F is not None and len(res_rf.F) > 0:
print(f"RF MOO Tuning found {len(res_rf.F)} solutions.")
    plt.figure(figsize=(10, 7))
    # Objective 1 is error (1-accuracy), Objective 2 is n_estimators
    # We can plot Accuracy vs n_estimators for better intuition
    accuracies = 1.0 - res_rf.F[:, 0]
    n_estimators_vals = res_rf.F[:, 1]

    plt.scatter(n_estimators_vals, accuracies, s=50, facecolors='none', edgecolors='purple')
    plt.title('Random Forest Tuning: Accuracy vs. Complexity (n_estimators)')
    plt.xlabel('Complexity (Number of Estimators)')
    plt.ylabel('Cross-Validated Accuracy')
    plt.grid(True)
    plt.savefig('rf_moo_tuning_front.png', dpi=300)
    plt.show()

    # Show some example solutions from the Pareto front
    print("\nExample solutions from Pareto Front (Complexity, Accuracy):")
    # Sort by complexity (n_estimators) for display
    sorted_indices = np.argsort(res_rf.F[:, 1])
    for k in range(min(5, len(sorted_indices))):
        idx = sorted_indices[k]
        n_est_sol = int(round(res_rf.X[idx, 0]))
        m_depth_sol = int(round(res_rf.X[idx, 1]))
        acc_sol = 1.0 - res_rf.F[idx, 0]
        comp_sol = res_rf.F[idx, 1] # This is n_estimators from the objective
        print(f"  Sol: n_est={n_est_sol}(obj:{comp_sol:.0f}), max_d={m_depth_sol} => Accuracy={acc_sol:.4f}")
else:
    print("RF MOO Tuning did not yield results.")
```

Listing 7.24 Python code for MOO of Random Forest hyperparameters

The Pareto front generated and shown in Fig. 7.8 has a trade-off: achieving higher accuracy often requires a more complex model (more estimators and higher depth). With this result in hand, data scientists can then choose a model from this front based on their specific needs. Sometimes, a slightly less accurate model is acceptable if it is significantly simpler and faster to train and predict. Or, depending on the situation, the team may opt for maximum accuracy instead.

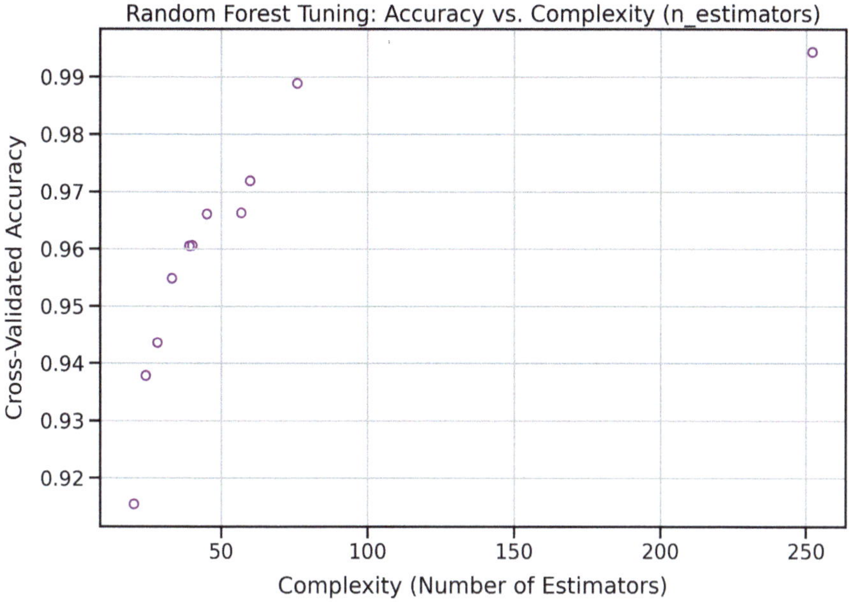

Fig. 7.8 The Pareto front approximations for the two-bar truss design problem through three different techniques. *Source* The author, 2025

7.3.3 Example 12: Portfolio Optimization

Let's assume that an investor wants to distribute capital across N available assets. In this case, the investor wants to maximize the expected portfolio return, minimize the portfolio risk (using the variance as a metric), and maximize the average liquidity score of the chosen portfolio. These three objectives should be optimized simultaneously.

We can rely on percentages to represent this asset distribution, where 0% means no capital allocated and 100% means that all capital is invested in an asset i. However, in programming code, we often represent the 0%-100% range as the interval [0, 1]. Therefore, our decision variables are the weights w_i representing the fraction of capital invested in asset i, where $\sum_{i=1}^{N} w_i = 1$ (that is, we should allocate 100% of the capital across all assets) and $w_i \geq 0$ (we should allocate either 0% or more than 0% to each asset).

7.3.3.1 Mathematical Formulation

As a starting point, we already know that we will be working with the asset weights, and there should be N weights in total. Therefore, let's first establish that $\mathbf{w} = [w_1, w_2, \ldots, w_N]^T$ is the vector of asset weights.

In the same way, we need to determine the expected returns and the liquidity scores for each asset. It means that we should have at least two additional vectors. The first one will be the vector of expected returns:

$$\mu = [\mu_1, \mu_2, \ldots, \mu_N]^T$$

And then, the vector of liquidity scores:

$$\mathbf{l} = [l_1, l_2, \ldots, l_N]^T$$

Also, we need to calculate the $N \times N$ covariance matrix Σ of asset returns.

Finally, remember we have three objectives: the expected portfolio return $f_1(\mathbf{w})$, the risk $f_2(\mathbf{w})$, and the liquidity $f_3(\mathbf{w})$. Since $f_1(\mathbf{w})$ and $f_3(\mathbf{w})$ are maximization problems, we need to convert them into minimization problems first by taking their negative. It means, in practice, that our problem can be mathematically defined as:

$$\begin{aligned} \min \mathbf{f}(H, A) &= \begin{bmatrix} f_1(\mathbf{w}) = -\mathbf{w}^T \mu \\ f_2(\mathbf{w}) = \mathbf{w}^T \Sigma \mathbf{w} \\ f_3(\mathbf{w}) = -\mathbf{w}^T \mathbf{l} \end{bmatrix} \\ \text{subject to } & \sum_{i=1}^{N} w_i - 1 = 0 \\ & w_i \geq 0 \quad \forall i = 1, \ldots, N \end{aligned} \tag{7.6}$$

7.3.3.2 Python Implementation with Pymoo

For this example, we will use pymoo again, especially because of its implementation of the NSGA-III algorithm, which is suitable for our scenario with three objectives. In this implementation, we will need to first create synthetic data for expected returns, covariance, and liquidity.

Then, as seen in the other examples, you will notice that some patterns will repeat: the definition of the class containing the Optimization problem according to pymoo's documentation, and the call to the optimizer is located in a single line of code. Finally, we dedicate the last 25 lines to generate a 3D scatterplot containing the Pareto front. This scatterplot is seen in Fig. 7.9.

```
import numpy as np
from pymoo.core.problem import Problem
from pymoo.algorithms.moo.nsga3 import NSGA3 # NSGA-III is good
    for many objectives
from pymoo.util.ref_dirs import get_reference_directions # For
    MOEA/D
from pymoo.optimize import minimize as pymoo_minimize
import matplotlib.pyplot as plt
from mpl_toolkits.mplot3d import Axes3D # For 3D scatter plot

# Creating synthetic data for 5 assets
```

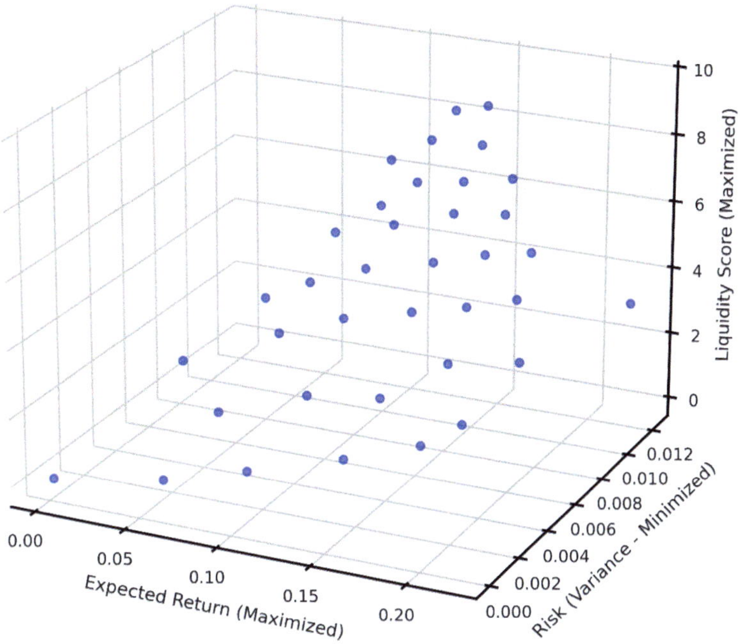

Fig. 7.9 The Pareto front approximation for the portfolio optimization problem. *Source* The author, 2025

```
np.random.seed(42)
n_assets = 5
expected_returns = np.random.rand(n_assets) * 0.2 + 0.05 # avg returns 5-25%
temp_cov = np.random.rand(n_assets, n_assets)
covariance_matrix = np.dot(temp_cov, temp_cov.transpose()) / 100 + np.eye(n_assets)*0.001
liquidity_scores = np.random.rand(n_assets) * 10 # Scores from 0 to 10

class PortfolioProblem3Obj(Problem):
    def __init__(self, n_assets, expected_returns, covariance_matrix, liquidity_scores):
        super().__init__(n_var=n_assets, n_obj=3, n_constr=1, xl=0.0, xu=1.0)
        self.expected_returns = expected_returns
        self.covariance_matrix = covariance_matrix
        self.liquidity_scores = liquidity_scores

    def _evaluate(self, W, out, *args, **kwargs): # W is a matrix of weight vectors
        obj1 = -np.sum(W * self.expected_returns, axis=1) # Maximize return
        obj2 = np.array([w @ self.covariance_matrix @ w for w in W]) # Minimize variance
        obj3 = -np.sum(W * self.liquidity_scores, axis=1) # Maximize liquidity
```

```python

        out["F"] = np.column_stack([obj1, obj2, obj3])

        # Constraint: sum of weights - 1 = 0
        out["G"] = np.sum(W, axis=1) - 1.0

# Instantiate problem
problem_portfolio = PortfolioProblem3Obj(n_assets, expected_returns, covariance_matrix, liquidity_scores)

# Create reference directions for 3 objectives
ref_dirs_3d = get_reference_directions("das-dennis", 3, n_partitions=12)
algorithm_portfolio = NSGA3(pop_size=100, ref_dirs=ref_dirs_3d)

print("\nOptimizing Portfolio (3 Objectives) with NSGA-III...")
res_portfolio = pymoo_minimize(problem_portfolio, algorithm_portfolio, termination=('n_gen', 100), seed=42, verbose=False)

# Plotting 3D Pareto Front
if res_portfolio.F is not None and len(res_portfolio.F) > 0:
    print(f"Portfolio MOO found {len(res_portfolio.F)} solutions.")
    fig = plt.figure(figsize=(10, 8))
    ax = fig.add_subplot(111, projection='3d')

    # Objectives were minimized: -Return, Variance, -Liquidity
    # So, for plotting: Return = -F[:,0], Variance = F[:,1], Liquidity = -F[:,2]
    ax.scatter(-res_portfolio.F[:,0], res_portfolio.F[:,1], -res_portfolio.F[:,2], s=30, c='blue', marker='o', alpha=0.6)

    ax.set_title('Portfolio Optimization: 3D Pareto Front')
    ax.set_xlabel('Expected Return (Maximized)')
    ax.set_ylabel('Risk (Variance - Minimized)')
    ax.set_zlabel('Liquidity Score (Maximized)')
    ax.view_init(elev=20., azim=-65) # Adjust view angle
    plt.grid(True)
    plt.savefig('portfolio_3d_front.png', dpi=300)
    plt.show()

    print("\nExample portfolio weights from Pareto Front (Return, Risk, Liquidity):")
    for k in range(min(3, len(res_portfolio.X))):
    weights = res_portfolio.X[k]
        ret = -res_portfolio.F[k,0]
        risk = res_portfolio.F[k,1]
        liq = -res_portfolio.F[k,2]
        print(f"  Sol {k+1}: Weights {np.round(weights,3)}, Return={ret:.3f}, Risk={risk:.4f}, Liquidity={liq:.2f}")
else:
    print("Portfolio MOO did not yield results.")
```

Listing 7.25 Python code for 3-objective portfolio optimization

In this problem, we used NSGA-III, as it is designed to handle many-objective problems by using reference directions to maintain diversity. In fact, you can see from the results that the resulting 3D Pareto front shows a range of diverse solutions considering the trade-offs between maximizing return, minimizing risk, and maximizing liquidity. On the other hand, notice a challenge in the visualization, as trying to compare the solutions in a three-dimensional static image is challenging. It is for that reason that techniques such as parallel coordinate plots are often used.

7.3.4 Example 13: Heat Exchanger Design

We have mentioned the adoption of Optimization problems toward heat exchangers in the book. Although in practice it is a complex engineering problem, we can simplify it as an example for learning purposes.

Therefore, let's consider the challenge of designing a simple shell-and-tube heat exchanger. In this problem, consider that we can modify the tube diameter (D_t), the number of tubes (N_t), and the tube length (L_t).

Here, we need to maximize the heat transfer rate (Q), minimize the pressure drop (ΔP) on the tube side, and minimize the manufacturing cost related to surface area A_{surf} and number of tubes N_t. All these objectives must be optimized simultaneously.

7.3.4.1 Mathematical Formulation

Let's focus on the mathematical formulation, as it happened in the last examples. Remember that this example is an oversimplification of a real example, though.

Having that said, the heat transfer rate could be formulated as:

$$Q = N_t D_t L_t$$

Remember that this objective should be converted to minimization for compatibility purposes with our algorithms. Therefore, the first objective f_1 can be determined as being $-Q$.

The second objective f_2 is the pressure drop ΔP (i.e., $f_2 = \Delta P$), which is defined as:

$$\Delta P = L_t/(N_t D_t^3)$$

The third objective f_3 is the cost C, which is defined as:

$$C = (N_t D_t L_t) * 5 + (N_t) * 2$$

Also, the constraints could be subject to real-world acceptable bounds, such as:

- The tube diameter D_t could be anything between 0.01 m and 0.05 m.
- The number of tubes N_t could be anything between 10 and 200.
- The tube length L_t could be anything between 0.5 m and 5.0 m.

Summarizing, we end up with:

$$
\begin{aligned}
\min \mathbf{f}(N_t, D_t, L_t) = &\begin{bmatrix} f_1(N_t, D_t, L_t) = -(N_t D_t L_t) \\ f_2(N_t, D_t, L_t) = L_t/(N_t D_t^3) \\ f_3(N_t, D_t, L_t) = (N_t D_t L_t) * 5 + (N_t) * 2 \end{bmatrix} \\
\text{subject to} \quad & 0.01 \leq D_t \leq 0.05 \\
& 10 \leq N_t \leq 200 \\
& 0.5 \leq L_t \leq 5.0
\end{aligned}
\tag{7.7}
$$

7.3.4.2 Python Implementation with Pymoo

Let us adapt the previous example, keeping basically the same structure. We will still use NSGA-III from `pymoo` to solve this problem. Note the class `HeatExchangerProblem` contains the implementation of our (simplified) problem. Here, we have three variables and three objectives. However, we have zero constraints (`n_constr=0`, in line 10). This is possible because `pymoo` enables us to inform the lower (`xl`) and upper (`xu`) bounds for each variable, and, as seen in our mathematical formulation, all constraints are the bounds for each variable, in practice.

```
1  import numpy as np
2  from pymoo.core.problem import Problem
3  from pymoo.algorithms.moo.nsga3 import NSGA3
4  from pymoo.optimize import minimize as pymoo_minimize
5  import matplotlib.pyplot as plt
6  from mpl_toolkits.mplot3d import Axes3D
7
8  class HeatExchangerProblem(Problem):
9      def __init__(self):
10         super().__init__(n_var=3, n_obj=3, n_constr=0,
11         xl=np.array([0.01, 10, 0.5]), # Min: Dt=0.01m, Nt=10, Lt
       =0.5m
12         xu=np.array([0.05, 200, 5.0])) # Max: Dt=0.05m, Nt=200, Lt
       =5m
13
14     def _evaluate(self, x, out, *args, **kwargs):
15         Dt = x[:, 0]
16         Nt = x[:, 1]
17         Lt = x[:, 2]
18
19         # Obj 1: Maximize Heat Transfer Q (so minimize -Q)
20         obj1_neg_Q = -(Nt * Dt * Lt)
21
22         # Obj 2: Minimize Pressure Drop dP
23         obj2_dP = Lt / (Nt * Dt**3 + 1e-9) # Added 1e-9 for
       stability
24
25         # Obj 3: Minimize Cost
26         obj3_Cost = (Nt * Dt * Lt) * 5 + Nt * 2 # Arbitrary cost
       factors
```

```
        out["F"] = np.column_stack([obj1_neg_Q, obj2_dP, obj3_Cost
    ])

# Instantiate problem
problem_hx = HeatExchangerProblem()

# Create reference directions for 3 objectives
ref_dirs_3d = get_reference_directions("das-dennis", 3,
    n_partitions=12)
algorithm_hx = NSGA3(pop_size=100, ref_dirs=ref_dirs_3d)
print("\nOptimizing Heat Exchanger (3 Objectives) with NSGA-III...
    ")
res_hx = pymoo_minimize(problem_hx, algorithm_hx, termination=('
    n_gen', 100), seed=42, verbose=False)

# Plotting 3D Pareto Front
if res_hx.F is not None and len(res_hx.F) > 0:
    print(f"Heat Exchanger MOO found {len(res_hx.F)} solutions.")
    fig = plt.figure(figsize=(10, 8))
    ax = fig.add_subplot(111, projection='3d')

    # Objectives: -Q, dP, Cost
    # Plotting: Q = -F[:,0], dP = F[:,1], Cost = F[:,2]
    ax.scatter(-res_hx.F[:,0], res_hx.F[:,1], res_hx.F[:,2], s=30,
     c='blue', marker='o', alpha=0.6)

    ax.set_title('Heat Exchanger Design')
    ax.set_xlabel('Heat Transfer Q (Maximized)')
    ax.set_ylabel('Pressure Drop dP (Minimized)')
    ax.set_zlabel('Manufacturing Cost (Minimized)')
    ax.view_init(elev=20., azim=-65)
    plt.grid(True)
    plt.savefig('heatexchanger_3d_front.png', dpi=300)
    plt.show()
else:
    print("Heat Exchanger MOO did not yield results.")
```

Listing 7.26 Python code for MOO of Heat Exchanger Design

In this case, we can see the Pareto front approximation in Fig. 7.10. When comparing against the previous example, note that we are applying the same Optimization concepts in different domains of knowledge: now, we are using NSGA-III toward Engineering design.

7.3.5 Example 14: Feature Selection for a Regression Problem

We can also rely on Optimization algorithms for feature selection problems in Data Science. As seasoned data scientists know, using too many input features can lead to increased model complexity, higher computational costs, and the curse of dimensionality.

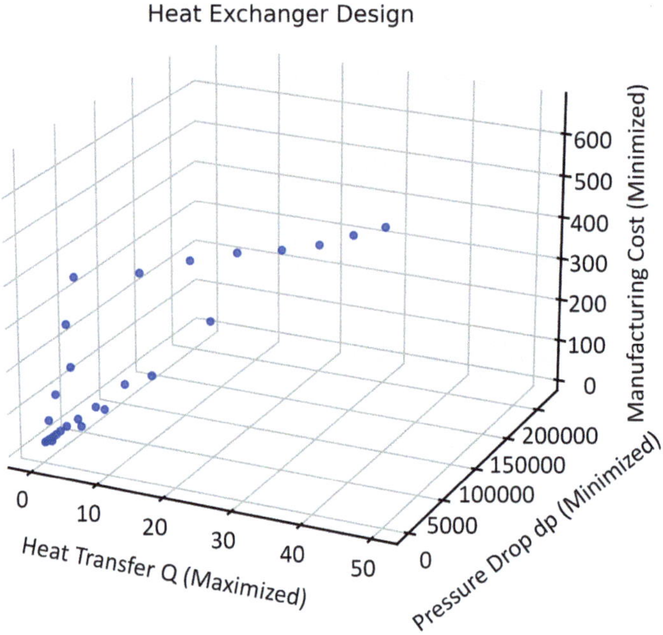

Fig. 7.10 The Pareto front approximation for the heat exchanger optimization problem. *Source* The author, 2025

Therefore, in this example we want to select a subset of features for a regression model that minimizes the number of selected features and minimizes the prediction error. Since it is a regression model, we could use the Root Mean Squared Error (RMSE) as a prediction error metric.

For testing purposes we will fit a linear regression model using the California housing dataset, another toy dataset for regression problems.

7.3.5.1 Mathematical Formulation

Since it is a feature selection, we need to consider that we will have N features, and our decision vector then has the form:

$$\mathbf{x} = [x_1, x_2, \ldots, x_N]$$

It is a binary vector, where $x_i = 1$ means that the feature i is selected, or $x_i = 0$ otherwise. In this case, we should minimize, at the same time, the cross-validation RMSE (f_1) of a model trained with the features as selected in $\mathbf{x}$, and the number of features selected in $\mathbf{x}$. These two objectives can then be formalized as:

$$
\begin{aligned}
\min \mathbf{f}(\mathbf{x}) &= \begin{bmatrix} f_1(\mathbf{x}) = \text{RMSE}(\mathbf{x}) \\ f_2(\mathbf{x}) = \sum_{j=1}^{N} x_j \end{bmatrix} \\
\text{subject to} \quad & x_i \in \{0, 1\} \quad \forall i = 1, \ldots, N
\end{aligned}
\tag{7.8}
$$

7.3.5.2 Python Implementation with Optuna

In this case, let us use Optuna for Multi-Objective Optimization. This is an interesting use case, as several data scientists use this and other similar libraries only to solve Single-Objective Optimization problems.

Here, we wanted you to notice a few interesting things: first, several data scientists use Optuna to optimize the hyperparameters of an ML model, but not the features. In this case, we are preparing Optuna for this second scenario. In the "objective_optuna_fs" we are creating a mask of selected features: a list containing the values 0 and 1, where 0 means that the feature was not selected, and 1 means that the feature is selected. For example:

$$[0, 0, 0, 1, 0]$$

means that only the fourth feature has been selected out of five features, while:

$$[1, 0, 1, 1, 0]$$

means that the first, third, and fourth features have been selected.

After understanding which features are selected, we will train a model (in our case, a simple linear regression model) and evaluate it, returning the RMSE for the model trained with that feature subset, as well as the amount of features selected.

Then, to optimize it we are using Optuna's `NSGAIISampler` (lines 50–52), which is a sampler based on the NSGA-II technique. After defining it, we will start the optimization with a limit of `n_trials` evaluations.

```
1 import numpy as np
2 from sklearn.linear_model import LinearRegression
3 from sklearn.model_selection import cross_val_score, KFold
4 from sklearn.datasets import fetch_california_housing
5 from sklearn.preprocessing import StandardScaler
6 import optuna
7 import matplotlib.pyplot as plt
8
9 # Load California Housing dataset
10 X_fs, y_fs = fetch_california_housing(return_X_y=True)
11 n_features_total = X_fs.shape[1]
12
13 # Scale features
14 scaler_fs = StandardScaler()
15 X_fs_scaled = scaler_fs.fit_transform(X_fs)
16
17 # Objective function for Optuna
18 def objective_optuna_fs(trial):
19     # Suggest a binary value (0 or 1) for each feature
```

```
    selected_features_mask = []
    for i in range(n_features_total):
        # Each selection is an independent choice
        selected_features_mask.append(trial.suggest_int(f'feature_{i}', 0, 1))

    selected_features_mask = np.array(selected_features_mask, dtype=bool)
    num_selected = np.sum(selected_features_mask)

    # No features selected: we want to force high values to disencourage.
    if num_selected == 0:
        return 1e10, n_features_total

    # Otherwise, continue. Training the model with the selected features only.
    X_subset = X_fs_scaled[:, selected_features_mask]

    model = LinearRegression()
    # Consistent KFold for each trial
    cv_kfold = KFold(n_splits=5, shuffle=True, random_state=42)
    neg_rmse_scores = cross_val_score(model, X_subset, y_fs, cv=cv_kfold, scoring='neg_root_mean_squared_error')
    rmse = -neg_rmse_scores.mean()

    # Optuna expects a tuple of objective values for MOO
    # This is why we are returning 2 values
    return rmse, num_selected

print("\nOptimizing Feature Selection with Optuna...")

# Define the sampler: NSGAIISampler is good for multi-objective problems
# population_size is similar to Pymoo's pop_size
# seed for reproducibility
sampler_fs = optuna.samplers.NSGAIISampler(
    population_size=50, # Aligning with Pymoo's pop_size
    seed=42
)

# Create a study object. For MOO, specify directions for each objective.
# We want to minimize RMSE and minimize the number of features.
study_fs = optuna.create_study(
    directions=['minimize', 'minimize'],
    sampler=sampler_fs,
    study_name='feature_selection_moo'
)

study_fs.optimize(objective_optuna_fs, n_trials=150)
pareto_front_trials = study_fs.best_trials

if pareto_front_trials:
    print(f"Feature Selection MOO (Optuna) found {len(pareto_front_trials)} solutions on the Pareto front.")

    # Extract objective values (F) and decision variables (X)
```

```python
    F_optuna = np.array([trial.values for trial in pareto_front_trials])

    X_optuna_params = [trial.params for trial in pareto_front_trials]
    X_optuna_bool_masks = []
    for params in X_optuna_params:
        mask = np.array([params[f'feature_{i}'] for i in range(n_features_total)], dtype=bool)
        X_optuna_bool_masks.append(mask)
    X_optuna_bool_masks = np.array(X_optuna_bool_masks)

    plt.figure(figsize=(10, 7))
    # F_optuna[:, 1] is num_selected, F_optuna[:, 0] is RMSE
    plt.scatter(F_optuna[:, 1], F_optuna[:, 0], s=50, facecolors='none', edgecolors='blue')
    plt.title('Optuna Feature Selection: RMSE vs. Number of Features')
    plt.xlabel('Number of Selected Features (Minimized)')
    plt.ylabel('Cross-Validated RMSE (Minimized)')
    plt.grid(True)
    plt.savefig('fs_moo_optuna_front.png', dpi=300)
    plt.show()

    print("\nExample solutions from Optuna's Pareto Front (Num Features, RMSE):")
    # Sort by number of features for display
    sorted_indices_optuna = np.argsort(F_optuna[:, 1])

    for k in range(min(5, len(sorted_indices_optuna))):
        idx = sorted_indices_optuna[k]
        num_feat_sol = F_optuna[idx, 1]
        rmse_sol = F_optuna[idx, 0]
        selected_indices = np.where(X_optuna_bool_masks[idx])[0]
        print(f"  Sol: NumFeat={num_feat_sol:.0f}, RMSE={rmse_sol:.4f}, Indices: {selected_indices}")
else:
    print("Feature Selection MOO (Optuna) did not yield results or no trials completed successfully.")
```

Listing 7.27 Python code for MOO Feature Selection

After finishing the Optimization, we will get the best individuals for our Pareto front approximation with the `study.best_trials` attribute, resulting in the examples below.

```
Example solutions from Optuna's Pareto Front (Num Features, RMSE):
Sol: NumFeat=1, RMSE=0.8373, Indices: [0]
Sol: NumFeat=2, RMSE=0.8085, Indices: [0 1]
Sol: NumFeat=2, RMSE=0.8085, Indices: [0 1]
Sol: NumFeat=2, RMSE=0.8085, Indices: [0 1]
Sol: NumFeat=3, RMSE=0.8018, Indices: [0 1 6]
```

Listing 7.28 Results for the Example 14

Also, we are able to generate a 3D scatterplot with the Pareto front approximation, as seen in Fig. 7.11.

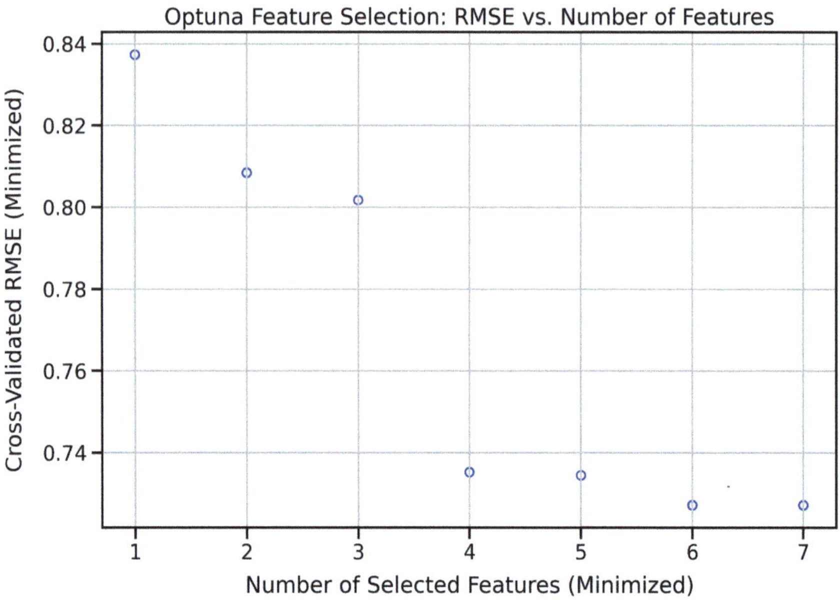

Fig. 7.11 The Pareto front approximation for the feature selection of a regression model. *Source* The author, 2025

7.4 Multi-Criteria Decision-Making (MCDM) for a Posteriori MOO

In the previous sections we also mentioned the relevance of using MCDM strategies, and how helpful it can be to assist decision-makers in selecting one or a small subset of solutions from the Pareto front approximation for implementation. Therefore, in this section we will explore how we can use MCDM techniques with a Pareto front approximation generated based on our previous MOO examples. As we have standardized our previous objectives to always be minimization problems, we will keep with this standard in this section.

Important
You will notice that we will demonstrate and compare multiple MCDM techniques for a given Pareto front approximation. However, in real-world scenarios you are not expected to do the same: often you might rely on a single visual technique, a single quantitative technique, or a combination of both. In a few exceptions you might explore one or two extra techniques, but not all of them.

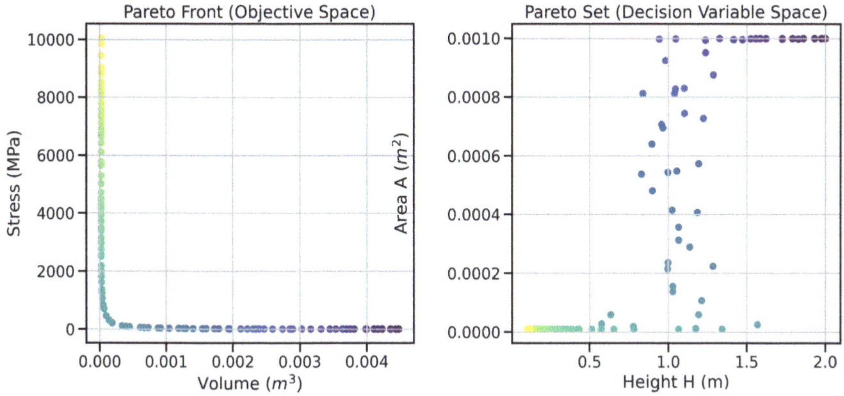

Fig. 7.12 2D scatterplots. On the left, the scatterplot for a Pareto front approximation. On the right, the corresponding Pareto set. The colors of the solutions in both plots are matching. *Source* The author, 2025

7.4.1 Visual Techniques

Therefore, let us first focus on the visual techniques in order to understand the trade-offs of our solutions and effectively see our Pareto front approximation.

7.4.1.1 2D and 3D Scatter Plots

Let us start with a 2D scatter plot using Example 11 (two-bar truss design) seen in Sect. 7.3.1. It is an MOO example with two objectives. In Fig. 7.12 we are showing the Pareto front approximation and the corresponding Pareto set found by the NSGA-II implementation in that example.

Notice that we were able to put both the front and set side-by-side to make the process of comparing the values easier. Also, we set a color scale to match both plots: a yellow dot in the Pareto front (high stress, low volume) can also be seen in the Pareto set (low height, low area). With that in mind, we are able to understand the relationships between the trade-offs in the variable coefficients against the objectives.

Now, what if we had three or more objectives? Let us use the portfolio optimization example in Sect. 7.3.3 to demonstrate this. It is an MOO example with three objectives.

For two objectives, a simple scatter plot as seen earlier is very effective. For three objectives (as in Sect. 7.3.3), 3D scatter plots can be used, though they can be harder to interpret fully without interactivity (e.g., zoom, hover, and scale dynamically). As an example, see the Pareto front approximation in Fig. 7.13. This is the same front, but seen in 9 different angles. Depending on the angle, we can have misleading interpretations about how close or far a given set of solutions are.

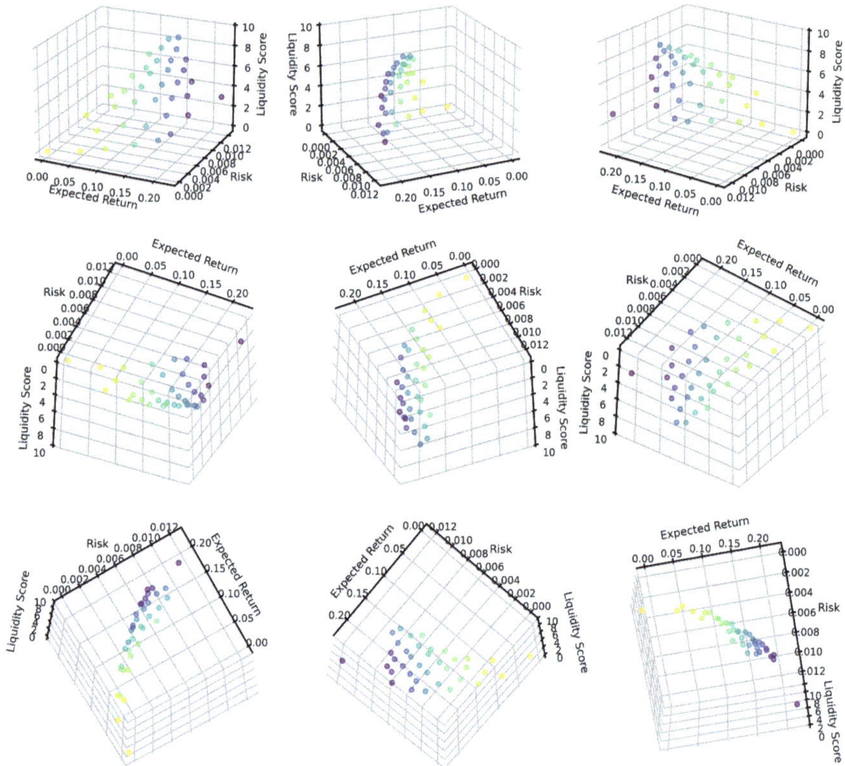

Fig. 7.13 3D scatterplots from the same Pareto front approximation, seen from different angles. *Source* The author, 2025

7.4.1.2 Parallel Coordinate Plots

So, the parallel coordinate plots can be quite useful for problems with three or more objectives. As we have mentioned in Sect. 5.5.1.3, each objective is represented by a vertical axis, and each solution is drawn as a line connecting its values across these axes. Solutions that are good in one objective but poor in another will show steep slopes, while balanced solutions might appear more horizontal.

Let us reuse Example 13 (portfolio optimization) to demonstrate it. As seen in the code below, we are first sorting all solutions in the Pareto front according to the first objective, and then we assign colors based on this order. In this case, lighter colors refer to lower values of the first objective, and darker colors to higher values.

```
import numpy as np
import pandas as pd
import matplotlib.pyplot as plt
from sklearn.preprocessing import MinMaxScaler
from pandas.plotting import parallel_coordinates

```

```
scaler_F = MinMaxScaler()
F = res_portfolio.F[sort_indices]
F[:,0] = -F[:,0]
F[:,2] = -F[:,2]
F_normalized = scaler_F.fit_transform(F)

obj_names = ['Expected Return', 'Risk', 'Liquidity Score']

for front, label in [(F, 'Objective Values'), (F_normalized, '
    Normalized Values')]:
    df_mcdm = pd.DataFrame(front, columns=obj_names).sort_values(
    by='Expected Return', ascending=False)
    df_mcdm['index'] = df_mcdm.index

    plt.figure(figsize=(12, 6))
    parallel_coordinates(df_mcdm, 'index', cols=obj_names, alpha
    =0.4, colormap='viridis')
    plt.title('Parallel Coordinate Plot of Pareto Solutions')
    plt.ylabel(label)
    plt.xticks(rotation=10)
    plt.grid(True)
    plt.legend().remove()
    plt.savefig(f'mcdm_parallel_{label.replace(" ", "_")}.png',
    dpi=300)
    plt.show()
```

Listing 7.29 Python code for Parallel Coordinate Plot of MOO solutions

This code generates two parallel coordinates. Figure 7.14 shows all values not normalized. Although some libraries scale them accordingly, this is not the case of this example: it is hard to compare the results of the first and second objectives. Now, compare with Fig. 7.15. It is now clearer to see how the different objectives interact—especially the first and second objectives.

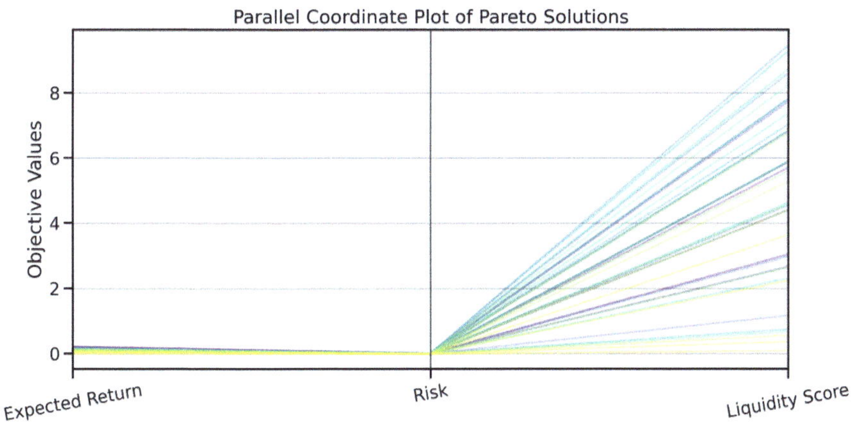

Fig. 7.14 Parallel coordinates for the portfolio optimization problem. All values are not normalized. *Source* The author, 2025

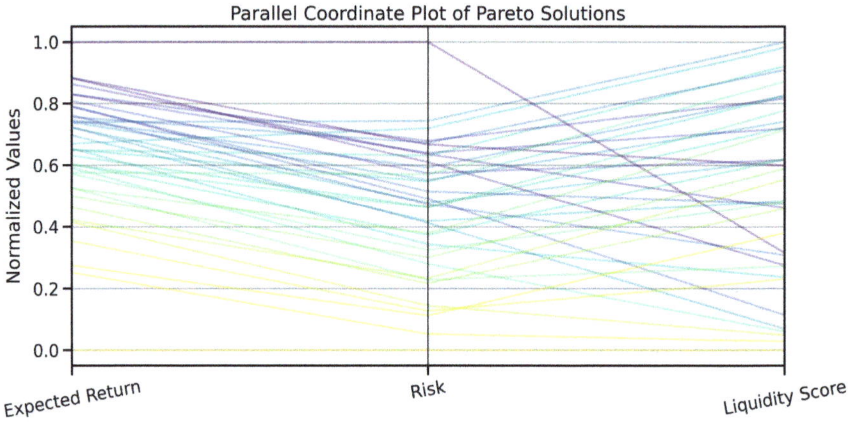

Fig. 7.15 Parallel coordinates for the portfolio optimization problem. All values are normalized. *Source* The author, 2025

7.4.1.3 Scatter Plot Matrices

Another way to visualize solutions for n-dimensional objective spaces where $n > 3$ is through the adoption of a scatter plot matrix, as mentioned in Sect. 5.5.1.1. With this visualization we can see every pairwise combination of objectives, and we can also see in the diagonal the histograms or density plots for each individual objective.

In Python, it is easy to create this visualization through the `Seaborn` library. Its implementation is named `pairplot`, as seen in the code below. The generated image is shown in Fig. 7.16.

```
import matplotlib.pyplot as plt
import seaborn as sns
import pandas as pd

F = res_portfolio.F
F[:,0] = -F[:,0]
F[:,2] = -F[:,2]
df_mcdm = pd.DataFrame(F, columns=['Expected Return', 'Risk', '
    Liquidity Score'])
sns.pairplot(df_mcdm)
plt.savefig('portfolio_pairplot.png', dpi=300)
plt.show()
```

Listing 7.30 Python code for Scatter Plot Matrix

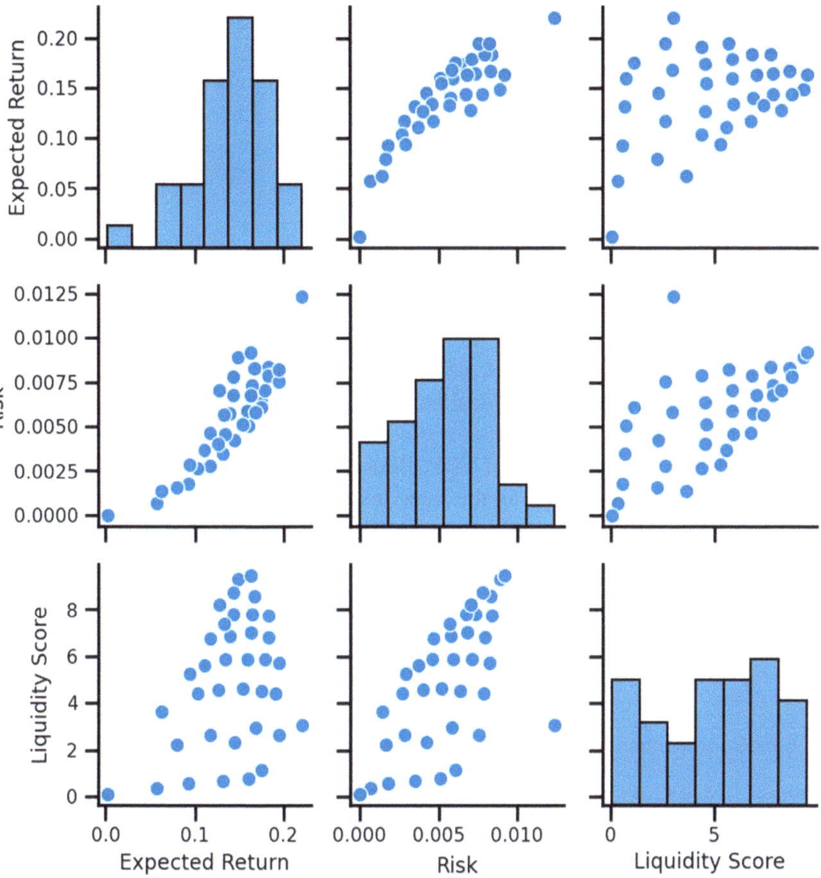

Fig. 7.16 Scatter plot matrix for the portfolio optimization problem. *Source* The author, 2025

7.4.2 *Quantitative Techniques*

Remember that visual techniques are useful for initial exploration, but they do not suggest which solution you should choose or prioritize. To this end, we can rely on quantitative techniques.

However, before applying any MCDM method it is strongly recommended to normalize our objective values, as MCDM often ranks and scores the solutions with distances and weighted sums. We already used a very similar code to generate a few plots earlier. Here, we will reuse parts of that code to generate a normalized Pareto front approximation for the portfolio optimization problem. Therefore, `df_mcdm` will be the normalized front we will use throughout this section.

```
import pandas as pd
from sklearn.preprocessing import MinMaxScaler

```

```
scaler_F = MinMaxScaler()
F = res_portfolio.F[sort_indices]
F_normalized = scaler_F.fit_transform(F)

obj_names = ['Expected Return', 'Risk', 'Liquidity Score']
df_mcdm = pd.DataFrame(F_normalized, columns=obj_names)
display(df_mcdm)
```

Listing 7.31 Python code for Min-Max Normalization

7.4.2.1 TOPSIS (Technique for Order of Preference by Similarity to Ideal Solution)

As you might remember from Sect. 5.5.2.1, this method ranks first the alternative that is simultaneously closest to the ideal solution and farthest from the negative-ideal (or anti-ideal) solution.

Check below a simple usage of TOPSIS using the `pyDecision` library. In this implementation we are assigning the same weights to all objectives, and informing that all of them were minimized (`'min'`). The first 5 results and last 5 results are shown in Table 7.2.

```
from pyDecision.algorithm import topsis_method

# Weights
weights =  [0.25, 0.25, 0.25]
criterion_type = ['min', 'min', 'min']
results = topsis_method(df_mcdm, weights, criterion_type, graph=
    False, verbose=False)

ranked_indices_topsis = np.argsort(results)[::-1] # Sort
    descending
display(df_mcdm.iloc[ranked_indices_topsis])
```

Listing 7.32 Python code for TOPSIS method

Table 7.2 Top 5 and bottom 5 results of the TOPSIS ranking. Each row represents a solution

Expected return	Risk	Liquidity score
0.254346	0.592529	0.174029
0.260616	0.552132	0.258665
0.170006	0.679179	0.184007
0.170295	0.640120	0.280815
0.277856	0.478696	0.382255
	...	
0.724784	0.112622	0.620158
0.645802	0.127952	0.770047
0.584006	0.144877	0.950485
0.747835	0.053276	0.970856
1.000000	0.000000	1.000000

7.4.2.2 ELECTRE (ELimination Et Choix Traduisant La RÉalité)

We can also benefit from already existing libraries such as `pyDecision` to use ELECTRE. The example below covers a test of ELECTRE I over our portfolio optimization Pareto front approximation. For illustrative purposes we have redistributed our weights, now giving more importance to the second objective.

One of the outputs of this implementation is the kernel of the best solutions according to this criterion. The implementation of this algorithm outputs the kernel solutions with an internal identification (i.e., `a1`, `a2`, `a3`, ...), instead of the table indices (i.e., 0, 1, 2, ...). This is why we fixed the indices first with `[int(i[1:])-1 for i in kernel]`.

```
import numpy as np
from pyDecision.algorithm import electre_i

# Weights
weights = [0.25, 0.50, 0.25]
c_hat = 0.70
d_hat = 0.50

_, _, _, kernel, _ = electre_i(df_mcdm.values, W=weights, c_hat=
    c_hat, d_hat=d_hat, graph=False)
kernel = [int(i[1:])-1 for i in kernel]
display(df_mcdm.iloc[kernel])
```

Listing 7.33 Python code for ELECTRE I method

We can see the objective values for the solutions in the kernel in Table 7.3. If we modify the weights as well as the discordance and concordance thresholds (`c_hat` and `d_hat`, respectively) you will notice the output kernel will be different.

Table 7.3 Kernel solutions, as found by the ELECTRE I ranking. Each row represents a solution

Expected return	Risk (min)	Liquidity score (max)
0.000000	1.000000	0.684008
0.584006	0.144877	0.950485
0.747835	0.053276	0.970856
1.000000	0.000000	1.000000
0.117682	0.610389	0.724453
0.208417	0.491401	0.885515
0.237591	0.473743	0.691460
0.276744	0.412790	0.929917
0.345940	0.343588	0.762430
0.407860	0.280302	0.938537
0.473691	0.228408	0.727453
0.535995	0.216546	0.540590
0.577442	0.233319	0.446565

7.5 Conclusion

This chapter served as a practical wrap-up, guiding you through implementing various Single-Objective and Multi-Objective Optimization problems using Python and its rich library ecosystem. We presented a sequence of examples, from classic operations research problems like production planning and the knapsack problem to complex non-linear engineering designs and advanced data science tasks like hyperparameter optimization and multi-objective classifier tuning. The intention was to help transform theoretical knowledge into practical, usable skills.

You can adapt these basic scripts for other problems beyond this book or use our examples as a starting point to test different libraries, algorithms, and parameters. By investigating the documentation of libraries like `SciPy`, `PuLP`, `Pymoo`, and `Optuna`, you'll better understand the nuances of various techniques and implementations, enabling you to create your own code.

Furthermore, remember that the ability to model, solve, and interpret optimization problems is an important hard skill for career progression. Understanding how these techniques work in practice allows you to propose more informed, data-driven decisions in your field. This translates into personal and professional growth.

Third-Party Material Attributions

The following third-party materials have been used in this work. We are grateful to the authors and creators for making their work available.

- Figure 1.2 by Axel Hindemith.
  License: Public domain.
- Figure 1.3 by Amanda Coutinho.
  License: Creative Commons Attribution 2.0 Generic.
- Figure 1.4 by NSYS Group.
  License: Unsplash License (Can be used for commercial use, no permission needed).
- Figure 2.2 by Andrey Matveev.
  License: Unsplash License (Can be used for commercial use, no permission needed).
- Figure 2.3 by Han Dora.
  License: Unsplash License (Can be used for commercial use, no permission needed).
- Figure 3.1 by Mikhail Nilov.
  License: Pexels License (Can be used for commercial use, no permission needed).
- Figure 3.6 by Pixabay.
  License: CC0 (Creative Commons Zero).
- Figure 3.9 by Magda Ehlers.
  License: Pexels License (Can be used for commercial use, no permission needed).
- Figure 4.1 by Clément Proust.
  License: Pexels License (Can be used for commercial use, no permission needed).

W. R. Monteiro and G. R. Meza, *A Practical Guide to Optimization in Engineering and Data Science*, https://doi.org/10.1007/978-3-032-04633-8

MIX
Papier aus verantwortungsvollen Quellen
Paper from responsible sources
FSC® C105338

If you have any concerns about our products,
you can contact us on
ProductSafety@springernature.com

In case Publisher is established outside the EU,
the EU authorized representative is:
Springer Nature Customer Service Center GmbH
Europaplatz 3, 69115 Heidelberg, Germany

Printed by Libri Plureos GmbH
in Hamburg, Germany